Water Resources Monograph 16

WATER: SCIENCE, POLICY, AND MANAGEMENT

Challenges and Opportunities

Richard G. Lawford
Denise D. Fort
Holly C. Hartmann
Susanna Eden
Editors

American Geophysical Union
Washington, DC

Library of Congress Cataloging-in-Publication Data
Water: science, policy, and management : challenges and opportunities / Richard G. Lawford ... [et al.].
p. cm. -- (Water resources monograph ; 16)
Includes bibliographical references.
ISBN 0-87590-320-7
1. Water resources development. 2. Hydrology. 3. Water-supply--Government policy. 4. Water-supply--Management. I. Lawford, Richard G. II. Series.

TC405.W36 2003
363.6'1--dc22

2003063034

ISBN 0-87590-320-7
ISSN 0170-9600

Printed in the United States of America.

CONTENTS

Preface

Although the Earth is blessed with an abundance of water, how we use (or misuse) water is becoming a central issue for sustaining life as we know it. Indeed, in the twenty-first century the availability and use of water, whether in public or private hands, involves local, regional, national and international values and needs.

The way in which we regard water also reflects our vision for the future. At the same time, regional differences in local climates, cultures and levels of economic development affect how we deal with water. Certainly, the way in which we manage water in a specific river basin is as much a product of all of the developments that have occurred in river basins since civilization dawned there, as the new means we have adopted to ensure sustainability.

In the United States, for example, policy makers need to consider how much water should be shared between basins and whether or not water development should be privatized or controlled by large government-funded initiatives. In Europe, the newly implemented European Community Water Framework Directive will introduce new standards for water and land management for streams and rivers that have not had strong environmental regulations. When water basins extend beyond the political jurisdiction of one to several nations, differences between environmental laws, water use patterns and technological developments are frequently a source of tension.

In arid and drought regions water has always been a pressing issue. During 2002, after several years of drought in the Midwestern U.S., the Secretary of Interior reduced the amount of water from the Colorado River for California in 2003 to that specified in the Colorado River compact. However, due to a successful court challenge by the Imperial Irrigation District in California, the final allocation is now being negotiated. Droughts and climate may be closing a critical spigot that supplies California's Central Valley agribusiness and its concomitant community growth and may be moving the management of the Colorado Compact into the courtroom. In Texas, proponents for the exploitation of groundwater reserves are developing plans and seeking approvals to pump these waters and sell them in thirsty urban areas. That this plan impacts local geological processes, such as ground subsidence, along with impacts on regional ecosystems and sustainable agriculture, is clear. Yet scientific research here is not being done. Water has also become a topic of discussion between President Bush and President Fox of Mexico. The dry riverbeds that no longer supply water to Mexican farmers have now become paths for illegal immigrants seeking to flee local drought conditions for a better life in the United States.

Fortunately the U.S. science community has begun to organize itself to more effectively address water issues. Under the U.S. Global Change Research Program and the more recent Climate Change Science Program, scientists have been advancing a water cycle agenda that includes a coordinating committee to draw together water resource research agencies with other agencies that make policy decisions. Although this effort primarily sustains a physical science focus, it is trying to bridge the gap with policy makers.

Internationally, there is a growing consensus that the years to come will bring us ever more difficult water problems to solve. In March 2003, ministers and water experts from around the world gathered in Kyoto at World Water Forum III to discuss such problems from many perspectives. A number of programs and studies had been set in motion to prepare for that event. The United Nation's World Water Assessment Project (WWAP) and the Dialogue on Water, to name two, presented their reports at the meeting. Funded by Japan and the Netherlands, respectively, these initiatives drew attention to the potential contributions that science can make in resolving water issues at the basin level. Other projects, such as Hydrology for Environment, Life and Policy (HELP), sponsored by UNESCO and WMO, have similar goals. Unfortunately, although the benefits of using water science more directly in the service of society are clear, and while these projects and their forums open or sustain dialogue between the ministerial and scientific communities, science has yet to find its voice at the table when major decisions are being made.

In this book, we recognize the need for new skills and understanding to deal with the interface between water science and policy, and the work required to enhance the dialogue between these communities. In the first section of the book, authors describe major issues that affect water availability and use, and the consequences of water shortages in sectors such as cities, and agriculture. In the second section, authors consider the role of institutional arrangements in constraining the management of water. As shown in the third section, previous efforts to link science and policy have sometimes had major successes, sometimes less success. Lessons to guide future interactions between the science and policy communities should be learned from these experiences. In the final section, authors explore the role of science and provide examples of how science interacts in different ways with the policy and management communities. The work is derived from a special session of the spring meeting of the American Geophysical Union held in Washington in May 2000, where many of the issues discussed here produced intense interest.

1

Introduction

Richard Lawford and Denise Fort

BACKGROUND

Water issues exist in nearly every part of the globe and touch every citizen on this planet. For example, inadequate water supplies and poor water quality give rise to health and equity issues, limit agricultural productivity and economic prosperity, and pose national security risks in some countries. Problems of this nature have been increasing in scope, frequency, and severity because the demands for water continue to grow while the supply of renewable water remains fixed. Furthermore, when water is returned to natural reservoirs after use it often bears the imprint of that use because of changes in the concentrations of chemicals, bacteria, pathogens and even heat in the water. The factors affecting water have their roots in more complex problems, including growing population and changing demographics, poorly planned or coordinated watershed management, and undisciplined industrial development. A number of governments have growing concerns about the state of water within and adjacent to their borders, and those concerns are reflected in recent statements by international organizations. In its millennium goals, for example, the United Nations has included a need to "halve the proportion of people with no access to safe drinking water or those who cannot afford it by 2015" (Millennium Target Number 3).

In May 2003, the G8 members agreed to undertake actions to resolve water issues throughout the world. This development followed statements from the United Nations, the World Summit on Sustainable Development, and the World Water Forum urging responsible parties to make clean water a human right and to ensure water sustainability for future generations. Clearly, however, it will take more than edicts by national or international bodies to address the problems associated with water.

Water: Science, Policy, and Management
Water Resources Monograph 16

10.1029/016WM01

Although this book describes the scope and nature of a number of water issues and the factors contributing to them, it is much more than a cataloguing of perspectives from the different communities involved. A central theme emerges that also invests some urgency to our future interactions: developing an integrated approach to solving water problems. The book thus offers examples and recommendations for new directions to bring scientists, engineers, policy makers, managers and the affected public together in meaningful ways to address water issues. It deals with knowledge generation, policy opportunities and constraints, and the conversion of knowledge into actions of benefit to society.

In the United States, significant water management issues loom. Aging water infrastructure in urban centers is reaching a critical point, with some large cities requesting federal assistance for infrastructure replacement. Water quality monitoring programs tend to be reactive to problems rather than preemptive; strengthened monitoring programs are needed to provide better direction for water quality programs. The quality of water in return flows from irrigation and their impacts on land salinity have become an issue as there is pressure to reuse water as many times as possible. The need for scientific understanding to support the conjunctive management of ground and surface waters is being accelerated by innovative projects that store excess streamflow from the spring freshet in aquifers, where it is kept for use during water shortages. Questions remain about the effects of rock formations and groundwater flows on the quality of water withdrawn from these aquifers. Allocating the true cost of water to the user is one way to encourage conservation and the utilization of technologies such as desalinization. Emerging in the western U.S. is a need to negotiate settlements of Native American water claims, so that equitable access is provided and states and water users gain a better understanding of the potential consequences of these settlements for future water availability and use patterns.

In addition to reviewing major water issues, this book assesses the status of the interactions between the scientific, policy, and operational communities involved in water management. While this book points to some directions for solving water problems, it deals more directly with the processes whereby these solutions can be developed. Among other issues, it addresses the process of knowledge generation to identify and delineate problems and options for solutions to water problems; reviews the legal infrastructure that controls water management decisions and discusses some conditions under which water managers have utilized information generated by the science community; discusses the engineering discipline, which plays a pivotal role in applying science for social benefit; and evaluates the potential of citizen groups to influence the implementation of solutions that optimize flexibility in dealing with water issues. These interactions are broad, and while some interactions have achieved success in terms of actual results, others are less mature with less success to their credit. To provide a range of options for the future, a number of ideas are presented on ways to improve the dialogue between those who generate knowledge related to water and those who use this knowledge to shape water policy and manage water systems.

SCIENCE: SOCIETY'S PRINCIPAL KNOWLEDGE GENERATION PROCESS

Research

Research has played a significant role in informing society on options for addressing water supply and management issues, and giving society the technologies needed for implementing these options. In the past, the research community made its greatest contribution to engineering solutions to water problems. More recently, science has shown that it can contribute to non-structural water management practices as well.

Although people have realized the value of water from the dawn of civilization, and used different means to control water, it was the development of reliable methods to measure changing water levels and volumetric streamflow, and the implementation of measurement networks, that facilitated the systematic development and control of water through the planning, design, construction, and operation of reservoir systems. As scientific research determined that water borne chemicals were associated with many health problems and as the capability to measure chemicals in water to parts per hundred thousand or even parts per million was developed in the 1960s and 1970s, the public became more aware of the range of substances in their water and in what concentrations they became harmful. In response, the public began to lobby for environmental programs to improve water quality. It is our belief that science is reaching a stage of development where it soon will make another incremental increase in the data and tools it can make available to society to enable water management. This new development will come from integrating our capabilities to measure water cycle variables at and near the surface and from space, together with our ability to model the Earth system in advanced assimilation and prediction systems. We can only hope that the water management and policy communities will directly benefit from these new predictive capabilities.

Scientists recognize the importance of water management in altering natural water systems. Yet scientists have tended to look at such impacts through discrete investigations involving two or three experts carrying out exploratory analyses, rather than by comprehensive reviews and assessments. Fortunately, large-scale assessments are now being carried out that deal with water management practices and their links to the environment. For instance, the United Nations World Water Assessment Program is currently developing its second World Water Development Report. The Global Water System Project has also been launched to assess the impacts of engineered water management systems on natural water distribution and climate on a global basis. Although we anticipate that these studies will be useful in identifying policy issues, more work will be needed before policy options can be presented.

Education

The transmission of knowledge regarding water science and water problems presents a very significant challenge due to the breadth of issues involved. Few university programs train "water scientists" on the many facets of science related to water and its multiple uses. Water touches so many aspects of our lives that many university departments offer different courses on some aspects of water with little to no coordination between them, even on the same campus. Civil engineering programs teach hydrological sciences and engineering applications and focus on the application of physical principals to water problems. Environmental chemistry and chemical engineering departments teach courses that consider water chemistry along with air chemistry and chemical oceanography. Meteorology departments teach the atmospheric aspects of the water cycle, including rain formation and climate prediction techniques. Water issues are taught in geography departments that often look at geomorphology and the movement of water, as well as some of the social and economic issues affecting water use. Water programs also exist in economic departments, law schools, social science programs, ecology programs and other departments. In many cases, these programs would have greater impact if they focused on water as a central disciplinary theme and considered the relationship of water management to specializations, such as the economics of water or the ecology of water. Developing specific programs of study to provide scientists and water managers dealing with water issues with a fully integrated overview has become a necessity.

POLICY: DEVELOPING SOCIETY'S CHANGE AGENT

In the context of this book, policy represents the wide range of instruments of governance that can be applied in water management. It includes high-level policy instruments such as legislation and legal agreements, as well as criteria for establishing priorities for daily management decisions. In some cases, the chosen policy approach has led to standards, guidelines, and regulations, sometimes with punitive implications for non-compliance. In other cases, achieving certain objectives may rely on incentives and establishing voluntary targets. Mechanisms for dispute resolution on issues, such as study boards and commissions, also can serve as policy instruments for the negotiated settlement of water conflicts. In cases where the knowledge base is inadequate, however, a research program may be needed before launching a new initiative.

Choosing Policy Instruments

The scale of the water problem generally influences the policy approach and information required. In some areas of jurisdiction, federal governments deal with policy on a national basis, while more basin-specific management is usually carried out at the state or county level, sometimes with federal oversight or with federal involvement or financial support. In the U.S., water allocation is a state-by-state matter with federal government intervention through the construction and control of infrastructure, pollution laws, and fiduciary relationship to tribal governments. Water law is thus characterized by diversity and leads to variances in the way some water issues are handled in different parts of the country. The reliance on riparian water law in the eastern states and on the prior appropriation doctrine in the western states has led to different regional perspectives on water. Conflicting federal law mandates can also lead to situations that highlight the lack of national agreement over policy goals. For example, on July 15, 2003, the U.S. Army Corps of Engineers (a federal agency) announced that it could not comply with a judge's order to reduce water levels on the Missouri River to protect endangered birds because the ruling conflicted with a federal ruling in 2002 that there must be enough water in Missouri for barges to navigate and power plants to operate.

Jurisdictional boundaries often partition basins along political boundaries, which highlights the need for approaches to policy making and program implementation that can accommodate several types of interaction, including: different legal frameworks, resource use patterns and policies; different levels of technological sophistication and information support; and different levels of basin development. For some transboundary issues, an international body vested with specified authority, such as the International Joint Commission (IJC) that deals with issues on the U.S.-Canada boundary, can effectively coordinate water discussions and resolve potential disputes without resorting to the courts. The kinds of information required to support decisions at these different levels also vary. Federal agencies may be well suited to maintain data systems and archives that are national in scope or where overlapping jurisdictions occur. Similarly, federal agencies can help coordinate local data systems to facilitate gathering of data from small scale to provide larger scale information.

Science-Policy Interactions

Internationally, water is recognized as a central resource issue and a growing environmental problem for which interdisciplinary discussion is urgently needed. Although constraints on water availability and management have been known for decades, many water problems, first evident as local inconveniences, have quickly grown into regional tensions. In some less developed countries, water issues are now listed with poverty and health as principal concerns related to survival.

Historically, the science and policy communities have not always worked well together. On occasion, the policy community has found its expectation for authoritative inputs from the science community frustrated by the variety of viewpoints scientists express, the inability of science to address specific problems in the short time frames associated with policy development, and the inability of some scientists to communicate their information in terms meaningful to policy makers. When approaching the science community, policy makers may also be confused by the plurality of opinions and capabilities available. This plurality can raise questions such as "who speaks for science?" Policy makers frequently have been compelled to address problems in relatively short time frames because of the public's demand for action.

For their part, many scientists believe that their task is to build a more complete body of knowledge, which can take years and even decades. The consequent tensions over the proper balance between long-term fundamental science and shorter-term problem driven investigation and problem solving has not been lost to science. Scientists have also expressed the view that their inputs are rarely or never openly solicited, and that the policy community has an over simplistic perception of problems and science's potential contributions to them. Scientists also point out that, when given scientific inputs, policy makers may misinterpret or ignore the information or give higher priority to non-scientific factors. Fortunately, scientists and policy makers both recognize the importance of improving their dialogue. Policy makers realize that decisions not founded in authoritative science will not be to the advantage of society, nor will they withstand public scrutiny as water availability decreases. On the other hand, scientists realize that their knowledge must contribute to, and be seen to contribute to, contemporary water issues to ensure that public support continues for water-related research.

It is clear from the discussions in this book that, in general, science alone will not solve major water problems - it can only help define them and provide options for solving them. However, policy operating in a vacuum, without inputs from science, would not have the potential to solve these problems either. Policy will be enriched and gain stronger public acceptance by relying more directly on scientific inputs.

Engineering approaches, which focus on structural solutions to water management problems, will continue to be important although the scope of implementation in many countries may be hampered by the lack of financial capital. The policy community often finds that it is easier to deal with the engineering community than the science community. Engineers develop solutions to problems that are based on specific measurements, reasonable assumptions, and statistical data analyses, including assessments of risk and the return period of extreme events. In contrast, scientists are direct about the limitations and gaps in their knowledge and are perceived as less practical in their approach to problem solving.

During the past two decades scientists and policy makers have successfully dealt with several major environmental issues including acid rain and the ozone hole. For example, the growth of the ozone hole has been successfully arrested as a result of negotiated international agreements controlling the emission of harmful chlorofluorocarbons. The need for ongoing dialogue has been highlighted by discussions about the environmental impacts of greenhouse gas emissions and possible mitigation and adaptation strategies. The current openness for dialogue between science and policy was advanced at the international level by the Third World Water Forum, which brought scientists and policy makers together in Kyoto, Japan. Clearly, the benefits already gained from informed dialogue about water policy need further support and development from concerned parties at the national level as well.

Lessons from Past Science-Policy Dialogue

Policy may set the criteria for scientific knowledge without a full appreciation of the limits of science or of the time and funding required to generate the supporting science. The need to control pollutants is a case in point, with government agencies often driving the policy process here. For example, Congress initially directed the Environmental Protection Agency (EPA) through environmental legislation (Clean Air Act and Clean Water Act) to address toxic chemicals using risk assessments and related health standards. Although this approach was logical, it did not account for the complexity of the assessment process, the challenges in attributing the results of specific health symptoms to a particular concentration of toxic chemicals, and the challenges in establishing an appropriate effluent standard for a specific toxic chemical. Furthermore, the resources allocated to advance this process were far from adequate to ensure an intensive research and review program. Consequently, with only limited progress to show after several years work, Congress changed its approach and gave EPA a list of chemicals to subject to regulation, and specified that technological feasibility should be a major factor in the regulation requirements and strategies. Although this approach led to many more toxic chemicals being regulated, it did not expand our knowledge of the full range of health effects and integrated risk arising from them.

Congress occasionally has provided some opportunities for scientific judgment in the regulation process. In a 1990 provision of the Clean Air Act, Congress gave more direction to EPA on how to set standards for certain toxic air pollutants (hazardous air) using risk based considerations. In this case EPA was asked to advise Congress each time it set a standard that would allow exposures that could lead to an increase of more than one in a million additional cancer cases. By this provision Congress gave some endorsement to what, in the science community, had evolved as a commonly used basis of determining acceptable risk.

In the end, however, there is little consensus on how we should approach the issue of toxic chemicals. Many lawmakers believe that risks have been exaggerated, and that statutory measures to mandate "good science" are required to correct the excesses of agency personnel. Science faces a challenge here: It must bring clarity to this complex policy process by defining what science can do while not having ready access to policy levels where legislation is formulated. For their part, industry lobbyists welcome the delays that result when agencies are required to meet further obligations related to terms as nebulous as "good science." New studies, that either increase or decrease estimates of risk, also lead to public confusion. In a democratic society, however, it is appropriate that the standards for regulation be set through an open process, and that policy makers take into account the costs and benefits of these standards.

Trends in Science-Policy Interactions

New approaches to water policy also are being considered. Efforts related to the management of risk in US environmental laws constitute a compelling example of the problematic intersection between policy and science. In broadest terms, the risk presented by toxic chemicals came to the forefront of lawmaking after the first generation of environmental laws was passed in the early 1970s. From a legal perspective, the uncertainties replete in risk assessments and risk management offer administrative agencies great discretion. In particular, the process of risk assessment offers multiple opportunities for the exercise of judgments by all parties. Science will need to define its role in this process. Judgments made by scientific experts may differ from one chemical to another, and because they are judgments, can be questioned by other scientists.

In addition to the uncertainties inherent in risk assessments, risk management requires difficult decisions about what levels of toxic chemicals should be prohibited, or permitted, in environmental exposures. At the root of these questions is a lack of societal consensus: our desire to be free from cancer, birth defects, and other illnesses brought on by exposure to toxic chemicals is moderated by our desire to continue with a lifestyle that is dependent on them. Thus some benzene in our gasoline tanks might be acceptable if it keeps the price of gasoline at a comfortable level.

COVERTING KNOWLEDGE INTO ACTION

While science is concerned with facts, policy must concern itself with choices. Society's tolerance to risk changes over time and varies according to the group being affected. In a democratic society the jockeying to control the agenda is on-going. Clearly many factors influence policy, but it is imperative that science be given a prominent place. Science can help to delineate options and provide guidance on which proposed approaches are most viable and which ones will not produce desired results.

The Need for Science in the Policy Arena

The need for science and policy to work jointly has been acknowledged in a number of fields. For example, the role of human activities in renewable resource issues and systems has been recognized and is now being addressed by research programs. Society may need to organize some of its policy and management responses differently to address environmental and resource shortage problems in new ways. Achieving this objective will require that science and policy work more closely together despite the different paradigms involved, and the inadvertent obstacles that that arise because of them. As the looming global water crisis takes hold, policy makers and politicians will increasingly be faced with issues such as drought losses, flood damages, destruction of ecosystems, water borne diseases, conflict between users of the same water sources, rate increases for water services, and so on. Resolving these issues will require meaningful on-going interactions between the science and policy communities.

Change leading to increased collaboration here has been, and will be, facilitated by developments in new technologies. Water managers' increased dependency on new techniques and scientific data may require extensive training programs to best enable them to use the tools available. Policy and legal instruments will need to be updated to ensure that they facilitate these transitions. In this regard, policy instruments that favor flexibility and accommodate learning should be encouraged. In the U.S., we have already seen a shift from exclusively structural solutions to problems to a mix of structural and non-structural solutions. In the developed world as a whole, change is being accelerated by new satellite observations, assimilation schemes, and prediction systems that provide improved information support to water managers with accuracies, resolutions and coverage never before available. While these systems provide global coverage, their primary use will involve current and predicted information for a particular watershed, and even a specific location. With suitable technology and training in the appropriate use of this information, such systems could support water management activities globally. New information technologies and the kind of understanding they prompt also can make the assessment process more precise, efficient, and credible.

At the same time, many global environmental programs are considering the global implication of the cumulative effects of local water management decisions and procedures. This grows in importance when considering how surface water and ground water are increasingly considered a single resource that needs to be managed under a unified legal and policy framework. Change is also coming from the private sector as it lobbies for a larger role in water delivery and resource exploitation, and as local communities gain more responsibility for managing the resource at the basin scale. However, many aspects of such revised approaches to water management need further research and understanding before societies fully commit themselves to a new course for water management. While the contributors to this book will not provide a consistent assessment of the adequacy of the knowledge base to address these changes, there is consensus that more science and policy research is necessary, particularly in the interface areas noted. It will also be necessary to advance this research agenda on a priority basis, specifically in response to the social factors fermenting change, which have grown more active as their proponents have grown more impatient.

The Challenge of Knowledge Transfer

The transfer of knowledge from the science community to other communities is a major challenge for science. When interacting with policy makers, scientists must find ways of converting the large data streams from satellite and in situ measurements into meaningful information that policy maker's can use when planning change. This means developing indices that can represent the net effect of a number of factors, including physical and non-physical factors, and through an analysis of trends in those indices indicating to policy makers if conditions are worsening or improving for a basin, a nation, or the globe. The transfer of scientific results to the engineering community can also be difficult because engineering practices are frequently entrenched in standards and regulations, making modification here difficult. As a result, the engineering community may be resistant to accepting changes in standards that can be inferred from the results of scientific research. At present, for example, the engineering community has been slow to adopt the changes that would be needed to deal with a non-stationary climate and the effects of changing water use, such as the requirements of ecosystems for minimum in-stream flows. Eventually, however, the engineering community may use climate change to rationalize structural solutions to problems within the framework of the debate over structural and non-structural solutions to water problems.

Researchers also are faced with the challenge of transferring knowledge to the operational water supply forecasting community, so that operational services can be improved and delivered more effectively. In cases where the operational community is a partner in the research program, adopting new findings and techniques generally proceeds quickly. Some of the strategies that work for technology transfer to operations may also be useful in solidifying the linkages between science and policy. Several international programs, such as the Hydrology for Environment Life and Policy (HELP), are beginning to explore this possibility. Regarding policy, there may be a need for information brokers who can translate the findings of the research community into terms that are meaningful to policy makers. The need for experts is particularly evident where water managers have access to a number of sophisticated but diverse hydrologic models for aiding in decision-making. Often the choice of model is dependent on climate, season, nature and scale of the problem being resolved, and geomorphological conditions: factors which the average water manager will have neither the time nor the expertise to analyze. A need also exists for expertise in determining when uncertainties in scientific results have significance for the interpretation of the information, and when these uncertainties make the contributions of science to policy premature.

Science and technology can promote the diffusion of technology and information services for more efficient and effective water management and contribute to meeting expectations of developing countries. Science programs have provided significant benefits for national hydrologic services by transferring new scientific understanding and models to operational forecast centers. The development of web-based information systems, such as that outlined by Hartmann et al. [this volume], exemplify practical innovations of use to water managers. In this case, the transfer of technology can be attributed to the investigators efforts to make their innovation known to the user community, and to the water managers who have seen the benefits of adopting a technology that will improve local decision-making. Cases also could be identified, however, where new technologies have not diffused into the user community, or have only been applied after a number of years following their development. The diffusion of policy innovations between jurisdictions is an even more challenging task. In fact, the policy framework is often most ready to look at new approaches immediately after a crisis or disaster such as a flood or a drought. Given this reality, it is important that the science and policy communities take advantage of opportunities presented by floods and droughts to introduce innovations related to water management when special budgets become available for such innovations.

One final word of caution must be sounded. There are risks in a full science-policy merger. On the one hand, a policy process held captive by scientists would not be healthy because it would not possess a capacity for negotiations and tradeoffs. On the other hand, science which had given up on a "search for the truth," and was open to radical utilitarianism, ready to endorse and justify any social fad, would degrade the authoritative voice of science and compromise its objectivity. To further develop the science-policy interface, each community needs to be open and willing to solicit and consider the concerns of the other community at each stage of the policy and knowledge development processes.

SUMMARY

This book grew from an evident need to bring the science and policy communities closer together and to initiate a dialogue to better understand how science could strengthen its links with policy. The dialogue is only in its preliminary stages and requires nurturing to reach fruition. Despite this caveat, this book documents some recent major gains in developing a common understanding of the problems and the potential contributions of the various communities in addressing water issues. As more experts become aware of the nature and opportunities inherent in this dialogue, they will engage the policy and science communities more vigorously. As such, we offer this book to both communities with a view to stimulating discussion and debate.

We have structured the book to introduce the reader to the extent and nature of water issues in the early chapters. After discussing the issues generally, a special focus is given to water for agriculture, and water for urban development. Groundwater, a special development issue, is also discussed. The second part of the book deals with the management of water through discussions about the role of policy in water management and the complex problems posed by transboundary basins. The third part of the book provides an assessment of the state of our scientific knowledge in many aspects of the sciences. The final section deals with the challenges and opportunities to bring science and policy together in new creative ways.

The perspectives and issues presented in this book provide a starting point for an ongoing comprehensive dialogue between scientists, policy makers, and water managers. To advance this dialogue interested experts will need a forum that will organize and promote these interactions. Such a forum would be most effective if it was associated with a scholarly society that would foster appropriate research, organize scientific conferences, facilitate publications, and promote education. Readers convinced of the importance of the issues raised in this book, and who wish to advance the necessary dialogue, are encouraged to discuss their interests and ideas with one of the editors.

Richard Lawford, NOAA Office of Global Programs, 1100 Wayne Avenue, Suite 1210, Silver Spring, MD 20910; richard.lawfor@noaa.gov

Denise D. Fort, University of New Mexico, School of Law, Albuquerque, NM 87131; fortde@libra.unm.edu

Introduction To Section 1

This section provides an overview of selected critical issues facing water managers during the next several decades. The scope of water issues affecting the globe are driven by water management practices, the use of land and other environmental components, climate, and socio-economic factors such as globalization, population growth, and related demographic trends. Because precipitation and water supplies are distributed unevenly over the earth's surface and levels of economic and water infrastructure development vary widely, water issues differ from nation to nation. Three of the major water issues facing us worldwide are outlined in greater depth, including: water for food security, exploiting fossil subsurface waters, and aging urban water systems. In each case, we consider possible policy issues and the potential contributions of science to problem resolution. Other central water issues are discussed briefly in the opening overview chapter. The present and future role of water in the energy sector, water and public health, and the two-way interactions between water systems and the global environment, for example, receive more extensive treatment elsewhere.

Arguably, there is no other topic where the contrast between development and long-term sustainability paradigms are as clearly seen as they are in the field of water. Given the growing interest in water both nationally and internationally, the many direct and indirect links that exist between water development and use and public expenditures, and the contrasting perspective between many who view water as a human right and some who view it as a commodity to be marketed, it is inevitable that science and policy will be required to deal with water issues in a more substantive way. This section highlights the issues that give urgency to the science-policy dialogue.

Water: Science, Policy, and Management
Water Resources Monograph 16

10.1029/016WM02

2

Using Science to Address a Growing Worldwide Water Dilemma for the 21st Century

Susanna Eden and Richard G. Lawford

INTRODUCTION

The challenge of ensuring usable water in sufficient quantities to meet the needs of humans and ecosystems is emerging as one of the primary issues of the 21st Century. Information on current trends in water use and availability indicates that stresses on the global supply of fresh water are manifold and growing. While pessimists suggest that it may be too late already to avert a crisis, the authors of this chapter believe that sustainable water resource use is attainable through concerted efforts by the research, management and policy communities in consultation with the affected public. However, they also contend that there is urgency attached to this issue. Sustainability must be sought and should be achieved now, so that major water shortages will be avoided in the future.

Water sustainability may require the redesign of major institutions and infrastructure for water distribution, management and use to incorporate advanced scientific understanding of the water cycle, ecosystems, and the behavior of social, economic and engineered systems. This chapter considers a number of the issues affecting society's ability to achieve sustainability. These include the needs of society for water and the role of water in development, improving conditions for the world's poor, the spread and prevention of disease, food production and food security, and the maintenance of ecosystems and the services they provide. In addition, it considers the issue of climate variability and change, which complicates water management for sustainability. Addressing these issues requires examination of potential economic controls on water use; alternative governance strategies for water management; alternative technologies and emerging opportunities to exploit improved hydroclimatological forecasts.

Water: Science, Policy, and Management
Water Resources Monograph 16

Published 2003 by the American Geophysical Union
10.1029/016WM03

In order to make useful contributions on policy and management issues, the science community must accommodate changes in water management paradigms. Progress in the development of US water resources has taken water managers from the years when infrastructure could be put in place to solve all water supply problems to the more recent years when water management has supplemented its solution set with non-structural strategies. In addition, the public has become much more involved in the decision process. These changes reflect the changing values that society has for the way in which the nation's water resources are being developed. This same pattern is evident in much of the developed world.

In the past, social well-being was considered to follow directly from economic development, thus any water use that contributed to this goal was welcomed. As people gained wealth, they became more concerned with the values of environment and the consequences of environmental degradation. These concerns have grown in importance globally, especially in countries with high standards of living. In these countries, institutions and infrastructure have evolved in response to such concerns. For example, in the developed world, new technologies, stimulated by regulations and consequent economic incentives, have enabled industry to reduce its use of water while continuing to increase its industrial production. In developing countries, the tension between traditional economic development and ecosystem values continues to manifest itself. Science must join law and policy to help developing nations achieve economic goals without sacrificing the environment.

This chapter introduces a number of issues that are dealt with more extensively in the following chapters of this book. It considers the implications of past and current practices as well as newer directions and potential futures for water resources. It examines issues involving natural constraints on water availability, water management, changing policy perspectives and the role of scientific research. Its goal is to introduce the key concerns and components of a science based sustainable water resource strategy.

WATER: A CRITICAL RESOURCE

Ready access to usable water is positively linked with economic development and social well-being. Although the number of options for water management is expected to increase as technology develops, regions that lack ready access to water will be challenged to ensure the viability and progress of their societies. In developed countries limited access to water is emerging as a constraint on development, and in developing countries it can be a significant determinant of survival.

Rivers, lakes and aquifers are needed to support transportation, power, recreation, ecosystems and environmental services. Society needs water from its natural streams and reservoirs for drinking and other domestic uses, including sanitation, and for commercial and industrial uses, which return water to the system, usually degraded in quality. People also take water for agriculture, which "consumes" a large percentage of the water through evapotranspiration. When clean, fresh water was plentiful, it made some sense to take from the most readily accessible points for whatever use and return wastewater wherever convenient, relying on the natural system to adapt and to recycle and redistribute the resource. As the resource is squeezed by growing demand and increasing waste loads, however, it is becoming necessary to reexamine and redesign water resource systems. New approaches based on advanced scientific understanding will be needed if the world is to meet its many needs for this critically important resource.

Water for Development

Throughout history water has been an important factor in the development and survival of societies. Ready access to water for domestic use, irrigation and transportation facilitated economic and social advances. Regional patterns of precipitation and rivers have given some countries better access than others to water and its associated natural advantages for economic development. Even today, an abundance of fresh water fuels economic growth. For example, Falkenmark [2002] has correlated water use with Gross National Product for different countries to show the strong positive correlation between water and economic activity.

Because water is central to economic development, communities eager for development have tended to elevate the economic uses of water above its other functions. In the U.S.A. for example, during the expansive first half of the twentieth century, this was the dominant attitude. The goal of water management activities was full resource development, which would produce healthy stable communities wherever its benefits extended. Large dams and channel modifications were built to tame rivers and irrigation was introduced to the semiarid American West. More recently it has been acknowledged that full resource development produced problems along with benefits. In addition, it is recognized that the environment itself is a resource whose values are poorly understood and too long under-appreciated. To achieve a sustainable balance between human water development and environmental health, a reexamination is needed of practices that affect the long term sustainability of water resources, including the appropriate role of dams and other infrastructure developments, the use of groundwater at rates in excess of natural discharge rates (groundwater mining), and planned or inadvertent use of rivers to transport wastes and chemicals.

Water for Health

Clean water is essential for health. In countries where water is scarce and/or treatment capabilities are inadequate, water-related diseases are major causes of death. According to World Health Organization (WHO) estimates, more than five million people die each year from diseases caused by unsafe drinking water, lack of water for hygiene, and inadequate sanitation [*Young et al.*, 1994] and up to 6,000 children die each day (2.2 million children/year) from water-borne diseases [*WHO*, 2003]. Communities that lack adequate water supplies and sanitation services are especially susceptible to disease. For example, ninety percent of cholera outbreaks worldwide can be traced to inadequate water for drinking and washing and inadequate sanitation. Around many of the world's largest and most rapidly expanding cities, where poor migrants from rural areas occupy undeveloped lands, human wastes pollute the surface water and aquifers that supply drinking water because population growth has overwhelmed the capacity of the infrastructure. Two of the United Nations millennium development targets relate to safe drinking water and the quality of life for the urban poor. Specifically, the Millennium targets are: "by 2015, reduce by half the proportion of people without access to safe drinking water" and "by 2020, achieve significant improvement in the lives of at least 100 million slum dwellers."

Water for Food

Water is essential for food security. Although the world's agricultural system currently can produce enough food for everyone, it is estimated that 777 million people cannot get enough food to live productive lives [*FAO*, 2003]. To meet the demands arising from future population growth, food production will have to increase. Estimates suggest that food production that will need to increase by 50 to 100 percent over the next 30 years. [*Gleick*, 1998]. Changing food habits arising from increased prosperity also contribute to increases in water demand for the food sector. The dramatic increase in food production that took place in the last quarter of the 20th century was aided by new, high-yield plant varieties, fertilizers and pest control; however expanded irrigation arguably was the most important factor [*Postel*, 1997]. While crop production is still rain fed on most of the world's agricultural land (83 percent), 40 percent of the total food product comes from irrigated agriculture [*Gleick*, 2000]. The demand for irrigated food production is expected to increase substantially in the developing world over the next few decades.

THE FINITE LIMITS OF THE WATER CYCLE

An important early step in developing a sustainable water resource strategy involves assessing the adequacy of the world's water resources, based on an inventory of how much usable water is cycling through the Earth system and hence available for human use, and to predict how that amount will vary with time. While observations exist for water stores and fluxes in some regions, the lack of adequate global databases is a significant source of uncertainty in these assessments.

The Cycling of Water

The water cycling through the Earth system satisfies human uses, supports the Earth's ecosystems and provides basic functions in the atmosphere's circulation and heat exchanges between the equator and the polar regions. This cycling, which is modulated by the march of the seasons and by shorter-term variations in the weather, ensures long-term continuity in the water supply.

The cycling of freshwater between Earth's major water reservoirs determines the renewable component of water resources. Water evaporates from the ocean (and to a lesser extent from the land surface) into the atmosphere. In areas of ascending atmospheric motion or suitable atmospheric instability, clouds form, precipitation develops within these clouds and this precipitation falls back to the Earth at some distance from the area of evaporation. In this cycle the oceans are a net supplier of moisture to the atmosphere, while the land surface receives more water than it looses. Once it is on the land surface, water flows over the surface as runoff, resides on or in the shallow soil, or infiltrates the deeper soil layers until it reaches groundwater reservoirs. According to Shiklomanov (1997), who assumed that over the long term the average change in soil water content is negligible, about 108,000 km^3 of water precipitates annually over land and about 60 percent (61,000 km^3) evaporates, leaving about 47,000 km^3 to flow toward the sea in river networks where it is accessible for human use, or to recharge underlying aquifers.

The World's Water Inventory

Water resource inventories need to consider two basic components: the total freshwater held in reserve and the amount that is renewable each year. The amount of precipitation that falls on land can be considered the renewable part of the resource and approximates the water that can be used without drawing down the water stored in reservoirs or aquifers. At present the ability to monitor precipitation and changes in surface and subsurface stores around the globe is limited. In the absence of good observations of water stores and fluxes, other methods for estimating these values through the use of models and data assimilation systems need to be developed.

The Earth's water is stored in three principal reservoirs, namely the oceans, atmosphere, and land (including surface and subsurface water storage), and is continuously cycling between these reservoirs. The amount of water held in the atmosphere is only a very small fraction of the total in the Earth system but this it plays a critical role in the climate. Atmospheric processes are critical in the transport and distillation of ocean water that later precipitates over land. According to current estimates, more than 97 percent of the water in the Earth system resides in oceans and seas and is too saline for most human uses. The remaining 3 percent of the Earth's water resides as freshwater on the land.

Two thirds of this fresh water exists in ice caps, glaciers, permafrost, swamps, and deep aquifers, where it is largely inaccessible.

As noted above, the annual renewable freshwater supply (precipitation – evaporation over land) is about 47,000 km^3. Of this it is estimated only 9,000 km^3 to 14,000 km^3 is potentially usable with technically, socially, environmentally, and economically feasible water development. At present, approximately 3,400 km^3, or about 30 percent is withdrawn for use on an annual basis [*Seckler*, 1998]. In areas where shortages exist, some research projects to develop and exploit technologies such as rain enhancement and desalinization of seawater are being undertaken. These efforts may be locally significant, but as yet have not made a significant impact globally.

According to WMO [1997], if the annual runoff were equally accessible to everyone, there would be approximately 7,700 m^3 of water per person per year. However, precipitation varies regionally consequently some continents have abundant runoff (South America has 12, 000 km^3) and can support large populations while others have limited annual runoff (such as Australia/Oceania, which has only 2,400 km^3). Much of the world's river flow occurs during seasonal floods with 60 to 70 percent of runoff occurring during the flood season. During flood periods much of the water flows unused to the oceans and water shortages then occur during the drier periods. In addition, year-to-year variability can be large, especially in arid regions.

Distribution Systems

Through the construction of reservoirs the periods of high and low flows have been smoothed out in many areas. Dams, groundwater wells, irrigation systems, and urban distribution systems permit communities to capture, store, and transport water so that it is available where and when it is needed. The level of infrastructure development is a major determinant of water resource availability.

These engineering innovations have been the major technological means of water development and management throughout the 20th century. Worldwide there are about 48,000 dams higher than 15 meters, with an average reservoir capacity of 269 million m^3 [*World Commission on Dams*, 2000]. In the U.S. there are 6,575 such dams [*Gleick*, 2002] with an estimated reservoir storage capacity of 1,060 million km^3 [*Frederick*, 1995]. Although the era of dam building has ended for the most part in the U.S.A., the role of man-made storage is likely to continue to increase in the developing world where economic growth and the hope of prosperity allow for it. According to the World Commission on Dams, "the services produced by dams are considerable—in the order of 12-16% of world food production, 19% of world electricity supply among others." Many existing dams were constructed before 1970. Since that time, their susceptibility to variations in flow and their potential environmental impacts have become better known. In addition, other problems have been identified. For example, the World

Commission on Dams found significant problems associated with their construction and maintenance, such as underperformance with respect to targets, loss of capacity due to sedimentation, increased risk of failure due to aging of structures, and increased flood damages. In the U.S.A. alone, it is estimated that an initial five-year investment of $2 billion would be needed "to begin to rehabilitate all the documented unsafe dams in the United States." [*ASCE*, 2001] In addition, retrofitting dams to new standards and expectations could be very expensive.

The value of large-scale water development projects including water conveyance systems for large population centers and large irrigation systems also are undergoing reassessments. Generally, these projects have contributed substantially to economic development and wellbeing. Greater Los Angeles, for example, depends heavily on such developed water infrastructure. For developing countries, large water projects still are attractive because near-term gains carry greater weight than long-term costs and risks. For example, Libya is justifying mining groundwater from an aquifer beneath the desert and transporting it through pipelines to coastal population centers based on the argument that "[s]ustainability can be assured if use of the water" builds the national capacity "to develop alternative supplies" [Saad al-Ghariani, quoted by de *Villiers*, 1999]. Recently, however, "the high financial and environmental costs associated with these projects have led to consideration of 'nonstructural alternatives', including conservation and recycling." [*Frederick*, 1995] It should be noted, however, that these non-structural solutions also may have high costs, both in terms of the development of the technologies and in economic and institutional restructuring needed to make the strategies effective. Both structural and non-structural solutions to water resource availability will rely on increased scientific understanding and technological innovation to make them more effective and affordable.

Changing water development attitudes is difficult, however, because the *status quo* is supported by a complex set of laws, regulations, agreements and perceptions. Guidance from the science community is needed now to identify where changes could improve water management. For example many of the smaller dams in the United States were built 50 or more years ago and their performance is impaired by aging or their functions are now less needed. Sediments that accumulated over the years have reduced storage capacities to near zero in some reservoirs. Removing the dams would release the sediments downstream, which could be extremely destructive. In addition, these sediments frequently contain toxic substances and other contaminants [*Baish et al.*, 2001]. Management is faced with the option of removing the dams or dredging the sediment, which both carry risks of releasing contaminants into the river and the environment. Scientific studies are needed to determine the criteria for choosing between these and other possible options.

Scientific research also is needed to understand the water needs of ecosystems so as to be able to set times and places of withdrawals and returns to minimize adverse environmental impact. A new infrastructure is needed that supports multiple, sequential water uses and only releases clean effluent to the environment; science is needed to establish principles of design for such a system. Such a system will need to be considered holistically, including appropriate extraction and recharge of groundwater, therefore greater scientific understanding of the dynamics of subsurface water also will be needed.

PROBLEMS: SYMPTOMS AND CAUSES

This section reviews the status of some major global water problems and identifies the trends that may exacerbate these conditions in the future.

Water and Health

Water-related diseases continue to be a major global health problem. Water-related diseases include those 1) contracted by ingesting or coming in contact with water contaminated by pathogens or their hosts; 2) that could be prevented by washing hands and food; and 3) spread by vectors that require water during their early stages of development. The first two categories includes most of the enteric and diarrheal diseases caused by bacteria, parasites, and viruses, such as cholera, giardia, and rotaviruses. Diseases like malaria are spread through insect vectors (e.g., mosquitoes). Other diseases carried by water related insect vectors include dengue fever, yellow fever, and West Nile virus.

Water management practices can have a significant impact on water-related disease. For example, construction of large dams has created conditions favorable for the snail hosts of schistosomiasis and led to major outbreaks of the disease. In the Sudan, Egypt, and West Africa, for example, construction of major dams led to the infection of nearly the entire nearby populations [*Gleick*, 1998]. On the other hand, experience has shown that the development of water supply and sanitation systems reduces the incidence of water borne and water washed diseases such as cholera. The statement of goals announced by the recent World Summit on Sustainable Development (WSSD) in Johannesburg includes a commitment by the ratifying nations to reduce the threat of water borne diseases through extension of water supply and sanitation services.

Efforts have been undertaken in the past to relieve water supply and sanitation problems. In 1980, the UN General Assembly proclaimed the period 1981 to 1990 the International Drinking Water Supply and Sanitation Decade. Although millions of people received these basic amenities for the first time as a result of these initiatives, according to WHO [2001], 17 percent of the world population still lacked access to safe water supply and 40 percent lacked adequate sanitation in 2000. Most of these people live in the developing world, where the systems that do exist are in need of repair and maintenance. More research is needed to provide the basis for actions capable of keeping these problems in check.

Exposure to chemical contaminants in water also is a public health concern worldwide because of the many associated chronic and acute health effects, including cancer, neurological and cardiovascular disease and reproductive problems, as well as acute toxicity [*Nash*, 1993]. Common sources of contamination include industrial effluent and runoff from farm fields and urban streets. In addition, water systems can create problems by transforming, concentrating and providing access to naturally occurring contaminants. Examples include trihalomethane production from reactions of chlorine disinfectants with organic matter and tube wells drawing water from formations rich in elemental arsenic.

Water and Environmental Quality

There are several factors that have contributed to water quality degradation and other development impacts on ecosystems. Population growth in developing countries and economic and social pressures on rural life leads to rural-to-urban migration and the rapid growth of "megacities" with their attendant water problems. In many cities the large influx of rural poor has overwhelmed the capacity of systems to cope. Many urban planners do not have the tools or information to properly assess the environmental consequences of their development plans. Irresponsible industries that operate in countries without environmental regulations or adequate enforcement produce widespread water contamination. The institutional and economic capacity to control these activities often is lacking.

Water pollution is a frequent by-product of agriculture and industrial development. Inadequately treated domestic sewage, inadequate controls on water use, ill-considered siting of industrial plants, deforestation, and poor agricultural practices often create major water quality problems for ecosystems. The majority of marine pollution is caused by human activities on land because rivers transport this pollution to coastal and marine ecosystems. River transport of nutrients to coastal waters has caused the formation and growth of hypoxia zones in the Gulf of Mexico with profound effects on aquatic ecosystems and fish populations. Atmospheric transport is also important for the deposition of many nitrogen and sulfur compounds and organic contaminants in aquatic ecosystems.

Sustainable development and management of global fresh water reserves requires recognition of the inter-connectedness of the components of the global water cycle. Aquatic and terrestrial ecosystems are affected by development of all kinds. Consequences of development that can be avoided through careful planning and implementation include pollution, erosion, sedimentation, deforestation and desertification. Sensitive development can minimize and mitigate many negative environmental effects when ecosystem health is considered in the planning. Ecosystem needs are not fully represented in many development decisions because impacts of development alternatives on ecosystems are not known in the quantitative terms needed for planning. In addition, ecosystem values, such as species diversity, cannot be accurately quantified in economic terms. Ecological benefits arising from ensuring adequate water levels, such as maintaining wetland habitat for waterfowl, are difficult to articulate in a meaningful way in decisions that are driven by engineering considerations or the need to act quickly in a crisis mode.

Science should provide a basis for careful assessment of decision options and their likely environmental consequences. The results of scientific inquiry must be communicated effectively to those who make decisions, along with explanations of the implications, uncertainties and limitations of the research. Some progress has been made in using indices of environmental quality to communicate environmental needs for water. However, most indices do not tell the whole story, and they must be supplemented by a more comprehensive set of indicators that signal the health of the ecosystem as a whole. These indicators should include physical, chemical, biological, radiological, and ecological parameters of the water and of the material it carries. [*Young et al.*, 1994]

Irrigation and Food Security

The amount of water withdrawn for irrigation has increased since 1960 by more than 60 percent in response to the world's demand for affordable food, primarily in developing countries [WMO, 1997]. Irrigation increases the amount of food produced per unit area of land because it allows farmers to regulate the timing and amount of water applied to crops. However, irrigation development is neither cost neutral nor environmentally benign. Some of the physical effects of irrigation that reduce its attractiveness are water pollution, water-logging and soil salinization.

During most of the 20th century, the amount of land under irrigation increased at a rate greater than population growth, but during the last two decades the rate of irrigation development has slowed. Since 1978, irrigated area *per capita* has actually decreased by 5 percent [*Tilman et al.*, 2002]. Reasons suggested for this change include increased irrigation efficiency and increased costs of irrigation development. In addition, most of the land suitable for irrigation has already been developed and irrigable land is being taken out of production due to soil degradation. About 20 percent of the 250 million hectares of irrigated land in the world is affected by salt to the extent that crop production is reduced significantly [WMO 1997]. In addition, other irrigated lands are lost each year to urban development.

As a result of this slow-down in the growth of irrigation, along with other factors, annual increases in yields from agriculture are slowing [*Kendall and Pimentel*, 1994]. However, the demand for food continues to increase at about 1.3 percent per year as a result of population growth and rising incomes [*WWC*, 2000]. Some traditional water development will be needed to meet this increasing demand for food; however, aggressive development of irrigation systems seems unlikely. Some irrigation system development is occurring through groundwater development for supplemental irrigation during drought periods in humid climates, but only about a 10 percent increase in water for irrigation is expected in the next 30 years from the construction of new dams [*Tilman, et al.*, 2002]. For food production to keep pace with population growth, innovations in water harvesting, water reuse, and water use efficiency also will be needed.

If water is scarce and/or water development is not an option for a country, food will have to be imported [*WWC*, 2000]. The water represented by imported food can be significant. For example it is estimated it would require 43.4 km^3 per year of water to provide sufficient irrigation to grow crops in Southeastern Asian equivalent to the amount of food that is imported from North America. In total it would take 255.53 km^3 of water to produce the crops exported from North America to the rest of the world. [*Oki*, 2003]. This line of reasoning suggests that agreements could be reached whereby water rich countries would be encouraged to become food providers. Means will have to be devised to pay for food imports and issues of national food security and self-sufficiency will have to be addressed.

Scientific understanding can provide a basis for better strategies to use available water supplies to meet future increasing requirements for food. Improving irrigation efficiency by traditional methods such as lining canals will produce some water savings. Additional water savings can be achieved by changing the crop mix and cropping schedule. However, greater water savings will likely require more technologically sophisticated approaches and a stronger commitment to their implementation [*Seckler et al.*, 1998].

Climate Change

Climate change is expected to affect water availability. Warmer temperatures are expected to result in higher evapotranspiration rates, changes in the amount, distribution, and variability of precipitation, and sea level rise, with attendant impacts on the availability of freshwater [*IPCC*, 2001]. Compared to the effects of other global factors, such as population growth and land use changes, impacts associated with gradual global warming may seem small, but their cumulative effects over time may be significant. In certain situations, small temperature changes can exacerbate conditions in overstressed areas. In addition, climate scientists have speculated that some changes, such as changes in the thermohaline circulation could be abrupt, surprising and large [*NRC*, 2002].

Current climate models project that temperatures globally will rise by 1.4° to 5.8°C over the next century [*IPCC*, 2001]. Global climate models are less consistent in projecting the direction and magnitude of changes in precipitation and soil moisture. Uncertainties increase as one moves from the global to the regional scale, especially when projecting socio-economic conditions.

Regional impacts will be highly variable, with some areas experiencing precipitation decreases while others experience increases. Changes in precipitation regimes could affect the timing and magnitude of floods and droughts and alter groundwater recharge characteristics. Changes in atmospheric stability also may lead to more variability in the intensity, severity, and number of major storms. It follows that runoff events on smaller watersheds may become more "flashy" and flood frequencies may also be affected. In addition, saltwater intrusions into coastal aquifers are projected to increase as a result of rising sea levels. Furthermore, because the components of the water cycle are interdependent, changes in temperature will affect evapotranspiration rates, cloud characteristics and soil moisture, leading to changes in vegetation patterns and in moist-process feedbacks to the atmosphere. In turn, changes in vegetation patterns will lead to changes in rates of evapotranspiration and thus, reservoir levels, river flow and groundwater recharge.

At the present time, estimates of climate change impacts on water resources are too uncertain to provide a basis for recommendations for specific irreversible infrastructure changes. Projections of impacts are very sensitive to the climate model used and the water demand scenario. For some variables and areas, predictions have greater consistency than for other. For example, Nash and Gleick [1993] concluded that climate change is likely to have the greatest adverse impact on countries around the Mediterranean, in the Middle East and southern Africa, areas that are among the least able to cope with additional water problems [*Gleick,* 1998].

Climate change impacts on water resources can be considered from several perspectives. Adaptive responses to changes in water resources could lead to more efficient designs of systems for water management and reuse. Also, they could provide increased flexibility in selecting alternatives for protecting environmental values where local environments are changing. On the other hand, many in the water resource community fear that the current climate change projections for runoff and soil moisture are not sufficiently reliable at the regional scale to justify action. Consequently some water managers believe they will be able to "muddle through" using the same strategies they use now to deal with climate variability. However, the design of reservoirs and associated infrastructure often is based on 30- or 50- year average flows and the assumption of a steady state climate. Infrastructure should be designed to account for possible changes in the average timing, number, duration and intensity of precipitation events. History has shown that dams can be overdesigned if steady state climate assumptions are applied in areas experiencing low frequency variations or change in the precipitation and runoff patterns. Water management may have to deal with changes in streamflow and water availability that render current practices and infrastructure ineffective or inadequate. An approach that uses the results of climate model projections and scenarios to develop policy options could make systems more robust to climate and other types of change.

WATER MANAGEMENT CHALLENGES

Physical Controls on Water Supply

Water availability is controlled by a complex web of both natural and anthropogenic factors. As noted in the previous section, anthropogenic factors with significant impacts on water availability include land-use changes, water system engineering, climate change, population growth and economic development. Land use changes lead to changes in the quantity and quality of runoff. Water system engineering can change rates of flow, infiltration and evaporation and is often used to transport water from its natural basin to another basin. Over-exploitation of groundwater resources dries rivers and wetlands and causes falling water tables and land subsidence. In addition, demands for water increase with growing populations and economic development, change with social values and preferences, and increase or decrease with technological innovations. Although these factors have both positive and negative influences on the supply and demand for water, their net long-term effect has been to increase demand and decrease supply. Climate change with its attendant changes in the amount and timing of precipitation, runoff and groundwater recharge will exacerbate these stresses.

The interactions of water management practices and the environment also can aggravate the environmental stresses that affect water management. Human engineering of water systems produces substantial benefits for water supply and water use but also has significant environmental impacts. For example, the St. Lawrence Seaway has been an outstanding economic success in opening up the central part of North America to commerce, but the construction of the Seaway dislocated many people, provided an outlet for pollution from industries on the shores of the Lakes and estuaries along the St. Lawrence River, and enabled the migration of exotic species such as Sea Lamprey into the Great Lakes where they destroyed a significant component of the commercial fisheries.

The Effects of Changing Demographics on Water Demand

Population growth is the most important long-term factor affecting water supplies. In general, water demand increases most rapidly in those areas where population increases most rapidly. The precise impact is difficult to project because estimates of both population growth and future water use *per capita* are uncertain. The world population was estimated at 6.1 billion people in 2000 [*US Census Bureau*, 2002]. Projections of population growth range from 7.3 billion to more than 10 billion people by 2050 [*WRI*, 2001; *Gleick*, 2000]. The increased population will place additional stresses on water resources, as the additional people demand goods and services that rely on water. The regional distribution of this population growth is a central issue because much of the growth will occur in regions where water availability is inadequate and economic prospects are poor.

People need some minimal amount of water for their domestic uses, including drinking, cooking, bathing, and cleaning. Gleick [1998] estimates this "basic human water requirement" at 50 liters per person per day. In addition to domestic uses, water also is needed for use in industry and agriculture. Water withdrawn from streams and reservoirs for all uses divided by population yields *per capita* water withdrawals, a rough measure of water use. In 2000, there were 32 countries, with an aggregate population of about 484 million, which withdrew less than 50 m^3 of water per person per year for use in all sectors [*Gleick*, 2000]. At the other end of the spectrum, 1,688 m^3 of water were withdrawn for each of the 278 million inhabitants of the U.S.A.

Between 1900 and 1995 (when the population tripled) global water withdrawals were estimated to have increased about 650% [*Gleick*, 1998]. This rapid growth in water withdrawals is attributed to the increase in *per capita* water use that goes hand in hand with economic development. Since 1980, however, global water withdrawals have increased more slowly, at about the same rate as population growth. In the developed countries, *per capita* water withdrawals have decreased substantially and consumptive water use has leveled out since 1980. In these countries, conservation and increases in water use efficiency have reduced demands.

Resolving Conflicts over Water

As the number of people requiring water for households, agriculture, and industry increases, the potential for conflict based on real or perceived inequities in access to water also increases. At the watershed scale, people with different interests all wish to influence water allocation decisions in their favor. Conflicts often arise from competition between upstream and downstream water users. Downstream users usually have developed their water uses first and have established patterns of use that may be threatened by subsequent upstream development. On the other hand, upstream users want to control and use the water originating within their jurisdictions. Conflict may arise among people in different areas or water use sectors within countries that lack institutions for resolving disputes. In the U.S.A. water rights in many rivers were fully allocated more than 50 years ago and interested parties have engaged in developing processes for reallocating water among users. Methods are needed that will work in an international context when conflicts arise between countries sharing a river. Many of the countries sharing transboundary rivers will have to explore innovative and possibly expensive options for meeting their competing demands if they are to avoid conflicts.

For example, the U.S. has experienced strained relations with its neighbors over the use of transboundary rivers. Disagreements over upstream-downstream right and responsibilities for quantity and quality of transboundary flows have led to the creation of binational treaties and institutions to oversee compliance. The International Boundary and Water Commission has been active on the U.S. Mexico border since the mid 19th Century, and the International

Joint Commission has facilitated the resolution of cross-border problems on the Canada/U.S. border since 1909. Collaborative planning and management mechanisms like these have been established in other regions of the world, but more are needed. Without them the world will see more upstream-downstream conflicts such as in the former Soviet states of central Asia, where disagreements over the use of water have slowed actions to mitigate the environmental disaster occurring in the Aral Sea.

Development and Industrial Water Use

Industrial uses account for 22 percent of water use worldwide. If current trends continue, industrial water use will have doubled between 1995 and 2025 [WMO, 1997]. Developing nations are expected to account for a major part of the growth in demand for industrial water. The amount of water used *per capita* in the industrial sector generally increases with the level of development to a point at which wealth and technological sophistication lead to high water use efficiency. In many developed countries, increased water use efficiency has caused *per capita* industrial water demand to plateau.

Water is used to produce energy through hydropower, to cool energy plants using fossil or nuclear fuel, and in many manufacturing processes. Industry uses (consumes) some water taken from watercourses by incorporating it into products, but it returns most water for subsequent use. However, this returned water is frequently altered, either thermally or chemically or both, in ways that degrade its further usefulness. For example, power plant cooling water returned to a river may be too warm to support coldwater fish species, or tannery process water may make a river too polluted for drinking or irrigation. Thus, industry tends to "force" the system in two ways, through consumption, and through pollution loading.

Pollution will have a growing impact on the availability of usable water. In the developed world, government regulation and technological innovations have reduced pollution loads. In the developing world, industrialization has led to pollution loads that are increasing more rapidly than population or GNP. The pollution load in these countries has increased 5-10 times during the same period that population has doubled. [*Fallkenmark*, 2000] A water system based at least in part on water reuse technologies would provide the dual benefit of reducing pollution loads and increasing water supplies.

Water Infrastructure Failure in Urban Environments

The unplanned movement of people from rural areas to cities continues to place major stress on aging or otherwise inadequate infrastructure in many cities.

In 1995, 37 percent of the world's people lived in urban areas. By 2025 the urban population is expected to reach 56 percent of total population [WMO, 1997]. Some large cities are experiencing explosive growth, and the concentration of large numbers of poor people is overwhelming urban infrastructure. In most cases, water for these people will have to be provided by reducing the amount available for irrigated agriculture. At the same time rural agriculture in these developing countries will have to progress beyond subsistence farming to feed the growing urban population.

It is important for urban planners to face water and waste disposal issues as early as possible during urban development. Without adequate urban planning, untreated sewerage can contaminate water distribution systems; development on floodplains can increase the risk of flooding; and stresses on existing water supplies can lead to over-exploitation of groundwater reserves at the same time paved surfaces are reducing groundwater recharge. Large concentrations of wastes from urban centers are a major source of pollution. Already, pollution from human waste is degrading water resources in many large cities, including Sao Paolo, Delhi, and Mexico City [*Falkenmark,* 2000]. In some cities, where groundwater pumping is causing ground subsidence, sewer pipes have broken and water losses are occurring due to leakage. In the U.S., the aging water infrastructure of large cities is becoming a major problem. In cities like New York, leaky infrastructure is causing significant water losses. At the same time climate variability and change are leading to changes in water availability. A substantial investment in infrastructure designed to account for the effects of climate variability and change, on top of underlying stresses, will be required to provide urban populations with adequate quantities of good quality water.

MANAGING THE WATER DILEMMA

Decision makers at all levels of government, along with the private sector and all water users should contribute to avoiding the global water dilemma. As outlined below, this will require participatory water management and flexible approaches to managing for sustainability.

Water Management Strategies for the Future

Water resource management is a process whose goal is to match supply and demand on a continuing basis. Risks are generally related to oversupply, such as flooding, and shortfalls, such as droughts. The decisions to open a valve, install a meter, or raise the height of water in a reservoir may be made by professional water managers, but the laws, regulations, expectations, and priorities that constrain the decision are most often beyond their control. In some situations, market opportunities and economic priorities determine how water is used; in others a political process governs the planning and implementation of water management, either through a centralized, top-down or a decentralized, bottom-up approach.

Many people believe that the free market is the best mechanism for determining the most efficient use of the water resource. The operation of a free market generally requires a system of private ownership, which tends to shift the balance between public and private sector influences on water management. For many other people, loss of public control over water is associated with the sacrifice of public values to private gain. Water markets have been tested on a limited basis in the western U.S.A., where they may be seen as a way to get around inflexible legal and institutional constraints; but states are not inclined to give up public control of water allocation. For example, entrepreneurs in Texas and California are developing plans to pump groundwater to sell to large urban communities, but both schemes are likely to encounter litigation before progressing beyond the planning phase.

The ideological support for water markets in general is distinct from arguments for privatizing water supply and sanitation systems. It is argued that private companies responding to market incentives will do a better job of expanding systems to meet growing needs, and of maintaining and operating systems more efficiently than public agencies. On the other hand, there are few market incentives for serving the poorest communities or ensuring water for ecosystems. Privatization of supply can lead to neglect of underrepresented and underserved communities, worsening of economic inequalities and lack of community participation, monitoring and control over local resources [*Gleick et al.*, 2002]. Furthermore, with less public involvement the risk would increase for less sensitivity to sustainability issues. However, there are many different ways that private business can be involved in delivery of water services, ranging from ownership of the resource to contracting the operation of specific components of a water system. Some of these forms are more prone than others to the problems associated with the commodification of water. Proponents maintain that privatization can be designed to alleviate potential adverse consequences. Experience suggests that successful water supply and sanitation management, whether public, private or hybrid, depends on specific local conditions.

Local control of local resources, small-scale development, and more involvement of all stakeholders in decisions are alternate approaches for improving water resource management. Implementing these recommendations in a fully informed way will require major efforts in capacity building and institutional design. In many instances legal and institutional frameworks may have to be created or changed. The European Union's (EU) has introduced a new paradigm for managing water, with its Integrated Water Management Directive and its requirements for management plans at the watershed scale. The EU has asked member nations to introduce basin-specific management strategies as early as 2003, implement programs by 2006, and achieve environmental objectives by 2015. A major challenge in this approach will be gaining commitment from the diverse national and subnational governance structures within the boundaries of each basin to achieve these objectives.

Managing for Sustainability

Managing for sustainability is a relatively new concept in professional water resources management, although precedents exist. As noted earlier two paradigms influence water management: a development paradigm whose goal is to put water to work for humans and another that advocates water for the benefit of all including ecosystems and the physical environment. Sustainability comprehends both the economic aspirations of the development paradigm and the attention to broader, longer-term issues of the alternative paradigm.

A coordinated effort to bring the development paradigm together with environmental concerns was initiated in the 1980's under the leadership of Norway's then Prime Minister, Gro Brundtland. The initiative introduced the concept of 'sustainable development' to an international audience and gave it its widely accepted definition as: "a form of development that meets the needs of the present without compromising the ability of future generations to meet their own needs." [WCED, 1987] The application of this concept to water resources was immediately apparent. Sustainable water management involves simultaneously meeting the needs of human society for water and maintaining the the integrity of the hydrological cycle and the ecological systems that depend on it. Because of the interdependencies of the Earth system, sustainability requires the integration of global issues with local issues and local actions with global consequences. Water related problems usually manifest themselves on the local level, but often require regional, national or even international solutions. The water cycle functions regardless of national borders, although what happens to rain after it reaches the surface often depends on national infrastructure, industrial development, and environmental policies.

Integrated Water Resource Management (IWRM) [*van den Heuval and Willemse*, 2001] was developed to implement the goals of sustainability. It incorporates scientific research into management practices; coordinates water, land use and related resource management; and encourages cooperation among separate jurisdictions within a single basin. The comprehensive sustainability approach integrates technical, social, economic and environmental issues and considers these issues at regional, national, and international levels. In addition, it incorporates the values of participatory democracy, and therefore includes capacity-building, participation, transparency, and institutional development among its objectives.

DEVELOPING SCIENTIFIC UNDERPINNINGS FOR SUSTAINABLE WATER DEVELOPMENT

The need to support sustainability on a global basis has placed demands on the science community for an improved understanding of the global water cycle. In particular there is a need to understand the connections between the various components of the water cycle; to develop a capability to make predictions of future precipitation and runoff patterns, and to understand the linkages between the water cycle and nutrient and carbon cycling, ecosystems and human activities [*Hornberger et al.*, 2001; *NRC*, 2001]. Improvements in the capability to predict critical water cycle variables (e.g., soil moisture and streamflow) will require the development and implementation of advanced prediction and observational systems. Because water cycle science is multidisciplinary (involving atmospheric sciences, hydrology and water chemistry among others) and covers a range of timescales (including weather, seasonal to interannual and decadal timescales), more interdisciplinary projects are needed to provide the comprehensive knowledge base needed to address these issues.

Global Water Cycle Science

A growing recognition of the potential for a global water crisis has motivated many individual countries and international organizations to support scientific research on key water science issues. Lawford *et al.* (this volume) describe the contributions of many of these international initiatives in addressing water issues. The World Climate Research Program (WCRP), with its focus on the prediction of precipitation, is addressing questions related to evaporation from the oceans, the movement of water through the atmosphere, the formation of precipitation and land-atmosphere interactions. Research related to El Nino events has contributed to prediction of seasonal precipitation patterns in some regions. The World Meteorological Organization (WMO) has supported research under its World Weather Research Program (WWRP) that contributes to rainfall prediction at daily to weekly timescales. However, to understand the partitioning of rain into soil moisture and runoff requires a different set of models and data sets, including the terrestrial hydrology of the global water cycle which, until recently, had received less integrated international attention. Within the Integrated Global Observing Strategy Partnership, efforts are underway that will facilitate the construction a global data set of hydro-meteorological information combining continental and regional scale process studies with improved observational data sets. Agencies like NASA and NOAA are using satellites in combination with sophisticated Land Data Assimilation Systems (LDAS) to provide guidance and advice to water managers in the southwestern U.S.A.

To achieve the developments listed above, there are a number of advances needed in the basic hydrological sciences. In many cases the knowledge does not exist to confidently apply the information or technologies that are available. For example, while we can measure radiances from space, the interpretation of these measurements in terms of conventional water cycle variables is not easy. A lack of understanding of subsurface hydrology limits us in using our knowledge of remotely sensed surface soil moisture, antecedent precipitation conditions and soil textures to monitor deep layer soil moisture and recharge. The absence of predictability studies reduces our ability to take full advantage of those areas and periods when predictive skill exists for the seasonal timescale. As a result we fail to take full advantage of our knowledge.

As elaborated by Endreny *et al.* (this volume) the lack of basic science also limits our ability to design safe natural watershed collection systems. A lack of knowledge of the diffusive properties of pathogens prevents us from determining if groundwater will be contaminated from the release of these substances to the environment. The absence of long-term epidemiological studies of health effects prevents us from putting in place early warning systems to detect critical levels of industrial pollution before communities are put at risk. To address these needs, a strong water resources research agenda must be developed. The NRC [2001] has made some excellent progress in developing such a plan for the U.S.A.

In addition, scientific progress is needed on water chemistry issues. Water is Earth's most "universal" solvent. Consequently, water has been relied upon to dissolve and transport all types of waste, which has led to serious water quality degradation. The various aspects of point and diffuse source pollution that affect water quality either directly or indirectly should be documented, including types, rates, sources, and distributions. The Global Environmental Monitoring System (GEMS) is the best source of water quality information currently available. However, GEMS is limited by the adequacy of the data collection programs in individual countries and their willingness to share data. Enhanced methodologies and observations are required to assess the quality of water in different regions of the world during different seasons and over different timescales. Research is needed to develop new measurement techniques that will provide more comprehensive assessments and to bring together water quantity and quality assessment capabilities. More priority needs to be given to developing runoff models with water quality components, plume models for groundwater, and statistical estimation techniques.

A global science system must also address water uses. There is a need to assess the impacts of water use patterns and water control infrastructure on water supply, distribution and water quality. Important topics for study include the potential for conservation technologies, reuse, efficiency improvements and demand management. In-stream and ecosystem requirements for fresh water also should be studied. For example, despite the negative impacts of floods and droughts, their cycles of natural variability are vital to many ecosystems. The global water cycle supports a complex environmental system on land and in the oceans, and many of its components and processes are inadequately understood. Integration is needed to make use of multiple scientific perspectives and to address complex interdependencies among the natural, physical and socioeconomic realms.

Water Supply Assessments

Within the last decade there has been growing pressure from national governments and international organizations to determine how much water will be available for future generations and to identify where shortages will produce the greatest hardship. As a result of the UN concerns in this area, a Comprehensive Assessment of the Freshwater Resources of the World [*WMO*, 1997] was undertaken to provide provided estimates of the total renewable water of each region. Many data collection and data integration problems were encountered, including a fundamental lack of observations in some areas. This has led to a commitment on the part of the World Water Assessment Program to an iterative program that develops and refines the Assessment with each iteration.

Indicators of water stress are frequently used to characterize the extent of global water problems and to provide early warnings of where new problems may emerge. These techniques are very useful in communicating information to policy makers, although there are significant challenges in trying to represent such complex processes by a single number [*Arnell*, 1999]. The 1997 Comprehensive Assessment [*WMO*, 1997] used the ratio of water withdrawn for use to the Annual Water Resource (AWR), on the theory that the higher the percentage of water withdrawn to AWR, the smaller the buffer a country has against water stress. (The AWR is the average annual amount of renewable fresh water within a country or region.) By this categorization, 460 million people, more than 8 percent of total world population, live in countries with highly stressed water supplies (ratio>40%). In addition, more than one-third of the people in the world's poorest countries already face medium-high to high water stress (ratio between 20 and 40%). Many of these countries are located in arid and semi-arid regions of Africa and Asia.

Another index commonly used internationally is *per capita* AWR. Shortages are local and rare if this index is 1,700 m^3 or greater. However, if it is below 1,000 m^3, lack of water begins to impair quality of life, and below 500 m^3, "water availability is a primary constraint to life" [*Falkenmark and Lindh,* 1976]. According to these indexes, countries will migrate to higher stress levels when rapid economic growth and/or population growth are taking place under conditions of limited supply.

Both of these indicators attempt to integrate the factors affecting water supply adequacy into a single measure, but both have serious deficiencies. A primary challenge in developing indicators involves accurately characterizing water availability. In many nations, the infrastructure for acquiring the necessary data on water demand is absent. In addition, most groundwater is excluded from the calculation of AWR, although a significant percentage of global water use is supplied by groundwater; and the AWR is based on average annual runoff and consequently does not account for seasonal and interannual variability. Where runoff is stable and/or water storage is substantial, more of the resource can be used than where flows are highly variable and/or storage is lacking. Yet both areas may produce the same amount of runoff and therefore have the same AWR [*Seckler et al.*, 1998; *Gleick et al.*, 2002].

Sociopolitical and economic conditions also are significant factors in water stress assessments. Wealthier regions have greater ability to cope with water supply limitations than poorer countries. In addition, political turmoil and social instability increase vulnerability to water supply problems. New indicators that account for these factors are being used in an attempt to map global water stress and predict where pressure on water resources is most likely to be an issue [*Gleick et al.*, 2002]. However, to improve the overall credibility and utility of these assessments, detailed evaluations at appropriate scales and the development of more sophisticated models and data sets to support these global analyses are also needed.

Water Technology

Technology provides options for enhancing water supplies, although exploitation of these technologies should be planned within the context of the sustainability paradigm. In addition to infrastructure for storing water, technologies and strategies are needed for shifting water between uses and enhancing the recycling of water. This may mean that fresh water is routed through a hierarchy of uses where uses with the lowest quality requirements are at the end of the chain.

Technology has frequently been used to decrease variability of supplies or redistribute supplies from areas with surpluses to areas with deficits without a full evaluation of the consequences. In the past, such solutions were applied before there was an adequate knowledge base to identify and quantify many other "unintended" consequences. For example, the use of untreated sewage water for irrigation applications can lead to the inadvertent transmission of pathogens in the foods grown from that water, On the other hand, these developments contributed significantly to the economic growth of their regions. In particular, technological innovations such as water conserving industrial processes are credited with reducing industrial water demand and reducing pollution loads in developed countries.

Promising technological developments for enhancing water supplies include desalination and water recycling. Desalination is becoming economically viable. At present, desalination has been applied to brackish water for industrial applications and is now being used to meet urban water requirements in a few cities. Conservation technologies are also an effective way of making more water available for more users. More efficient irrigation systems that deliver water to the plant roots or nearer to the plant canopy so that the evaporation losses are reduced have resulted in significant reductions in demands for irrigation water. In urban settings many techniques have been identified to enable more efficient water use including "low" technology solutions, such as watering lawns and golf courses at night when evaporative demand is a minimum, low-flow toilets, etc. Recently, at the World Water Forum III in Kyoto, results were presented that showed the value of applying science to problems at all scales. On the small scale, new techniques for the home treatment of polluted water were demonstrated. These techniques allow people in the developing world to have access to clean drinking water for only a few pennies per day.

New technologies such as satellite data combined with hydrometeorological models can be used to predict water supply and availability, thus allowing water managers to make better resource decisions. In addition, satellites are making it possible to monitor changes in the water available regionally for the entire globe. The recent GRACE mission provides measurements of gravitational anomalies (admittedly at rather coarse resolution) that will enable us to estimate the water in the ground (both on and below the surface). When combined with other information it should allow for improved estimates of changes in groundwater amounts. Passive radiometric measurements have been used to detect moisture in the atmosphere and in the upper layers of the soil. The Tropical Rainfall Measurement Mission (TRMM) has been very successful in providing three-dimensional information on precipitation distribution. There are now plans to expand this system so this type of information can be acquired for the entire earth every 3 hours (or less). These innovations combined with improved distributed hydrological models and models capable of producing reliable seasonal to inter-annual predictions will greatly improve the information that will be available for local water managers.

Techniques are being developed for prospecting for groundwater remotely by aircraft. Tools for mapping sources of groundwater are being developed, along with means for extracting water more effectively from the ground. Also, technologies are available to monitor water use and the success of conservation programs.

Communications Between Science and Policy

Ultimately the responsibility for decisions lies with governments, the private sector, and above all, people. A great deal of knowledge exists that can facilitate better decisions and avert a potential global water crisis. Science should be used to ensure that options are realistic and that all likely consequences are considered. While good scientific information is essential to the decision making process, science alone cannot guarantee good decisions. It must be communicated effectively to a broad public, and those who make decisions must understand its content, implications, uncertainties and limitations. There is a need to make more effective use of information science to develop techniques and incentives that increase understanding and use of scientific information. The concerns and constraints perceived by the public are essential inputs to focusing scientific work on important issues. (Jacobs and Pulwarty (this volume) demonstrate the consequences of implementing projects when engineers, scientists and policy makers are not communicating effectively their needs and knowledge.) Furthermore, there must be a clear procedure for feedbacks and interactive learning between scientists and science users. Substantial efforts are underway to incorporate science into decision support systems, both in the design of these systems and in the flow and interpretation of data for these systems.

SUMMARY AND RECOMMENDATIONS

The world is facing a growing dilemma over water: a finite resource with multiple uses, a growing population and a system of increasingly complex interdependencies. The developing world is seeking increasing amounts of water to realize its economic growth potential and to enjoy the fruits of its economic growth. The goal of the international community to provide access to the millions of people living without adequate water supply and sanitation adds an additional dimension to the challenge. Pressures are growing for traditional supply augmentation schemes, such as groundwater mining and interbasin transfers, even if they are not sustainable or involve significant uncertainties. To address these developments the world needs new technologies to help it to use its water more efficiently. Where these new technologies are not available now, society needs temporary measures that can be applied now. Development of sustainable water management strategies, based on the latest scientific understanding, is essential if the world is to resolve its water dilemma.

Although there will be many issues that divert public attention from water, the following fundamental issues will need to be resolved in order to attain water sustainability in the future. Sustainable water management in the future will require new integrated strategies that take the complex interactions of the global water system and the physical environment into account. These strategies may require significant changes in accepted practices and established institutions. Any such change must be supported by scientific information that is both comprehensive in scope and focused on solving real problems. To meet this challenge, science will rely on greater cooperation among disciplines and greater integration of scientific observations, models and knowledge. In addition, scientists, policy and management decision makers and the public will have to become collaborators in projects over a range of scales.

As first steps toward implementing this vision, the authors recommend that attention be placed on the following areas:

1. Research

a) A concerted effort is needed to improving the quality of predictions available to water managers. The effort will need to include process studies, better observational and assimilation systems and improved global and regional prediction models. The focus of this effort should be improved seasonal predictions of water cycle variability at all scales.

b) Better integration of observation campaigns with field studies, across disciplines is needed, along with institutional as well as scientific innovations in managing data from multiple sources for greater access.

c) A systematic research program is needed to understand and address water and health issues at regional and global scales.

2. Assessments

a) An assessment framework involving individual nations and the United Nations is needed to assess the status of the world's water resources and the actions being taken nationally to conserve water.

b) The consequences for water resources arising from climate variability and change should be examined for different water basins in different climate regimes.

3. New technologies

a) As one of the largest users of water, new effective methods of irrigation should be developed and alternatives to irrigation should be sought to meet the growing demand for food.

b) New technologies must be developed that are affordable, easily emplaced and maintained, and ecologically benign to rapidly increase access to safe freshwater supplies and sanitation.

4. Education and Capacity Building

a) Capacity building is fundamental; from the watershed to the regional level there is a need for development of the skill base and institutional infrastructure for integrated water resource management.

b) Through the efforts of major Space Agencies and international organizations, an initiative should be developed whereby these agencies would organize the transfer to countries in the developing world of the technology and training needed to exploit satellite data and related products in water and land management.

Most of these initiatives would involve both science and policy and a practical program for implementation, bringing them together to address the issues effectively. The following chapters look at the needs and opportunities from both the science and policy perspectives and review past experiences in applying them to existing problems.

REFERENCES

ASCE, Statement of the AMERICAN SOCIETY OF CIVIL ENGINEERS on U.S. Water Infrastructure Needs Before the Subcommittee on Water Resources and Environment,

Committee on Transportation and Infrastructure, U.S. House of Representatives, March 28, 2001.

Arnell, Nigel, Climate change and global water resources, *Global Environmental Change*, 9: S31-S49, 1999.

Baish, Sarah K., Sheila D. David and William L. Graf, The complex decisionmaking process for removing dams, *Environment*, 44/4: 21-25, 2001.

De Villiers, *Water, the fate of our most precious resource*, Mariner Books, New Houghton Mifflin Co., New York, 1999.

Falkenmark, Malin, in cooperation with the Stockholm Water Symposium Scientific Program Committee, No Freshwater Security Without Major Shift in Thinking, ten-year message from the Stockholm Water Symposium, Stockholm International Water Institute, 2000.

Falkenmark, Malin, Human dimensions of freshwater: time to overcome water blindness and bridge the science/policy gap, *IHDP Update*, Newsletter of the International Human Dimensions Programme on Global Environmental Change, 1/01, http://www.uni-bonn.de/ihdp/IHDPUpdate0101/article1.html, 2002.

Falkenmark, Malin and Gunnar Lindh, *Water for a Starving World.* Westview Press, Boulder, 1976.

FAO (Food and Agriculture Organization), Securing food for a growing world population," in *Water for People*, 2003.

Frederick, Kenneth D., America's Water Supply: status and prospects for the future, *Consequences*, 1/1, Resources for the Future, Washington D.C., 1995.

Gleick, Peter H., *The World's Water 1998-1999, Biennial Report on Freshwater Resources*, Island Press, 1998.

Gleick, Peter H., *The World's Water 2000-2001, Biennial Report on Freshwater Resources*, Island Press, 2000.

Gleick, Peter H., with William C.G. Burns, Elizabeth L. Chalecki, Michael Cohen, Katherine Kao Cushing, Amar S. Mann, Rachel Reyes, Gary H. Wolff, Arlene K. Wong, *The World's Water 2002-2003, Biennial Report on Freshwater Resources*, Island Press, 2002.

Gleick, Peter H., Elizabeth L. Chalecki and Arlene Wong, Measuring Water Well Being: Water Indicators and Indices, chapter 4 in *The World's Water 2002-2003*.

Gleick, Peter H., Gary Wolff, Elizabeth L Chalecki and Rachel Reyes, The New Economy of Water: the risks and benefits of globalization and privatization of fresh water, Pacific Institute for Studies in Development, Environment, and Security, 2002.

Hornberger, G.M., J.D. Aber, J. Bahr, R.C.Bales, K.Bevan, E. Foufoula-Georgiou, G.Katulo, J.L.KinterIII, R.D.Koster, D.P.Lettenmaier, D.McKnight, K. Miller, K.Mitchell, J.O.Roads, B.R.Scanlon and E. Smith, *A Plan for a New Science Initiative on the Global Water Cycle*, U.S. Water Cycle Study Group of the U.S. Global Change Research Program (USGCRP), USGCRP Program Office, Washington, DC, 2001.

IPCC (Intergovernmental Panel on Climate Change), *Climate Change 2001: The Scientific Basis*, Cambridge University Press, 2001.

Kendall, Henry W. and David Pimentel, Constraints on the expansion of the global food supply, *Ambio*, 23/3 (http://dieoff.org/page36.htm). 1994.

Nash, Linda, Water quality and health, chapter 2 in *Water in Crisis, a guide to the world's fresh water resources*, P. Gleick, ed., Pacific Institute and SEI, Oxford University Press, 1993.

NRC (National Research Council) Committee on Abrupt Climate Change, *Inevitable*

Surprises. National Academy Press, Washington DC., 2002.

NRC (National Research Council) Water Science and Technology Board, *Envisioning the Agenda for Water Resources Research in the 21st Century,* National Academy Press, Washington DC., 2001.

Oki, Taikan, M. Sato, A, Kowamura, M. Musiake, S. Kanae and K. Musiake, Virtual water trade to Japan and in the world, Virtual Water Trade: Proceedings of the International Expert Meeting on Virtual Water Trade, A.Y. Hoekstr, ed., Institute for Hydrology and Environment (IHE), Delft, 2003.

Postel, Sandra, *Last Oasis, Facing Water Scarcity*, 2nd Edition, The Worldwatch Environmental Alert Series, L. Starke, Series ed., W.W. Norton & Co., New York, 1997.

Shiklomanov, I., ed., Assessment of water resources and water availability in the world, background document for the Comprehenssive Assessement of Freshwater Resources of the World, UN System/ Swedish Environmental Instutute, Stockholm, 1997.

Seckler, David, Upali Amarasighe, David Molden, Radhika de Silva and Randolph Barker, World Water Demand and Supply, 1990 to 2025: scenarios and issues, Research Report #19, International Water Management Institute, 1998.

Tilman, D., K.G. Cassman, P.A. Matson, R. Naylor and S. Polasky, Agricultural sustainability and intensive production practices, *Nature,* 418: 671-677, 2002.

UN (United Nations) World Water Assessment Program, Water for People, Water for Life, the United Nations World Water Development Report, UNESCO Publishing/Berghahn Books, Paris, 2003.

U.S. Census, Total Midyear Population for the World: 1950-2050, www.census.gov/ipc/; www/worldpop.html, 2002.

van den Heuvel, Marcel, and Edu Willemse, Achieving Water Security: making water everybody's business, *Sustainable Development International, strategies and technologies for Agenda 21 implementation*, 2001.

World Commission on Dams, Dams and Development: A New Framework for Decision-Making, an overview, November 16, 2000, 2001.

World Commission on Sustainable Development, *Our Common Future* (Brundtland Report), Oxford University Press, New York, 1987.

WHO (World Health Organization), Global Water Supply and Sanitation Assessment 2000 Report. 2001.

WHO (World Health Organization), Basic needs and the right to health in *Water for the People*, 2003.

WMO (World Meteorological Organization), Comprehensive Assessment of the Freshwater Resources of the World, Stockholm Environment Institute, 1997.

WRI (World Resource Institute), Pilot Analysis of Global Ecosystems (PAGE): freshwater systems, http://www.wri.org/wr2000/freshwater_page.html, October 2000.

World Water Council, Water for Food and Rural Development, Global Vision Draft, Vision on Water, Life and Environment in the 21st Century, February 2000.

Young, Gordon J., James C. I. Dooge, and John C. Rodda, *Global Water Resource Issues*, Cambridge University Press, 1994.

Susanna Eden, USGCRP/CCSP Global Water Cycle Program, 1717 Pennsylvania Avenue, Suite 250, Washington, D.C. 20006

Richard G. Lawford, NOAA Office of Global Programs, 1100 Wayne Avenue, Suite 1210, Silver Spring, MD 20910

3

Urban Water Issues – An International Perspective

José Alberto Tejada-Guibert and Cedo Maksimovic

Improving freshwater management in the urban environment is currently seen with a sense of urgency. Urban water problems are mounting throughout the world. Widespread mismanagement of water resources, growing competition for the use of freshwater, degraded sources—sometimes by pollutants of unpredictable effects —only heighten the acuteness of the problems. The situation is further exacerbated by an explosive growth of urbanization, particularly in the developing world, which exhibits as its most visible expression the emergence of megacities that often obey a massive internal migration toward the cities, thus feeding an uncontrolled, unplanned expansion. Over half of humanity will live in cities by year 2010.

Cities in industrialized countries face the problems of aging infrastructure causing underperformance of the urban water systems and often exhibit a reluctance to try innovative approaches. In the developing world, a major urban problem often is securing access to water supply and sanitation. These problems are of a complex nature and can only be properly addressed through a concerted effort involving scientific, social and institutional approaches. New paradigms are emerging to counter the urban water problems, reflecting an integrated management of all components, with emphasis on water reuse and water conservation approaches as well as on demand management practices.

In this chapter, a fresh look is taken at traditional water-borne sanitation, with consideration being given to less water-intensive alternatives. As well, we review the situation and the main reports on current management approaches and research efforts.

Global Urban Water Setting

The world is becoming increasingly urbanized. Between 1990 and 2000, the global population increased by 15% (from 5.27 to 6.06 billion), and the urban population grew by 24% (from 2.29 to 2.85 billion), meaning that the percentage

Water: Science, Policy, and Management
Water Resources Monograph 16

10.1029/016WM04

Table 1. The Twelve Largest Urban Agglomerations Ranked By Population (in millions): 1980, 2000 and 2015[a]

Rank	1980		2000		2015	
	City	Population (millions)	City	Population (millions)	City	Population (millions)
1	Tokyo	21.9	Tokyo	26.4	Tokyo	26.4
2	New York	15.6	Mexico City	18.1	Bombay	26.1
3	Mexico City	13.9	Bombay	18.1	Lagos	23.2
4	Sao Paulo	12,5	Sao Paulo	17.8	Dhaka	21.1
5	Shanghai	11.7	New York	16.6	Sao Paulo	20.4
6	Osaka	10.0	Lagos	13.4	Karachi	19.2
7	Buenos Aires	9.9	Los Angeles	13.1	Mexico City	19.2
8	Los Angeles	9.5	Calcutta	12.9	New York	17.4
9	Calcutta	9.0	Shanghai	12.9	Jakarta	17.3
10	Beijing	9.0	Buenos Aires	12.6	Calcutta	17.3
11	Paris	8.9	Dhaka	12.3	Delhi	16.8
12	Rio de Janeiro	8.7	Karachi	11.8	Metro Manila	14.8

[a]Adapted from United Nations, 2001b.

of people living in cities went from 43.5% in 1990 to 47% by 2000 [World Health Organization (WHO) and United Nations Children's Fund (UNICEF), 2000]. In the more developed regions of the world, the urban percentage in 2000 had reached 76%, while in less developed regions it was 39.9%, but growing rapidly [United Nations, 2001a]. The latest projections show that by 2010 over half of all humanity will live in urban centers. The locations of the world's largest megacities appear increasingly in developing countries. Table 1 shows the world's twelve largest cities in 1980, 2000 and makes an estimation for 2015. Tokyo is shown as the largest throughout this period, but while it was comfortably in the lead in 1980, several developing country cities will close the gap by 2015. In 1980, five of the twelve most rapidly growing cities were in industrialized countries. However, by 2015 this total is projected to go down to only two.

The ever-increasing urban population tends to be concentrated in larger cities. The number of cities in the world with over one million inhabitants grew from 234 in 1980, to 411 in 2000, and it is projected that the total will reach 564 in 2015. It is estimated that the percentage of the world's urban population in cities with over one million inhabitants grew from 35.2% in 1980 to 39.3% in 2000 and is expected to increase to 42.6% by 2015 [United Nations, 2001a]. This trend is similar for both developed and developing regions, but is comparatively more pronounced in the latter.

The much larger urban expansion, particularly in developing countries, goes far beyond a numbers game, because, as the United Nations Centre for Human Settlements [2001] states:

"Burdened with all the problems of growth, cities are increasingly subject to dramatic crises, especially in developing countries. Unemployment, environmental degradation, lack of urban services, deterioration of existing infrastructure and lack of access to land, finance and adequate shelter are among the main areas of concern."

An issue of serious concern is the deprived periphery (peri-urban areas or informal settlements) of many developing world cities, where the majority of population growth occurs. A survey by WHO and UNICEF [2000] showed that 27% of the population of the surveyed cities in Africa and 18% of those in Asia lived in these impoverished settlements. There are indications that the actual global figures might be even higher.

Table 2 shows overall figures for urban water supply and sanitation coverage in 1990 and 2000 for the world. Globally, despite an appreciable increase in the total urban population served, the number of people lacking water supply in cities has actually grown from 113 million to 173 million and the number of those lacking sanitation has stayed more or less constant at about 400 million. The percentage of population served in cities has roughly stayed constant throughout the various continents, with the exception of sanitation coverage in Asia. Here the percentage went from 67% in 1990 to 78% in 2000, the number of Asians served having increased more than 50% in 10 years (from 690 to 1055 million) – indeed a remarkable achievement! It should be

Table 2. Urban Water Supply and Sanitation Coverage by Region, 1990 and 2000[a]

Region	1990 Population (million)				2000 Population (million)			
	Total urban	Served	Unserved	% Served	Total urban	Served	Unserved	% Served
GLOBAL	2 292				2 845			
Urban water supply		2 179	113	95		2 672	173	94
Urban sanitation		1 877	415	82		2 442	403	86
AFRICA	197				297			
Urban water supply		166	31	84		253	44	85
Urban sanitation		167	30	85		251	46	84
ASIA	1 029				1 352			
Urban water supply		972	57	94		1 254	98	93
Urban sanitation		690	339	67		1 055	297	78
LATIN AMERICA AND THE CARIBBEAN	313				391			
Urban water supply		287	26	92		362	29	93
Urban sanitation		267	46	85		340	51	87
OCEANIA	18				21			
Urban water supply		18	0	100		21	0	98
Urban sanitation		18	0	99		21	0	99
EUROPE	522				545			
Urban water supply		522	0	100		542	3	100
Urban sanitation		522	0	100		538	8	99
NORTHERN AMERICA	213				239			
Urban water supply		213	0	100		239	0	100
Urban sanitation		213	0	100		239	0	100

[a]Adapted from World Health Organization and UNICEF, 2000.

pointed out that, including rural population, 1.1 billion people in the world still lack access to safe water and 2.4 billion are without access to improved sanitation—a dire situation that points to a factor affecting rural-urban migration.

The high water supply and sanitation coverage of Europe and North America seems to indicate that there are not pressing urban water problems in these regions. However, there are certainly difficulties linked to failing infrastructure. Some of the major problems are:

- Aging of the underground utilities, causing deterioration of water quality in distribution systems and leakage of wastewater from aged sewers causing adverse trends in the groundwater table;
- The insufficient capacity to cope with increased loads, causing frequent spills of combined sewer overflows, jeopardizing the quality of receiving water bodies;
- Flooding of settlements built-up in floodplains;
- stormwater pollution (ignored in the past);
- Reluctance or slowness in gaining acceptance of innovative solutions based on source control (industrial and stormwater);
- Low priority or lack of care for urban streams and other urban water features; and
- Inappropriate institutional framework for efficient integrated urban water management.

The process of globalization manifests itself in changing patterns of consumption, production, distribution and investment. Such changes can lead to greater economic efficiency, but all the same they contribute to an increasing pressure on resources. Thus, globalization goes beyond economics; it is a social, cultural and technological phenomenon with evident environmental implications. Urban areas concentrate population and industries, and thus require increasing amounts of energy and other resources to function, drawing them from a much larger area than the city occupies. The ecological footprint of a city could be defined as: "the area of land/water required to produce the resources consumed, and to assimilate the wastes generated by that population on a continuous basis, wherever on Earth that land may be located" [Rees, 1999]. Under this concept, the ecological footprints of typical residents of high-income countries range as high as five to ten hectares per capita, resulting in footprint areas that easily are hundreds of times greater than the physical areas of the cities! This is usually the case for large city water supply sources, as well as their disposal of wastewater and solid waste.

Urban Water Issues: Developed Versus Developing World

The water-related environmental problems that cities are facing can be grouped broadly into four overlapping and interacting categories: (i) access to water and sanitation infrastructure and services; (ii) urban wastewater pollution; (iii) resource degradation; and (iv) water-related hazards. In developed countries, categories (i) and (iv) are either under control or less prevalent categories [Tejada-Guibert and Maksimovic, 2001]. Cities in developing countries and lower-income urban areas often face all four sets of problems (intensified in many cases, as they tend to occur simultaneously and with higher intensity over longer periods of time).

Developed countries have gone through a period of industrial growth that has often left a persisting imprint of environmental damage. This damage is often so profound, that expensive environmental mitigation can take several generations to succeed. Cities in developed countries also have a massive legacy of existing infrastructure built on the basis of what was once an acceptable technology when these works were implemented. This can be a very heavy load when age, technological obsolescence and new conditions point to a change. Thus, cities in more developed countries typically experience infrastructural aging that calls for overhaul and modernization, and is often held back by new environmental concerns and standards.

The explosive population growth occurring in urban areas of less developed countries causes many of them to experience enormous difficulties in supplying running water to sprawling peri-urban locations, illegal settlements and the decaying city center, besides facing inadequate and overloaded sewer systems. The problems are great, but developing countries have the chance to learn from the mistakes of the industrialized world and can therefore avoid future costs and problems.

Urban water management facilitates the sustainable provision of specific services, including water supply, flood protection and drainage, wastewater treatment and disposal, and maintenance of water-based amenities. While in developing countries there is a need for enlarging the supply capacity, in developed countries the focus tends to be on reducing further growth of withdrawals by demand management, water saving, and wastewater reuse and recycling. Urban water management has to "accommodate" the relatively new function of protecting receiving waters as a recreational, aesthetic, environmental and ecological amenity.

Approaches that may have been successfully applied in urban areas of developed countries, might not work as well or not at all in cities of the developing world. The conditions are often considerably different: less human and financial resources, an incipient or inappropriate institutional structure and greater political instability. Some of the differences are more subtle: time-related dynamics (cities in developed countries and their current systems have evolved over many decades or even centuries), cultural sensitivity to certain solutions, climatic differences (technologically appropriate solutions for temperate climates may not work in the humid tropics or arid climates). Often authors, when discussing the problems of developed countries, tend to lump cities of these countries into one category—but there are striking differences between many of these cities that go beyond climate and culture, and depend, to a large extent, on the overall developmental context.

At the country level, the United Nations General Assembly has classified 48 LDCs (least developed countries), out of which 33 are located in Africa, nine in Asia, five in Oceania (small island states) and one in Latin America and the Caribbean.

The rapidly expanding urbanization in developing countries affects dramatically the water resources both in the cities and surrounding areas because of combined physical, chemical and biological impacts. The need for integrated management becomes particularly evident as these water resources become depleted or degraded. Beyond the physical and environmental aspects, improvement requires social involvement, directly or indirectly, in the quest for acceptable and lasting solutions. Other key issues include: governance and policy issues, interaction with other competing sectors, responsive institutional arrangements, participatory decision-making, economical and socially effective water rates and a host of other considerations. Shortcomings exist in developed and developing countries but they are indeed deeper, wider and more entrenched in the developing world.

Need for new Approaches in Times of Change

The principles issuing from the International Conference on Water and the Environment (ICWE), held in Dublin in 1992, have had a very strong influence on current debate and design of water resources management strategies. The Dublin Principles highlight the finiteness and vulnerability of water, the necessity of a participatory approach in the water development and management, the acknowledgment of the central role of women in water management, and the recognition of water as an economic good. The call for integrated planning and management of water and for the involvement of all stakeholders in the process [Lundqvist et al., 2001] of water management is evident.

One of the recurrent central concepts in international fora nowadays is that new approaches, even a new mentality, are urgently required to successfully face the current urban water challenges, especially in the developing countries. The Paris Statement [UNESCO and Académie de l'Eau, 1998] recognized this need and called for action on the:

- Implementation of demand management measures, recognizing the economic value of water and incorporating social sensitivity;
- Alleviation of competing needs for water in rural and urban areas, considering the safe supply of drained urban stormwater and wastewater for agriculture;
- Integrated management of surface and groundwater in urban areas;
- Timely consideration of environmentally sound projects that will increase the availability of water where it is needed;
- Development of appropriate approaches for urban drainage according to climatic differences and consideration of special problems of developing countries;

- Investigation of novel approaches, among others: dry sanitation as an alternative to traditional water-borne sanitation, consideration of stormwater as a resource, local treatment of stormwater using biological systems, and development of technologies to recycle nutrients from wastewater;
- Adoption of non-structural means for urban flood mitigation, since structural solutions have proved, at best, to be partial solutions;
- Active involvement and participation of all stakeholders and a willingness to invert the traditional top-down decision making process;
- Specific and urgent problems of the peri-urban, less privileged areas in the outskirts of cities in developing countries; and
- Conception and application of new solutions and systems when existing ones have proved inadequate and/or perhaps infeasible, considering the unique character and circumstances of each city.

Other approaches recently developed, such as the Household-Centered Environmental Sanitation [Kalbermatten et al., 1999] contain similar propositions. Patorni [2000] calls for learning and participation as catalysts for change. The high social and institutional responsibility attached to the proposed measures is evident and thus solutions must be grounded in sound science.

OUTLOOK FOR THE FUTURE

Integrated Urban Water Management Concepts and Obstacles

Integrated water management can be understood as the application of a long-term policy for developing water resource uses in an environmentally and socially sustainable manner. An implementation strategy must be developed, recognizing the existing legislative, administrative and other constraining and conditioning factors. Urban areas, with their specific management problems and solutions, represent only sub-elements of river basin management and must comply with and support the overall basin planning. The planning and implementation of integrated urban water management require a full spectrum of tools and procedures, ranging from communication with, and empowerment of, the public, to sophisticated computer tools and techniques. It is increasingly recognized that successful management of urban waters cannot be accomplished with conventional prescriptive water management. A better chance of success lies with adaptive management, which proceeds along the line of the envisaged best solution while monitoring system performance and making adjustments as required. This requires much better coordination and open communication channels between the relevant players.

According to Vlachos and Braga [2001], four broad evolving areas pertaining to water resources that need to be considered in urban planning and management are: (i) *conceptual breakthroughs,* including shifting paradigms in terms of ecosystem approaches, sustainability, recognition of non-linear and non-hierarchical systems, complexity, uncertainty and interdependence of surrounding environments; (ii) *methodological advances,* especially multipurpose/multiobjective approaches, Decision Support Systems, Risk Analysis, and the implications of rapidly expanding computational power and access to new information and communication technologies; (iii) *organizational mobilization*, in terms of new administrative mechanisms, institutional arrangements, participatory processes, renewed interest in river basin contingency planning, alternative dispute resolutions etc.; and (iv) *contextual changes*, signifying the full range of ongoing and future quantity and quality problems, new areas of concern, shifting priorities, potential socio-political intervention mechanisms and comprehensive resource policies.

Emerging Paradigms vs. Conventional Practices

The ongoing degradation of the urban environment is a threat to the sustainability of the entire urban system. The objective of sustainability should be rooted in the principle of total management of the water cycle, thus including especially the provision of healthy drinking water and elementary sanitation to all. However, to meet the long-run needs of society, new approaches will have to be developed, tested and applied in the following domains:

a) Integrated management of wastewater originating from stormwater, water supply and industrial uses;
b) Appropriate techniques for reuse of treated wastewater (for the disposal of pollutants and using it in the sub-potable water supply); used-water recycling in industry and
c) Water conservation-based approaches which include more efficient use of water, and substitution of traditional sources with recycled water for industry and other uses.

While in different cases one may aim at the same goal, the methods employed will differ. Access to appropriate sanitation is still a major concern in developing countries. Thus, the emphasis in developing countries may be placed on increasing the availability of proper sanitation, while in developed countries, the main goals regarding sanitation would be to improve the efficiency and sustainability of wastewater collection and treatment systems; in both instances, through changes in environmental behavior as well as technological advances.

A new urban water paradigm to handle the types of problems described above is emerging. It calls for the implementation of demand management practices, resource recycling, and environment-friendly solutions in planning, development and management of urban water. The concept of demand management implies encouraging a change in life styles with an emphasis on recycling, increasing energy efficiency, waste minimization (leakage reduction),

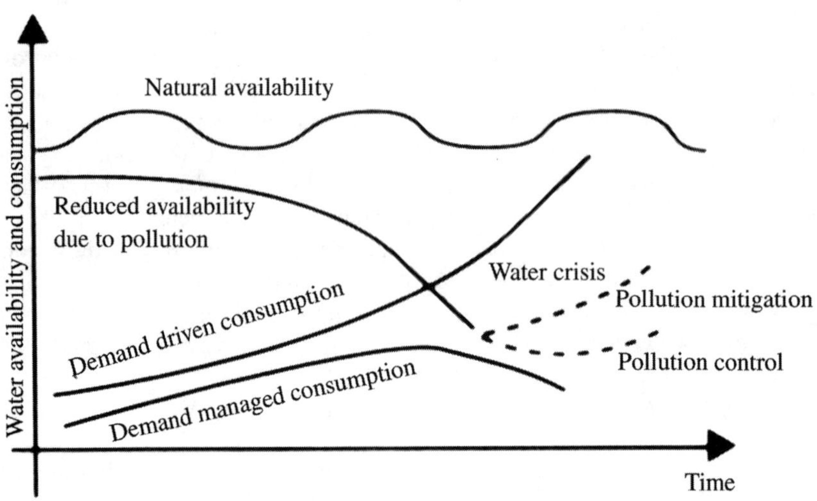

Figure 1. Demand driven vs. demand managed systems in an urban setting. Source: Butler and Maksimovic [2001].

reuse and reclamation, which ultimately may be conducive to diminishing demands for water rather than continuously increasing supply (Figure 1). This change would be coupled with the use of potential technological breakthroughs associated with the "post-industrial city" that can alter the form, structure and operation of urban aggregations and, thus, the shape of solutions for urban water problems. This idea is applicable in both developing and developed countries.

Technologies already developed and proven to work for potable, wastewater and stormwater systems are being applied elsewhere with some success in developed countries. These technologies should be further tested, implemented on a broader scale and made more sustainable. These include: advanced treatments, more efficient demand-driven systems, data acquisition and processing systems at lower costs and increased reliability, methodologies for fault detection and description of the state of the system, improvement of water quality in the distribution systems, and outputs from wastewater treatment plants and from stormwater disposal facilities. In the case of developing countries, a proper balance between affordability, technical competence of the society, and sustainability of the improvement should be sought. Developing countries may have the possibility of implementing innovative solutions from the start in cases where urban water systems are only incipient (or nonexistent).

It should not be assumed that all the high-tech solutions adopted in the developed world can be readily exported to developing countries, nor is it desirable to aim indiscriminately for this. There are documented cases where the flow in technology has reversed directions, i.e. new approaches developed in the South for local conditions that result in technologies or approaches which can be applicable in the North, as in the case of the concept of simplified sewerage developed in Brazil [Mara and Guimarães, 1999]. Simplified sewerage is widely used in Brazil and elsewhere in Latin America. In Brazil, it is serving more than one million people; in the capital Brasilia, more than 400,000 clients representing all social and income levels are served. The system is increasingly being used in Africa and Asia. In Europe, the cost saving potential of this system has been demonstrated in Greece [CEHA, 2000].

PRACTICE AND RESEARCH NEEDS: DATA, METHODOLOGIES, AND TOOLS

The contribution of science will be critical in the development of new approaches to urban water management. A number of relevant topics are outlined below. The thematic treatment shown below has benefited from the summaries by Niemczynowicz [1999] and by Butler and Maksimovic [1999].

Urban Hydrological Data

The need for organizing data-gathering hydrological networks at a regional or basin-scale for planning and management is readily understood and accepted. Standards have normally been long established. However, the requirements of hydrological data of urban areas demand a specific treatment. Factors such as the growing imperviousness due to pavements and buildings, the lack of natural water ways in large sections of the city, the complexity of the hydraulics of the stormwater evacuation system, the latent (or actual) interaction with sewerage and the potential consequences of an extreme event on the urban environment pose particular challenges concerning the types of data to be gathered and the necessary spatial and temporal resolution.

Rainfall measurement is clearly a fundamental hydrological process for runoff computation and modeling. In many cases, available rainfall data in the urban environment has fallen short of requirements for an adequate spatial and temporal representation. Water enters an urban area from numerous sources besides local rainfall: for instance, inflows from adjacent rural and urban areas, nearby rivers and streams flowing at extremely high levels that enter the city above ground or through the sewer system, and from rising groundwater levels. Discharge measurements within the city are essential to compute the water balance (diagnostics of the existing systems), to calibrate runoff models and to monitor various conveyance elements of the city system, particularly for real time operational purposes. More details on rainfall and discharge measurement may be found in Niemczynowicz [1996] and Maksimovic [1996].

The preservation of water quality requires consistent monitoring, as there are numerous polluting activities in urban areas, which may impose adverse effects on receiving bodies of water. Urban streams are thus an important asset; their water quality, hydrological and ecological status must be observed and assessed continuously. The acquisition of this type of data can be costly, thus intensive campaigns may be launched in experimental urban catchments. The observed parameters, whether chemical or bacteriological, necessitate careful selection in response to the specific environmentally impacting activities of the city and potential effects on urban dwellers and those downstream from the city [Marsalek, 1996].

Sediments in sewers and drains have a quite different genesis and composition from those in natural streams and certainly affect the behavior and efficiency of these structures. Measurement technology and modeling of urban drainage sediments is an evolving discipline [Ashley et al., 2000; Delleur, 2001].

The water balance in the distribution system requires special attention, particularly in cases where the quantification of uncertainty is of concern. Not only do the three major components (production, consumption and losses) call for quantification, but also the level of uncertainty involved in each of the three components should be qualified. One of the possible methodologies for addressing uncertainty and its spatial and temporal distribution is proposed by Jankovic-Nisic et al. [2002]. The art of quantifying uncertainty and its implications in the urban water context is yet to be mastered.

The water losses from the city system might actually raise the groundwater levels under it. For example, the Saudi desert city of Riyadh has had serious problems with a high water table resulting from excessive infiltration! It may also be that local abstraction is lowering the groundwater table under the city, or that sewage leaks are contaminating the groundwater. Thus monitoring groundwater processes is often quite important.

Planning and Management

Planning of the rehabilitation (upgrading) of existing urban water systems, and "ventional planning methodology practiced in the past, based on the personal knowledge and reputation of the chief planner-engineer, is gradually being replaced by a new one based on the extensive use of information systems support, stronger environmental considerations and the incorporation of other relevant factors including socio-economic interactions.

This approach can be described by the five "I's", principles highlighted by Butler and Maksimovic [1999] and expanded to seven by Vlachos and Braga, [2001]. They include the following: *Integration* (development of coordinated, integrated urban water models); *Interaction* (search for the additive, cumulative synergistic effects of complex urban systems); *Interfacing* (especially the public with the environment); *Instrumentation* (data reliability, quantification of uncertainty, real time control, sensors, non-invasive techniques); *Intelligence* (expansion of data, information, and knowledge through GIS [Geographic Information Systems], and hydro-informatics); *Interpretation* (complementarity of data and judgment, a wise combination of structured reasoning and disciplined imagination); and finally, *Implementation* (capacity for true and concerted action and the transformation of policies into reasonable practice that combine corrective and proactive strategies and tactics). At the same time, planning of urban water systems has to be well coordinated with water resources management at the river basin scale.

Drinking Water Supply

An ongoing debate in the international water supply field focuses on what constitutes the minimum level of service for domestic use. The right of access to water to meet basic water needs for all people has been reaffirmed at a number of high-level intergovernmental conferences, and implicit in this formulation is that a minimum quantity of resource requirements for human and ecological uses should be considered in the allocation of the resource.

Basic water requirements for humans should include drinking water plus water necessary for human waste disposal. The distinct health advantages of access to adequate sanitation facilities, and the protection of drinking water from pathogenic bacteria and other agents of disease, have been clearly established. While waste-disposal options that require no water are available, it has been recognized that there are additional health benefits when about 20 liters per capita per day (l/p/d) are provided for this purpose. This figure would suggest, as a basic water requirement, 25 l/p/d of clean water for drinking and sanitation. The supplementary needs for basic hygiene (washing, bathing) and food preparation have been estimated at about 25 l/p/d. Thus, adding the latter amount raises the basic water requirements to 50 l/p/d, as proposed by Gleick [1998], "as a new standard for meeting the four domestic basic needs: drinking, sanitation, bathing and cooking, independent of climate, technology, and culture."

However, the needs and current practices based on water-borne sanitation in modern urban settlements far exceed this amount. The actual figure is highly dependent on the size of the city, its level of development, and climatic conditions. More conservative approaches in which no demand management is exercised assume in general that 150-250 l/p/d are required to satisfy all personal requirements in a typical modern city, and an additional 150-200 l/p/d for municipal purposes, excluding industrial uses and losses by leakage [Shiklomanov, 1997]. Introducing demand management, leakage reduction and better overall management can significantly reduce these figures. The urban demand per capita has increased several-fold during the last century due to improved standards of living, industrialization and level of development. In industrially developed countries in Europe and North America, urban water use often falls in the 400-600 l/p/d range [Shiklomanov, 1997]. In cases of large cities in industrialized countries, where water is considered to be an abundant resource with little concern for demand management and where a warm climate may be a conditioning factor, figures may reach 1,000 l/p/d. It must be kept in mind, that in many cases, when a detailed system diagnosis is carried out, usually a high percentage of leakage is revealed, sometimes as high as 70–80% [Maksimovic et al., 2001].

The dynamics of demography are such that even though population growth rate is slowing all over the world, and cities will expand at a lower rate than during the past half-century, the increase of population will still be substantial (32% for the total population and 72% for the urban population growth foreseen between 2000 and 2030). Thus water, a finite resource, will need to be provided to a continuously increasing urban population for domestic and other uses. Solutions for dealing with the increasing population will need to be multisectoral (irrigated agriculture, industry, ecological maintenance, etc.), within an integrated framework, in order to maximize the well-being of society in the present and future.

Currently, as there is normally only one water supply network per city, the water distributed is of uniform quality, that necessary for drinking. In the urban environment, several approaches have been identified to enhance the chances of satisfying future needs—the most demanding use and thus the costliest due to the required treatment. Recognizing that the different uses of domestic water do not require a uniform quality, dual systems have sometimes been installed without fully satisfactory results, because of either high construction and maintenance costs or health safety concerns (accidental cross-connection of two types of water). However, innovative solutions based either on water recycling or harvested stormwater used for toilet flushing and washing is beginning to find its place at an individual and small community level. Herrmann and Schmida [1999] assert that in 1999, there were already over 100 commercial manufacturers competing in the rainwater usage market in Germany.

Water-sensitive urban planning (WSUP) [Shamir and Howard, 2000] encourages the use of urban runoff and harvested rainfall as a resource. Furthermore, they indicate point-of-use treatment as a technically feasible option for drinking water supply, and that large-scale, widespread point-of-use treatment deserves thorough consideration, as it increases water supply management flexibility and the potential to use untapped sources. They also point to the rise of bottled water usage all over the world, which in developed countries rarely is due to health concerns and runs counter to the paradigm of centralized collection and distribution of urban water supplies.

Optimization techniques have been applied to water distribution network design. However, Walski, one of the pioneers in this field, has sounded the alarm on why a real (even relative) optimum is not guaranteed [Walski, 2001], given that the algorithms are based on a cost minimization formulation, in which real expected or potential benefits of the system do not enter into play. Many factors, including uncertainties about future demand, make it difficult to pose a more realistic formulation based on the maximization of net benefits (benefits-costs). He asks practitioners to acknowledge this limitation and to search for ways to move beyond the cost minimization paradigm. It should be emphasized that optimization tools only serve as an aid to decision-making, and not as a substitute! Multiobjective optimization approaches, in contrast to single-objective optimization (i.e. cost minimization), are more complex but more realistic, and should be considered (see, for instance, Halhal et al. [1997]).

Rehabilitation of water networks is of paramount importance, particularly for cities in developed countries, as the existing networks can be up to 150 years old in the older sections and were built under standards and practices that are now outdated. However, the enormous cost makes it economically infeasible to replace older parts of the system, thus their maintenance and performance upgrading will continue to be important issues. A European group of experts [Saegrov et al., 1999] considered structural and functional deterioration of water pipes, describing the state-of-the-art situation in Europe and North America, as well as current research needs. Concerning inspection and control of water mains, they contrast the reactive approach of locating leaks (using acoustic leak detection systems) with the proactive approach, oriented toward detection of thinning pipe walls (due to corrosion) by means of an electromagnetic inspection technique. Concern over the structural deterioration of water pipes has resulted in the development of several models to forecast water main breaks, as well as several decision support systems. New innovative techniques for leakage detection and quantification are being developed. For example, Covas et al. [2001] present an interesting approach based on the application of the inverse transient method.

Figure 2 illustrates a framework to explore rehabilitation options. Saegrov et al. [1999] point to the following needs: (i) an improved understanding of static and dynamic factors affecting pipes, including statistical methods for analyses of pipe damage, relationship between pipe condition and pipeline performance and dynamics, advanced techniques for field inspection; (ii) better understanding of operational factors, addressing factors such as data collection and management, understanding the mechanisms of operational and seasonal variability in pipe burst rates; (iii) development of integrated network management strategies; (iv) new rehabilitation solutions, making use of trenchless techniques and emphasizing the exchange of information; and (v) establishment of appropriate procedural frameworks and incentives for technology transfer. Trenchless or "no-dig" techniques allow pipes to be installed or rehabilitated below the ground without open cuts. These techniques can be used to install replacement pipes along a new route (directional drilling and microtunnelling), install them along the existing route by destroying the old pipe (pipe bursting), or install a smaller diameter fully structural pipe within the old (slip lining) [Heavens, n/d]. Trenchless renovation strictly includes processes of pipeline rehabilitation that make use of the existing pipe structure such as cured-in-place lining (e.g., the epoxy version of Instituform), and coating by epoxy resin lining and cement mortar lining. The trenchless techniques can minimize traffic disruption and possible interference with other underground services (gas, electricity, etc).

In addition to conventional sources used to supply drinking water (such as abstraction, treatment of fresh surface and groundwater), desalination of brackish and salt water has become an affordable option in the cases of shortage of other sources—with the price of treatment falling below 1 $US/m^3.

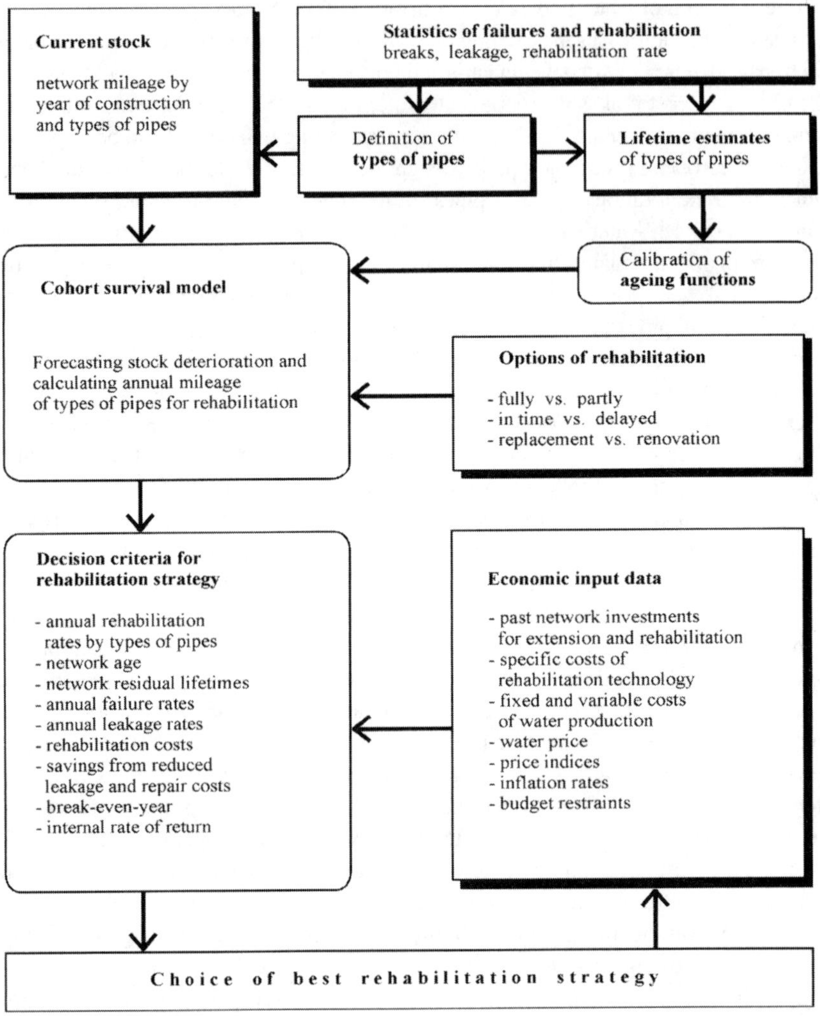

Current stock

network mileage by year of construction and types of pipes

Statistics of failures and rehabilitation
breaks, leakage, rehabilitation rate

Definition of **types of pipes**

Lifetime estimates of types of pipes

Calibration of **ageing functions**

Cohort survival model

Forecasting stock deterioration and calculating annual mileage of types of pipes for rehabilitation

Options of rehabilitation

- fully vs. partly
- in time vs. delayed
- replacement vs. renovation

Decision criteria for rehabilitation strategy

- annual rehabilitation rates by types of pipes
- network age
- network residual lifetimes
- annual failure rates
- annual leakage rates
- rehabilitation costs
- savings from reduced leakage and repair costs
- break-even-year
- internal rate of return

Economic input data

- past network investments for extension and rehabilitation
- specific costs of rehabilitation technology
- fixed and variable costs of water production
- water price
- price indices
- inflation rates
- budget restraints

Choice of best rehabilitation strategy

Figure 2. Framework for exploring water networks rehabilitation needs and strategies. Source: Saegrov et al. [1999].

Stormwater Management

During storms the impervious and relatively smooth surfaces of the metropolitan environment can cause quick and high concentrations of runoff. In order to allow city life to continue unimpeded, the excess water traditionally was (and in many cases still is) removed from inhabited and transited areas and disposed of into the nearest surface water body. This process of urban stormwater handling is evolving. Two centuries ago, even in large and leading industrialized cities, man-made

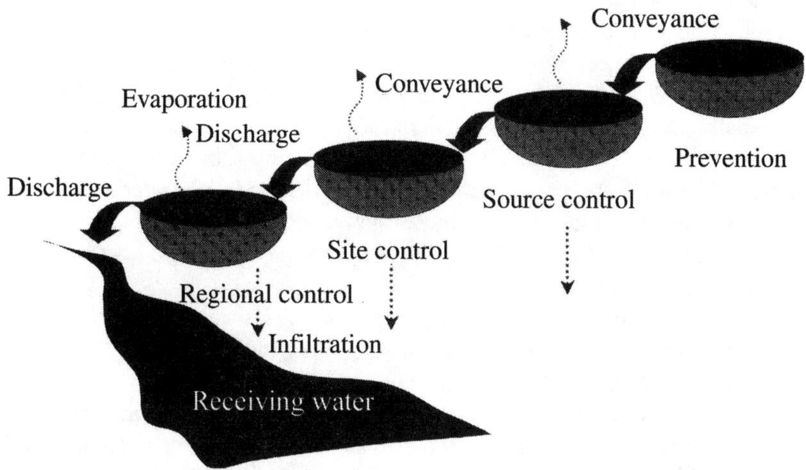

Figure 3. Treatment train for control if surface runoff and improvement of quality.

ditches, open channels and sewers were commonly used for draining surface runoff. Later, combined sewer systems, with conduits carrying both stormwater and sewage came into being; of course the resulting overflows and discharges were a health risk. Treatment facilities for the collected discharge were implemented to reduce the risks before they flowed into receiving waters.

It was later realized that fully separate sewer systems (for wastewater and for runoff) were a vast improvement over the combined sewers, thus they were adopted in most large cities (many countries still have large numbers of operating combined sewers, although they are rarely constructed nowadays). Normally, stormwater is carried away and disposed of outside the city.

During the last few decades there has been a shift in the thinking on how to manage stormwater by introducingso-called BMP (Best Management Practices) in the United States ("A device, practice or method for removing, reducing, retarding, or preventing targeted stormwater runoff quantity, constituents, pollutants and contaminants from reaching receiving waters." [Strecker et al., 2001]). The term SUDS (Sustainable Urban Drainage Systems) is accepted in the UK and in some other countries. Both SUDS and BMP essentially work on reducing peak flows and improving runoff quality at the source and throughout the runoff process by means of a "treatment train" (Figure 3). These include a range of measures, starting with better control (i.e. prevention), of wash-off of heavily polluting waste from individual properties and industrial sites. This is followed by the control of runoff at individual lots by detaining and infiltrating stormwater, and continued by the use of grass swales, infiltration facilities, site control measures, detention/retention ponds, infiltration basins, and other measures. Some infiltration and collection techniques, used in the past but largely foregone in the recent urbanizing wave, are making a comeback and being refined. This trend brings about several positive consequences: avoidance of large works, decentralization of the system, attenuation of flow peaks and turning stormwater into a resource, since it is used to augment local recharge to groundwater and/or to enhance the community in an ecologically harmonious way. Storm runoff control facilities have an added advantage in that they can easily be incorporated into urban amenity features.

Despite the wide acceptance and success of BMPs/SUDS, criteria and information necessary to rate their overall effectiveness and their effect on downstream aquatic environment are still receiving the attention of researchers [Roesner et al., 2001].

Thus, new solutions tend to be source-oriented in order to mitigate the formation of urban runoff and pollution, but, according to Niemczynowicz [1999], there is still not enough knowledge about potential environmental, economic and social effects, which may vary when applications move from small-scale to a more generalized use. The interactions between the river basins and aquifers, with regard to long-term potential changes in the release of nutrients and of recharge mechanisms, need to be investigated and documented. The problem of polluted sludge sedimentation in open systems, such as constructed wetlands, needs to be solved. The issue of avoiding, at the outset, the mix of pollutants with clean rainwater and the subsequent short—and long-run effects on the design and complexity of stormwater systems is a problem that must be addressed. Even though a number of promising approaches and technologies are being developed, the decisions of where and how to implement them will remain, leading to questions such as: only in new settlements? Under what circumstances is the replacement of old systems viable? Methodologies to handle these types of dilemmas, in a physically and socially integrated manner, are being developed and implemented [CIRIA, 2001]. An important factor for the success of these techniques in their practice is the acceptance (sense of ownership) and support by the local population, enhancing their chance for long-term sustainability.

The uniqueness of individual urban settings, concerning climate, culture and socio-economic development cannot be overemphasized. Many of the approaches and standards developed correspond to temperate climates, typical of North America and Europe; the development of specific approaches for other climates (such as arid, humid, tropical and cold ones) is an ongoing effort [Tucci, 2001; Nouh, 2001; Saegrov, et al. 2001]. Methodologies that take into account the socio-economic realities of less developed countries are receiving attention, for instance in the South Asian humid tropics [Kolsky, 1998; Parikh, 2001].

Climatic differences and the management of street cleaning (sweeping) are significant factors affecting the amount of solids caught by the surface runoff and transported down the drains. This process is important, as heavy metals are normally associated with small diameter particles found in storm runoff. Traditionally ignored as a source of pollution, this is nowadays being recognized as a significant contributor to diffuse pollution. Deletic and Maksimovic [1998] describe the modeling of this process. Control of diffuse pollution caused by storm drainage in the future will need to be based on better modeling of the process and more reliable assessment of the effects of management practices such as street sweeping, which a few decades back was considered inefficient. However, the improvements in sweeper designs are leading to greater efficiency in picking up and retaining small-diameter particles. Thus, these techniques will have to be re-examined for possible inclusion in pollution management.

Sedimentation in combined sewers, while largely ignored in the not too distant past, is under increasing study. The diversity of the ubiquitous inputs has a negative affect on the hydraulic performance of the system and the surrounding environment. As Delleur [2001] aptly describes in his excellent review, "problems that arise include blockage, surcharge, early overflows, large pollutant discharges, and costly removal. The erosion in sewers can release pollutants in concentrations that exceed the contributing sources of sediments and pollutants. Solid particles constitute the support on which other pollutants are adsorbed, among them heavy metals and hydrocarbons." Sediment-laden discharges from combined sewer overflows constitute a major pollution source in urban receiving waters. The cohesive nature of sediments due to tars, greases, chemical cementation and biological processes creates an added difficulty. Delleur identified several topics on sediment movement in urban drainage requiring further research, including incorporation of unsteadiness of flow, effects of conduit shape, understanding of movement of solids in sewers, effect of sediment cohesiveness, development of integrated models (sewer system, wastewater treatment plant and receiving waters), and improvement of field techniques. Ashley et al. [2000] also decry the current state-of-the-art understanding of the behavior of solids in sewers, and make a strong appeal for developing efficient and proactive sewer sediment management approaches, including better predictive models. Most of the current research effort is being carried out in Europe.

The U.S. Environmental Protection Agency (EPA) identified a number of research directions for urban wet-weather flow (WWF) management and pollution control [Field et al., 1998]. The flows in question are: combined sewer overflows, stormwater from separate stormwater drainage systems, and sanitary sewer-overflow. Control of WWF is a top cleanup priority for EPA, but the current high cost, which must be absorbed by municipalities make low-cost alternatives a research and development priority. The areas that EPA singles out are: (i) characterization and problem assessment, including improvement of monitoring methodologies, and determination of WWF receiving water impacts; (ii) watershed management research related to protecting ecological resources and potable water source areas; (iii) toxic pollutant impacts and control, oriented toward prevention and source control, and toward removal of toxic pollutants, particularly to mitigate their effect on receiving waters; (iv) WWF control technologies adopting an integrated risk management approach combining prevention and treatment; going from land management strategies to advanced collection systems and to the use of natural and created wetlands; and (v) infrastructural improvement, focusing on the effects of infiltration/inflow problems in sewers, especially as they are associated with aging and deteriorating urban infrastructure. Tafuri and Selvakumar [2002] address the wastewater collection research needs in the USA in three major areas: (a) assessment of system integrity; (b) operation, maintenance and rehabilitation; and (c) new construction. The nature of these problems illustrates clearly the concerns of cities in industrialized countries. Cities in developing countries can also benefit from implementing selected measures based on source control and pollution management, especially if they turn out to be less costly to construct and more affordable to maintain by less skilled personnel.

New Sanitation and Recovery of Wastewater Nutrients

Centralized conventional sewage treatment works are used widely today for domestic wastewater management. They consist of various stages: primary treatment usually involves separation grits and other solids by settling and removal of fats, oils and grease; secondary stages use alternative well known methods (activated sludges, trickling filters, oxidation ponds, etc.); tertiary stages exist where there is a sensitive environmental consideration or where advanced treatment is needed. These systems produce considerable amounts of sludge and can be energy intensive. A number of non-conventional alternatives for wastewater treatment including non-biological and biological methods ranging to aquaculture are described and compared by Burkhard et al. [2000]. A source book and training manual on these technologies is being issued by UNEP-IETC [2002]. There are good examples of successful applications of some of these techniques on small scale in a rural part of China [Hua, 2001].

Nonetheless, the existing conventional systems represent enormous investments and involve long-lasting infrastructure, thus the disposal of the out-flowing wastewater will have to be dealt with for a long time into the future. Wastewater reclamation and reuse must be seriously considered given the growing scarcity of new water sources.

With the emergence of megacities all over the world, this becomes a large-scale issue. Asano [2000] assessed the status of wastewater use in terms of large cities, describing the various categories of wastewater reuse and reporting on various recent experiences. He concludes that considerable progress has been achieved in making reclaimed wastewater a reliable and quality resource, but noting that continued research and demonstration efforts are needed in key research areas, such as: assessment of health risks associated with trace contaminants in reclaimed water; improved monitoring techniques to evaluate microbiological quality; application of membrane processes in the production of reclaimed water; and evaluation of the fate of microbiological, chemical, and organic contaminants in reclaimed water.

On the other hand, water for direct human consumption (hygiene, cooking and drinking, all requiring a high quality standard) is a relatively minor part of the total domestic water supply used. A significant amount of water consumed in a household is used for water-borne sanitation, such as in flush toilets. A major contradiction with respect to resource conservation is that, virtually with no exceptions, there is only one quality of water in an urban water system: costly water—of drinking water standard. Such water supply, for all uses, is unavoidable because there is only one water network. The outgoing wastewater then requires treatment before being released into the environment as described above. The performance of the wastewater treatment plant might not be enough to protect the receiving water, especially if no measures are applied at the source to prevent severe pollution or to make the treatment of water and sludge more efficient. For instance, nitrogen and phosphorus need to be removed to control eutrophication of lakes, reservoirs and bays. Other pollutants requiring attention are: heavy metals, persistent organic pollutants, endocrine—disrupting chemicals and pathogens [Matsui, 2001]. Pretreatment of industrial wastewater on the site is important in order to prevent heavy metals from reaching sewers and ending up in the sludge, making it unsuitable for use as fertilizer in agriculture.

An important and logical effort is now going into the development of less water-intensive forms of sanitation and into creating a resource out of "waste". Different kinds of toilets with low or no dilution and/or urine diversion are becoming important [Otterpohl, 2001]. The classic mixed wastewater (toilet, bath, kitchen, wash) will undoubtedly be a subject of increasing interest in the foreseeable future. There are also new approaches, emphasizing source control and using the following (color coded) wastewater "palette": black (toilet), gray (bath, kitchen, wash), light gray (bath, wash), yellow (urine), and brown (feces); locally drained stormwater would be "blue" (Figure 4). In addition, alternative technologies to be applied centrally (i.e. on the site or close to wastewater treatment plants) are also being developed.

There are a number of pilot projects testing water-saving technologies at household and community scales, especially in Northern Europe. An important element of these systems is urine separation toilets. Table 3 summarizes some results in water savings.

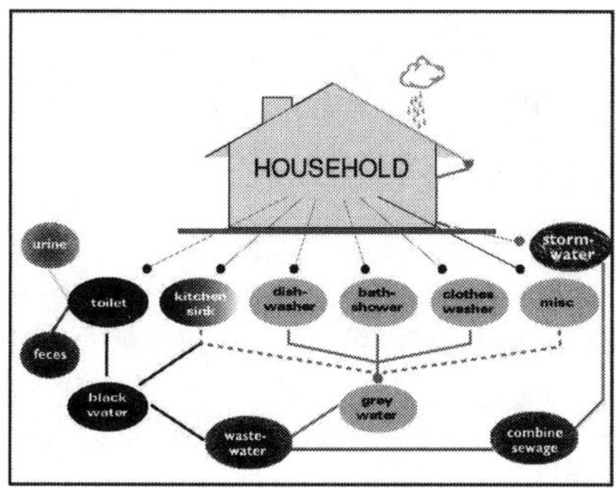

Figure 4. Wastewater separation at a household level. Source: UNEP-IETC [2002].

Water savings is a fundamental objective, and resource and nutrient recovery is an important complement to this approach. Table 4 shows typical volumes and components of wastewater and potential recovery. The fertilizer potential of urine can be quite significant with respect to the demand in many countries. Appropriate techniques exist for its use—among them the simplest is storage for 6 months to diminish pathogens and downgrade medical residues prior to application to the soil. There are other feasible ways to concentrate this form of fertilizer for easier storage, transportation and use. Low-diluted feces without urine (brown water) can be treated by anaerobic digestion, composting

Table 3. Fractionated Water Consumption. Values from Northern Europe [a](in liters per capita per day)

Use	Current water consumption	With savings (new installations, sewer rehabilitation)	Primary water consumption with savings and use of secondary water (for toilets and laundry)
Toilet	50	25/0	0
Bath	50	25	25
Kitchen	50	25	25
Laundry	10	5	1
Infiltration	80	25	-
Total	240	105	51

[a] Source: Matsui et al., 2001.

Table 4. Loads in Fraction of Household Wastewater and Source Control[a]

Yearly loads kg/capita/year	Gray water 25,000-100,000 l/capita/year	Urine ~ 500 l/capita/year	Feces ~ 50 l/capita/year (option: add biowaste)
Nitrogen (N) ~ 4-5	~ 3%	~ 87%	~ 10%
Phosphorus (P) ~ 0.75	~ 10%	~ 50%	~ 40%
Potassium (K) ~ 1.8	~ 34%	~ 54%	~ 12%
COD ~ 30	~ 41%	~ 12%	~ 47%
	Treatment ↓ Reuse/water cycle	Treatment ↓ Fertilizer	Biogas-plant composting ↓ Soil-conditioner

[a] Source: Matsui et al., 2001. See also Otterpohl, 2001, for a European pilot project figures.

and, in warm climates, desiccation. Technologies for treating gray water are also being developed. This topic will require considerable research in the future, both hi-tech and low-tech, because it promises to bring about a revolutionary change in sanitation solutions. Certainly cultural differences and social acceptability will need to be carefully considered. These efforts are closely allied with the concept of "ecological sanitation". The organizers of a recent international workshop in Mexico, dealing with ecological sanitation, reported matter of factly that "the outcome of the workshop represents a shift in the way people think about and act upon human excreta" [Esrey et al., 2001] and linked this approach to the attainment of food security.

Urban Groundwater

Groundwater is intensively exploited for private, domestic and industrial use in many cities, and as a result the subsurface in these cities has become the receptor of industrial and domestic wastewater. Urban groundwater has recently gained the status of a legitimate field of study. Lerner [1996] points to a number of distinguishing features that justify treating it as a specific subdomain:

- Recharges, affected by extensive sealing of surfaces, leaking water mains, sewers, and stormwater recharge, are often larger than under rural areas and exhibit large spatial variability.
- Geotechnical interactions such as flow interference due to deep basements, tunnels and pilings; subsidence stemming from excessive abstraction and rising groundwater levels because of excessive recharge.
- Groundwater quality, affected by point, multipoint and linear inputs of chemicals from the complex, heterogeneous features of a city, such as residential and industrial sources, wastewater systems, parks and landfills.
- Complex conditions for groundwater protection management, due to demand for the groundwater and the multiple sources of nearby pollution and rapidly urbanizing areas.
- Distinctive investigation problems due to heterogeneity of conditions, difficulties of access to private and other land and interference with field methods such as surface geophysics.

The complexity of the urban groundwater interactions is illustrated in Figure 5. Foster et al. [1998] classified urban groundwater problems as arising from inadequately controlled groundwater abstraction, from excessive subsurface contaminant load and from excess urban infiltration and identified the corresponding management need as shown also in Table 5. The problems can arise in cities in developing as well as developed countries, but normally the acuteness and urgency of the problem is greater in urban areas of developing countries because of unplanned urban growth, often haphazard wastewater disposal and

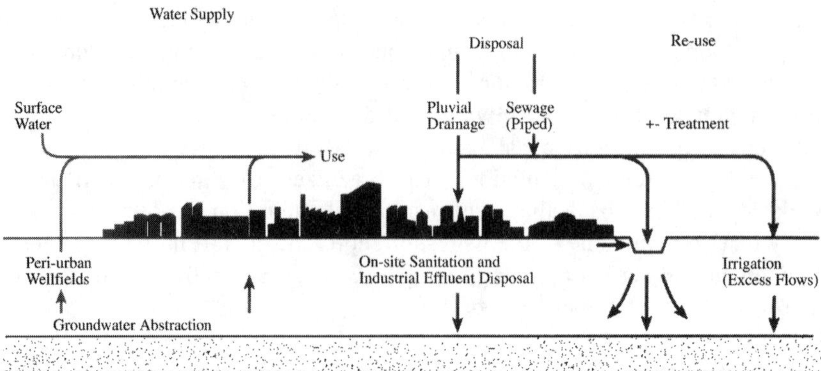

Figure 5. Interaction of groundwater supply and wastewater supply disposal. Source: Foster et al. [1998].

unregulated well water supply. The formulation of effective strategies and techniques for handling urban groundwater-related problems, geared to achieve healthy and sustainable results, will undoubtedly grow in importance in the coming years.

Peri-Urban Areas and Urban Agriculture

Peri-urban areas lie at the interface between the rural and the urban environment. In its normally accepted connotation, the term peri-urban does not refer to the relatively high-income, low density, green suburbs of many first world cities, but to the high-growth, precarious settlements springing up around large cities of developing countries. A number of definitions have been proposed. According to the United Nations Centre for Human Settlements (Habitat)-UNEP [1996], peri-urban areas include "settlements that are marginal to the physical regulatory boundaries of the formal city." Typically the peri-urban settlements have uncertain or illegal land-tenure, are low income and high density, and lack formal recognition. Many times they are situated within fragile ecosystems: low lying, flood-prone, plains or unstable mountain slopes, deemed undesirable for the formal expansion of the city. Evolving peri-urban areas may eventually become formally recognized peripheral communities or they may be absorbed by the formal cities forming low-income pockets.

Most the peri-urban areas in developing countries remain unserviced by piped water supply. They depend primarily on vendors who provide water of doubtful quality and exorbitant prices (typically 10-100 times higher than the formal water rates!). Sanitation coverage is even poorer than that of piped water, a problem that is often compounded by the occurrence of frequent flooding with inadequate drainage, with an increase of urban-vector diseases.

Table 5. Urban Groundwater Problems and Management Requirements

Underlying cause	Resultant groundwater problems	Management requirements
Inadequately controlled groundwater abstraction	Overabstraction of good quality resource within city limits	Reserve good, deeper ground water for sensitive uses and encourage use of shallow, poor groundwater for nonsensitive uses
	Overabstraction of good quality resource around city periphery (competition between urban supply and agricultural irrigation)	Reserve good groundwater for potable water supply and substitute treated wastewater or shallow, poor groundwater for irrigation
Excessive subsurface contaminant load	Contamination of municipal water supply boreholes/wellfields	Define source protection zones for priority control of surface contaminant load
	General widespread contamination of groundwater	Reduce contaminant load in selective areas, especially where aquifer is highly vulnerable, by appropriate planning provisions or mitigation measures
		Plan wastewater treatment/disposal taking account of groundwater interests and impacts
Excess urban infiltration	Rising water table beneath city causing: • Basement flooding • Malfunction of on-site sanitation units • Reversal of aquifer flow directions (with • contamination of periurban wellfields by • polluted urban groundwater)	Reduce urban infiltration by • Control of mains leakage • Reducing seepage from on-site • sanitation unit by mains • sewerage installation • Increase abstraction of shallow (polluted) groundwater for nonsensitive uses

[a]Source: Foster et al., 1998.

The high population density, accompanied by a lack of basic services and exposure to industrial wastes and effluents, can lead to disastrous sanitary conditions.

The challenge posed by peri-urban areas can be justifiably classified as social and institutional rather than technological, and requires an integrated solution. However, among the necessary prerequisites are the appropriate water supply, sanitation, treatment and reuse methods in order to reduce water-related diseases, and the means for closing the water and materials cycle, such as reuse of solid and liquid wastes as much as possible after adequate treatment and careful control [Bahri, 1999]. Decentralized water and wastewater management strategies appear to be more promising than an end-of-pipe to a source approach. Decentralized systems provide great flexibility for applicable treatment processes and reuse options, resulting in smaller wastewater outflows, less energy used and greater reductions in the amounts of sludge produced. Furthermore, they avoid the high infrastructural investment costs of connection to centralized networks, and management can be kept at the local scale, maximizing the participatory component in the decision-making process.

Innovative partnerships need to be encouraged. The private sector is taking an active interest in providing solutions to these underprivileged areas. For instance, Lyonnaise des Eaux [1998] equates the public sector to a high-cost, limited centralized approach ("all for some"), while indicating that private concessionaires can provide appropriate services following a customer-based approach ("some for all"). They analyze current conditions based on case studies, and propose some solutions. Regarding water supply they describe alternatives based on simplified distribution and storage schemes, but still connected to the network (for instance a 200-liter tank on the roof of a house with a small diameter pipe connection, and a water bailiff each day opening control valves limiting the consumption to 200 l/day per household). For sanitation, they conclude that local solutions such as technically well-designed latrines are usually the most appropriate. This type of development is encouraging, but still far from the concept of integrated, decentralized, recycling solutions.

Surprisingly, a growing economic activity in peri-urban areas is urban agriculture, occupying derelict tracts of land, producing food for local consumption and for sale in the city. This is a significant activity, which implies a greater self-reliance in food production, greater employment opportunities, beneficial recycling of urban waste, and enhancement of the environment. Most notably this potential is being exploited in a number of Asian cities (for example, 18% of China's largest cities are 90% self-sufficient in vegetables; 60% of metropolitan Bangkok is used for urban agriculture.).

Urban agriculture has historically formed part of the survival strategy of urban dwellers all over the world. In a wider sense urban agriculture refers not only to food crops and fruit trees grown in cities but encompasses different kinds of livestock, medicinal plants and ornamentals for other purposes world [Quon, 1999].

Table 6. Dimensions of Sustainability of Urban Agriculture[a].

	Environmental	Economic	Social
Synergy	plant nutrients in urban waste & sewage	access to inputs & markets	meeting human needs for green (recreational) urban space
		amenity	
	health aspect in context of 'urban greening'	employment & poverty alleviation	
Conflict	urban pollutants in agricultural produce	competition from urban land use systems	negative perceptions of (peri-) urban farming
	agrochemicals in urban environment	vandalism & theft	
	urban greening & nature-borne diseases		

[a]Source: van den Berg, 2000.

There is scope for expanding this activity by improved water management practices—as a matter of fact urban agriculture can legitimately form part of urban water management strategies. The expansion of sewer networks and wastewater treatment plants has increased availability of reclaimed water; in addition, sewage sludge, urban composts and other effluents can be judiciously and beneficially used as part of "closing the loop" in a recycling urban society. The challenge is to integrate this approach meaningfully in urban water strategies. Furthermore, urban agriculture is important to not only developing but to developed countries' cities, in a different but valid context. Van den Berg [2000] identified topics for research within the environmental, economic and social dimensions of urban sustainability, grouping them as either synergistic or conflictive, as shown in Table 6.

Urban Irrigation

In addition to urban agriculture per se, water used to support growth of urban greenery, vegetation in parks, planted trees along avenues etc. represents another source of urban water demand. Ancient civilizations (especially in arid areas) used to have separate systems of open ditches for surface irrigation of urban plants, some of which are still in use (Central Asia, Middle East). In modern cities, drinking water usually taken from the same municipal distribution system that serves household is also used for watering lawns, shrubbery and trees [Yevjevich, 1996]. If separate systems are used for irrigation, water quality can be lower; however, for public health reasons, it is advisable that water be chlorinated. The design of urban irrigation systems is subject to the principles used in agricultural irrigation. Given the high rate of evapotranspiration in irrigated fields, water demand is very much dependent on local climate conditions especially in arid zones.

Urban Streams

Most modern cities have been founded and developed either on the banks of a river (urban stream) or on the seacoast. Urban streams have often shaped the nature and the identity of the urban areas served, undergoing transformations throughout the development of the settlements. Many urban streams are the symbols and landmarks of the cities. The various aspects of the development of urban streams are illustrated in Figure 6.

Rivers in both rural and urban settings are complex, multifunctional ecosystems that have developed their own self-sustaining balance. Modification of a particular function over another may cause an imbalance that, in the case where it persists, may eventually lead to degradation of the aquatic environment and ecology. Historically, the development of watercourses has been undertaken for a variety of reasons such as flood control, water supply or navigation. There is a great diversity in the ways that rivers have been modified depending upon

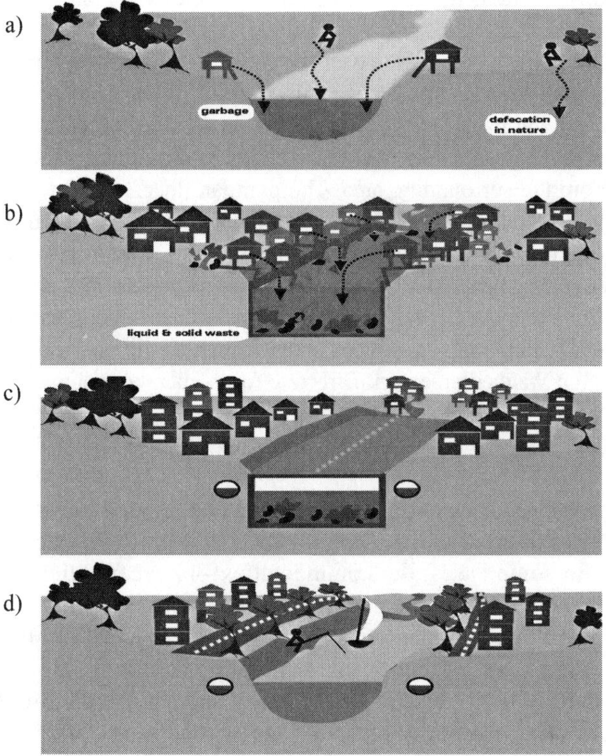

Figure 6. Development of urban streams. Source: UNEP-IETC [2002].

the needs of their adjacent communities. Unfortunately, throughout history many urban streams have gradually been turned into recipients of urban waste (solid and liquid), becoming foul and dirty. Many of them have been covered, turned into sewers and "forgotten."

With the recent increase in environmental awareness and concern, urban streams are being revisited and their aesthetic and environmental values appreciated. Thus rehabilitation and restoration projects are often being carried out nowadays. It is important to draw the distinction between rehabilitation and restoration of urban rivers and watercourses. Rehabilitation seeks to improve the state of the river in terms of physical characteristics, chemical quality, ecological diversity and aesthetic appearance, whereas restoration is directed at recreating the pristine physical, chemical and ecological state. Rehabilitation is a realistic objective in many urban areas, leading to broader social and economic benefits to the community, while in most cases true restoration does not constitute a practical option. The main concern associated with rehabilitating heavily modified and artificial watercourses is achieving a balance of functions consistent with social, economic and political demands upon the natural system.

In the past, urban watercourses have been confined to narrow river corridors with the natural channels turned into canals with concrete and other man-made materials forming the bed and banks of the river. Many urban streams have been converted into closed conduit sewers, and now receive both storm drainage and raw or diluted sewage from the surrounding area. The pollutant load frequently leads to poor water quality; indeed this adverse impact of urbanization often extends to the watercourses downstream of the urban area. In some cases, urban streams represent a severe threat to public health. The result is that many urban watercourses have virtually no aesthetic or amenity value, support only a limited range of ecosystems, and do not meet the water quality standards or objectives, such as the ones prescribed by the European Union Water Framework Directive for its member countries.

CLOSING REMARKS

The foregoing sections of this chapter have endeavored to provide an international perspective on urban water issues, highlighting especially the contrasts—and, in some cases, the commonalties—between cities in industrialized countries and those in the developing ones. While in a broad context the division of developed vs. developing world cities can aid in classifying the types of problems and concerns and of possible solutions, it is necessary to emphasize that each city has its own identity and set of conditions. Cities in developing countries actually present a very broad range of socio-economic and developmental settings.

There are winds of change in the way the urban water planning and management will be approached in the future, and the associated technological and scientific challenges are great. The paradigm shifts in this area go beyond the institutional and decision-making spheres, affecting the directions of research and technological development. For instance, further consideration should be given to the expansion of the possible range of solutions for sanitation to include less water-intensive (quantity and quality) options, to the inclusion of stormwater as a resource (and not a nuisance to be evacuated), to integrated urban water management approaches incorporating river basin planning, urban groundwater aspects, urban stream rehabilitation, urban/peri-urban agriculture etc. A number of these possibilities and concerns are pointed out in the text.

However, solving the urban water problems on the long-term basis requires much broader actions including not only water, but also other material or substances that are used and usually discarded in urban areas and that directly or indirectly interact with drinking or wastewater. Their negative effect on availability and quality of urban water is either fully eliminated or significantly reduced by dealing with them at the source. Some of these are:

- *Interactions of solid wastes*—in many developing countries solid waste is not properly managed: urban streams serve as ultimate recipients, solid waste is not separated at the source, leachate from landfills pollutes surface and groundwater, and hazardous wastes (including endocrine disruptors) are discarded into the environment;
- *Interactions with other material flows in urban areas*—such as materials used in industry, urban agriculture, transportation etc; and
- *Interactions with air pollutants*—the latter reach the urban streams via storm runoff.

Furthermore, the shift of approaches in urban water problem-solving requires going beyond the creation of sustainable technologies, to the consideration of all aspects affecting the situation in practice. This will entail the testing and application of such technologies within the realms of actual social, ethical, political and organizational conditions. These factors call for well-conceived and integrated multidisciplinary and participatory approaches for effective planning and implementation, particularly in this highly dynamic era of globalization.

North America and Europe exercise the leadership in research and technological development in urban water topics, but the needs differ in other parts of the world due to social dynamics, developmental levels and climatic differences. Thus the IGOs (international governmental organizations), such as those of the UN system, and the international scientific community and the evermore present NGOs (non-governmental organizations) should support the establishment of international and regional centers of excellence on urban water management and network-building efforts. The approaches and technologies that result from such a distributed effort will allow a more efficient and realistic handling of the critical situations that many cities are facing. And while recognizing that science and technology are essential to reaching integral and effective solutions, they must remain at the service of the objectives of sustainability and social equity.

REFERENCES

Asano, T., Wastewater management and reuse in mega-cities, in *Water for Urban Areas: Challenges and Perspectives*, edited by J.I. Uitto and A.K. Biswas, pp 135-155, United Nations University Press, Tokyo, 2000.

Ashley, R.M., A. Fraser, R. Burrows, and J. Blanksby, The management of sediment in combined sewers, *Urban Water*, Elsevier, 2 (4), 263-275, 2000.

Bahri, A., Alternative water management approaches to ensure environmentally sound urban and peri-urban water related relationships—closing the links, in *Proceedings Urban Stability Through Integrated Water-related management—The 9th Stockholm Water Symposium*, p. 137, Stockholm International Water Institute, Stockholm, Sweden, 1999.

Burkhard, R., and A. Deletic, and A. Craig, Techniques for water and wastewater management: a review of techniques and their integration in planning, *Urban Water*,

Elsevier, 2 (3), 197-221, 2000.

Butler, D. and C. Maksimovic, Urban water management—challenges for the third millennium, *Progress in Environmental Science*, 1(3), 213-225, 1999.

Butler D. and C. Maksimovic, Chapter 3: Interactions with the environment, in *Frontiers in Urban Water Management—Deadlock or Hope*, edited by C. Maksimovic and J.A. Tejada-Guibert, pp. 85-142, IWA Publishing, London, England, 2001.

CEHA (Regional Centre for Environmental Health Activities, Regional Office for the Eastern Mediterranean, World Health Organization), Technical Note: *Sanitation and Wastewater Management for Small Communities in Eastern Mediterranean Region Countries*, Amman, Jordan, 2000.

CIRIA (Construction Industry Research and Information Association), *Sustainable Urban Drainage Systems, Best Practices Manual forEngland, Scotland, Wales and Northern Ireland*, ISBN 0 86017 523 5, London, 2001.

Covas, D., I. Stoianov, D. Butler, C. Maksimovic, N. Graham, and H. Ramos, Leakage Detection in pipeline systems by inverse transient analysis—from theory to practice, Proc. CCWI Conference, Leicester in Water *Software Systems: theory and applications*, edited by B. Ulanicki, B. Coulbeck. J. Rance, Publ. Research Studies Pres, pp 3-16, 2001.

Deletic, A., and C. Maksimovic, Evaluation of Water Quality Factors in Storm Runoff from Paved Areas, *J. Env. Eng. Div.*, ASCE, Vol. 124 (9), 869-879, 1998.

Delleur, J.W., New results and research needs on sediment movement in urban drainage, *J. Water Resour. Plng. and Mgmt. Div.*, ASCE, 127 (3), 186-193, 2001.

Esrey, S.A., I. Andersson, A. Hillers, and R. Sawyer, Closing the loop—Ecological sanitation for food security, Publications on Water Resources No. 18, UNDP/Swedish International Development Cooperation Agency, Mexico, 2001.

Field, R., M. Borst, T.P. O'Connor, M.K. Stinson, C-Y Fan, J.M. Perdek, and D. Sullivan, Urban wet weather flow management: Research directions, *J. Water Resour. Plng. and Mgmt. Div.*, ASCE, 124 (3), 168-180, 1998.

Foster, S., A. Lawrence, and B. Morris, Groundwater in urban development: Assessing management needs and formulating policy strategies, *World Bank Technical Paper No. 390*, The World Bank, Washington D.C., USA, 1998.

Gleick P.H., The consequences of water scarcity: measures of human well-being, Proceedings of the International Conference on World Water Resources at the Beginning of the 21st Century—Water: a looming crisis? *Addendum to IHP-V Technical Documents in Hydrology No. 18*, UNESCO, Paris. 1998.

Halhal, D., G.A. Walters, D. Ouazar, and D.A. Savic, Multi-Objective Improvement of Water Distribution Systems Using a Structured Messy Genetic Algorithm Approach, *J. Water Resour. Plng. and Mgmt. Div.*, ASCE, Vol. 123 (3), 137-146, 1997.

Heavens, J.W., *Trenchless Renovation of Potable Water Pipelines,* Insituform® Technologies, Inc., 12 pp. 1997.

Herrmann, T., and U. Schmida, Rainwater utilisation in Germany: efficiency, dimensioning, hydraulic and environmental aspects, *Urban Water*, Elsevier, 1 (4), 307-316, 1999.

Hua, M., Towards a Recycling Society—A Case Study on the Successful Implementation of the Pilot Project in Dalu Village, China, in Proceedings Symposium Frontiers in urban water management: Deadlock or hope?, edited by J.A. Tejada-Guibert and C. Maksimovic, *Technical Documents in Hydrology No.45,* pp 161-167 UNESCO, Paris, 2001.

Jankovic-Nisic, B., C. Maksimovic, D. Butler, and N. J. D. Graham, *Use of Flow Meters for Managing Water Supply Networks,* accepted for publication, *J. Water Resour. Plng. and Mgmt. Div.*, ASCE, 2002.

Kalbermatten, J.M., R. Middleton, and R. Schertenleib, Household-Centred Environmental Sanitation. EAW-SANI, Swiss Federal Institute for Environmental Science and Technology, 1999.

Kolsky, P., *Storm Drainage—An engineering guide to the low-cost evaluation of system performance*, Intermediate Technology Publications, London, 1998.

Lerner, D., Guest editor's preface, Theme issue: Urban Groundwater, *Hydrogeology Journal*, 4 (1), 4-5, 1996.

Lundqvist, J., S. Narain, and A. Turton, Chapter 8: Social, institutional and regulatory issues, in *Frontiers in Urban Water Management—Deadlock or Hope*, edited by C. Maksimovic and J.A. Tejada-Guibert, pp 344-398, IWA Publishing, London, England, 2001.

Lyonnaise des Eaux, *Alternative solutions for water supply and sanitation in areas with limited financial resources*, Nanterre, France, 1998.

Maksimovic, C., Measurement of water quantity in urban areas, in Rain and floods in our cities—gauging the problem, edited C. Maksimovic, World Meteorological Organization, *Technical Reports in Hydrology and Water Resources No. 53*, WMO/TD No. 741, Geneva, Switzerland, 1996.

Maksimovic, C. M. Ivetic, D. Prodanovic, D. Pavlovic, N. Jacimovic, M. Milicevic, S. Ragibovic, V. Mauclert, *Elements of Sustainability in Water Distribution Systems—Case Study Laktasi.* Proc. of the LIFE Conference "Sustainability of Water & Environmental Systems Rehabilitation, edited by C. Maksimovic and B. Jaksic, ISBN 86-7440-013-2, pp 186-195, UZRS, Banja Luka, 2001.

Mara, D.D. and A.S.P. Guimarães, Simplified Sewerage: Potential applicability in industrialized countries, *Urban Water*, Elsevier, 1 (3), 257-259, 1999.

Marsalek, J. Urban runoff quality measurements, in Rain and floods in our cities—gauging the problem, edited C. Maksimovic, World Meteorological Organization, *Technical Reports in Hydrology and Water Resources No. 53*, WMO/TD No. 741, Geneva, Switzerland, 1996.

Matsui, S., M. Henze, G. Ho and R. Otterpohl, Chapter 5: Emerging Paradigms in water supply and sanitation, in *Frontiers in Urban Water Management—Deadlock or Hope*, edited by C. Maksimovic and J.A. Tejada-Guibert, pp. 229-263, IWA Publishing, London, England, 2001.

Niemczynowicz, J., Rainfall measurement, processing and application in urban hydrology, in Rain and floods in our cities—gauging the problem, edited C. Maksimovic, World Meteorological Organization, *Technical Reports in Hydrology and Water Resources No. 53*, WMO/TD No. 741, Geneva, Switzerland, 1996.

Niemczynowicz, J., Urban hydrology and water management—present and future challenges, *Urban Water, Elsevier*, 1 (1), 1-14, 1999.

Nouh, M., editor, Technical *Document in Hydrology N°40, Urban drainage in specific climates.* Vol. III. Urban drainage in arid and semiarid climates. UNESCO/IRTCUD, 2001.

Otterpohl, R., Black, brown, yellow, grey—the new colours of sanitation, *Water 21*, International Water Association, October 2001.

Parikh, H.H., Slum networking—Using slums to save cities, *Technical Document in*

Hydrology N°45: Proceedings of the International Symposium Frontiers in urban water management: Deadlock or hope?, Marseille, France, UNESCO, 2001.

Patorni, F.-M., Water Demand management strategies in cities, City Development Strategies—The Journal Online issue September 2000, The World Bank, Washington, D.C., 2000.

Quon S., Planning for urban agriculture: a review of tools and strategies for urban planners, *IDRC Cities Feeding People Report Series No. 28*, 1999.

Rees, W.E., The built environment and the ecosphere: a global perspective, *Building Research and Information*, 27 (4/5), 206-220, 1999.

Roesner, L.A., B.P. Bledsoe and R.W. Brashear, Are Best-Management-Practice criteria really environmentally friendly?, *J. Water Resour. Plng. and Mgmt. Div.*, ASCE, 127 (3), 150-154, 2001.

Saegrov, S., J.F.M. Baptista, P. Conroy, R.K. Herz, P. LeGauffre, G. Moss, J.E. Oddevald, B. Rajani and M. Schiatti, Rehabilitation of water networks. Survey of research needs and on-going efforts, *Urban Water*, Elsevier, 1 (1), 15-22, 1999.

Saegrov, S., J. Milina and T. Thorolfsson, editors, Technical *Document in Hydrology N°40,Urban drainage in specific climates.* Vol. II. Urban drainage in cold climates, UNESCO/IRTCUD, 2001.

Shamir, U., and C.D.D. Howard, Management of urban water: Introduction, *J. Water Resour. Plng. and Mgmt. Div.*, ASCE, 126 (3), 114-117, 2000.

Shiklomanov, I., editor, Assessment of water resources and water availability in the world, background document for the *Comprehensive assessment of the freshwater resources of the world*, UN System/Swedish Environmental Institute, Stockholm, 1997.

Strecker, E.W. M.M. Quigley, B.T. Urbonas, J.E. Jones and J.K. Cleary, Determining urban stormwater BMP effectiveness, *J. Water Resour. Plng. and Mgmt. Div.*, ASCE, 127 (3), 144-149, 2001.

Tafuri, A.N., and A. Selvakumar, Wastewater collection system infrastructure research need in the USA, *Urban Water*, Elsevier, 4 (1), 21-30, 2002.

Tejada-Guibert, J.A. and C. Maksimovic, Chapter 9: Outlook for the 21st Century, in *Frontiers in Urban Water Management—Deadlock or Hope*, edited by C. Maksimovic and J.A. Tejada-Guibert, pp 399-409, IWA Publishing, London, England, 2001.

Tucci, C., editor, *Technical Document in Hydrology N°40,Urban drainage in specific climates,* Vol. I. Urban drainage in humid tropics. UNESCO/IRTCUD, 2001.

UNESCO & Académie de l'Eau, *Proceedings Symposium Water, the City and Urban Planning, Paris 10-11 April 1997*, UNESCO Paris, France, 1998.

United Nations Centre for Human Settlements (Habitat) and United Nations Environment Programme (UNEP), Proposal for a Multi-Agency Programme for Integrated Water Resources Management in Peri-Urban Areas, a discussion paper prepared for the UN ACC Subcommittee on Water Resources, 1996.

United Nations Centre for Human Settlements (Habitat), *State of the Worlds Cities Report 2001*, Nairobi, 2001.

United Nations, Department of Economics and Social Affairs, *World Urbanization Prospects—The 1999 Revision*, ST/ESA/SER.A/194, United Nations, New York, 2001a.

United Nations, Department of Economics and Social Affairs and United Nations Centre for Human Settlements (Habitat), *Compendium of Human Settlements Statistics 2001—Sixth Issue*, United Nations, New York, 2001b.

UNEP-IETC (International Environmental Technological Centre), *Sustainable Wastewater and Stormwater Management, Source Book and Training Manual*, in press, 2002.

van den Berg, L., Urban Agriculture as the Combination of Two 'Impossible' though Sustainable Trends, ALTERRA Green World Research, Wageningen UR, IGU—Commission On Sustainable Rural Systems, Pusan Conference, Korea, 2000.

Vlachos, E. and B. Braga, Chapter 1: The challenge of urban water management, in *Frontiers in Urban Water Management—Deadlock or Hope*, edited by C. Maksimovic and J.A. Tejada-Guibert, pp 1-36, IWA Publishing, London, England, 2001.

Yevjevich, V., New concepts for water supply systems—what kind of water for what purpose in the future, in Water *Supply Systems: New Technologies*, NATO ASI Series, Vol. 15, edited by C. Maksimovic, F. Calomino and J. Snoxell, Springer, 43-57, 1996.

Walski, T.M., The wrong paradigm—why water distribution optimization doesn't work, *J. Water Resour. Plng. and Mgmt. Div.*, ASCE, 127 (4), 203-205, 2001.

World Health Organization and United Nations Children's Fund (UNICEF), *Global Water Supply and Sanitation Assessment 2000 Report*, Office of Publications, World Health Organization, Geneva, Switzerland/Editorial and Publications Section, UNICEF, New York, USA, 2000.

José Alberto Tejada-Guibert, Division of Water Sciences, UNESCO, 1 rue Miollis, 75732 Paris Cedex 15, France.

Cedo Maksimovic, Department of Civil and Environmental Engineering, Imperial College of Science, Technology and Medicine, London SW7 2BU, United Kingdom.

4

Groundwater Issues

David P. Ahlfeld and Weston R. Dripps

INTRODUCTION

This chapter introduces the reader to some of the challenges and issues involved in the use of groundwater. It begins with a review of the basic concepts of groundwater hydrology and the definition of some key terms and then continues with sections on the importance of groundwater to human needs, threats to its use, and current issues in groundwater policy.

Groundwater is the water present in the interconnected void spaces between geologic materials within the subsurface. It resides in the pores within granular materials, such as sand and gravel, and in fractures that permeate the rock matrix. Groundwater is the largest reservoir of liquid freshwater on the planet. Only 2.5% of the world's total water is fresh water; 68.7% of which is solidified in glaciers and permanent snow cover [*Shiklomanov*, 1993]. Of the remaining 31.3%, 96.2% is groundwater, with surface water, soil moisture, atmospheric water, permafrost, and biological water accounting for the other 3.8% [*Shiklomanov*, 1993].

Water enters the subsurface through infiltrating precipitation and through infiltration from surface water bodies. Once in the groundwater system, water moves through the subsurface, at varying rates, in response to pressure and elevation differences and the permeability of the geologic materials through which it travels. Water leaves the groundwater system via evapotranspiration, extraction by groundwater wells, and discharge to springs, wetlands, streams, lakes, and oceans.

Groundwater is present at depth in nearly all geologic formations. Those formations which have relatively high permeability are called aquifers. Aquifers can occur in a variety of geologic media and over a wide range of spatial scales. Aquifers can be classified as confined or unconfined. Confined aquifers are bounded by low permeability geologic formations and can exist under artesian conditions. Unconfined aquifers have no confining geologic formation at their upper surface. Figure 1 depicts a block section of a land form with an unconfined

Water: Science, Policy, and Management
Water Resources Monograph 16

10.1029/016WM05

aquifer. The water table represents the approximate boundary between sediments that are fully saturated with water (zone of saturation) and sediments that are unsaturated (unsaturated zone). The elevation of the water table, along with other dimensions and properties of the aquifer, indicate the volume of water stored in an unconfined aquifer. Groundwater often supplies flow to streams which can be critical for maintaining riparian, wetland, and in-stream ecosystems. Such a case is depicted in Figure 1 where the water table is at a higher elevation than the stream producing groundwater flow to the stream.

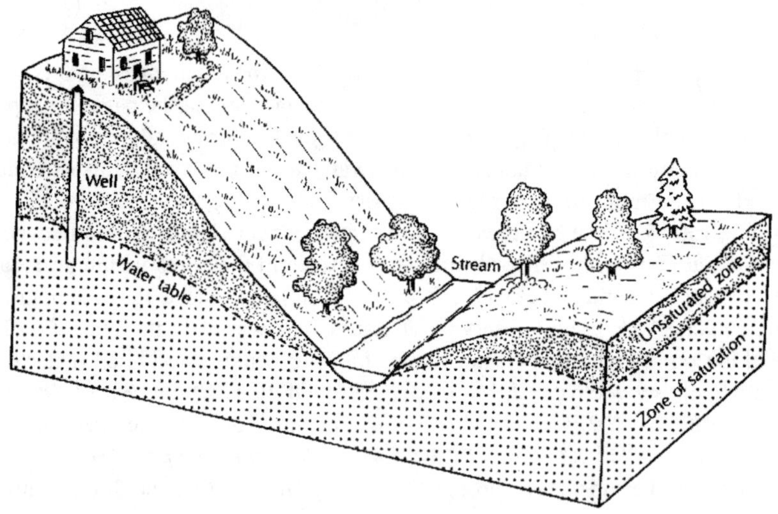

Figure 1. Water table aquifer showing saturated and unsaturated zones and groundwater connection with stream (from Fetter, 2001, by permission).

In many areas, aquifers contain large volumes of water and are highly valued for water resource management. Groundwater is often used to meet human and natural demands, particularly during extended periods of low precipitation. Groundwater flow into a stream maintains aquatic ecosystems, and can constitute a significant proportion of a stream's total flow, particularly during periods of extended low precipitation. Groundwater can also be used to meet human demands for consumption, especially when surface water sources are stressed or depleted. Removal of groundwater from an aquifer will reduce the volume of water stored in the aquifer. This "lost" water will be replenished naturally or recharged during periods of substantial precipitation.

GROUNDWATER USAGE

The volume of renewable fresh water available to support the world's population of over 6 billion people is less than 1% of the world's total water supply. This renewable volume translates to over 7,000 cubic meters per capita potentially available annually for human use [*World Resources Institute*, 2000]. According to Malin Falkenmark's "water stress index", an area whose per capita annual renewable fresh water availability exceeds 1,700 cubic meters per year is considered to be low water stress. Consequently, on a global average there should be adequate water supplies to meet human needs; however, spatial and temporal imbalances in the demand and the occurrence of both surface and groundwater resources produce localized non-sustainable use of water resources on every continent except for Antarctica [*Gleick*, 1998].

There have been a number of global assessments [e.g., *L'vovich*, 1974; *Falkenmark and Lindh*, 1974a; *Baumgartner and Reichel*, 1975; *De Mare*, 1976; *Korzoun*, 1978; *Ambroggi*, 1980; *Gleick*, 1993, 1998; *Postel et al.*, 1996; *Alcamo et al.*, 1997; *Raskin et al.*, 1997; *Gardner—Outlaw and Engelman*, 1997; *Seckler et al.*, 1998; *Shiklomanov*, 1998, 2000; *Vörösmarty et al.*, 2000; *WRI*, 2000] of renewable water resources and current water use, but to our knowledge, there has been no comprehensive detailed global study that isolates and accurately quantifies the groundwater recharge component. The World Resources Institute [2000] compiled and presented groundwater recharge and withdrawal information on a country-by-country basis, but no estimates were provided for most African, Central American, and South American countries due to a lack of available data. Current global water studies typically lump the renewable surface water and groundwater components together, and report total water annually available and total water annually used.

The annual groundwater recharge constitutes a very small percentage of the total global groundwater reserves which Engelman and LeRoy [1993] estimated at somewhere between 4 million and 60 million cubic kilometers. This is a vast amount at either bound although much of this groundwater lies too deep or remote to be considered reasonably accessible. Use beyond the annual groundwater recharge rate is inherently unsustainable.

Data on groundwater levels, aquifer storage capacities, and recharge rates are nonexistent for most countries of the world and are often incomplete or unreliable for those countries with data. Current data are simply insufficient to make an accurate assessment of the global groundwater recharge and the total groundwater resource. Similarly, a lack of records, instrumentation, and standard accounting procedures for most areas of the world prevent realistic estimates of global groundwater use. Existing data sets are of varying quality, scope, and reliability making aggregation problematic. Furthermore, groundwater's role in indirect uses, such as supporting ecological systems, has yet to be considered in global water use assessments. Only recently have water requirements for aquatic ecosystems begun to be placed on an equal level with agricultural, urban, and industrial demands, particularly in the Western United States where laws have been changed or passed that legally allocate water to help satisfy environmental water needs. From the limited data available, *Shah et al.* [2000] estimated annual global groundwater use at somewhere between 750 to 800 cubic kilometers.

Importance of Groundwater for Meeting Future Water Needs

Surface water is usually more readily accessible than groundwater, and consequently is typically utilized first to meet water needs, but in many areas of the world, surface water resources are nonexistent, already appropriated, or of too poor quality for the desired use. Furthermore, degradation of natural ecosystems by dams and surface water withdrawals are making additional appropriations of surface water increasingly less favorable in the public eye. Surface water demands have become so high along a number of the world's rivers, like the US's Colorado and Rio Grande Rivers, the Middle East's Jordan River, and China's Yellow River, that flow volumes have been reduced to a trickle at their outlets in given years [*Postel,* 1999a; *De Villiers*, 2000]. The Aral Sea, once the fourth largest inland body of water, has been reduced to one-third its historical volume due to the reduction of inflow from the Amu Dar'ya and Syr Dar'ya rivers as a result of heavy withdrawals for upstream irrigation [*Gleick*, 1993]. In China, 80% of the country's 50,000 major river kilometers are so degraded that they can no longer support any fish [*De Villiers*, 2000].

If water demands continue to rise and surface water sources become further stressed as most current projections predict [*Gleick*, 2000], humans will grow increasingly dependent on groundwater reserves to meet future water needs. In many areas of the world this increased dependence is already evident. In Taiwan, for instance, the percent of total freshwater supplied by groundwater increased from 21 to 40 percent from 1983 to 1991 [*Sampat*, 2000]. For many regions, groundwater has become more appealing than its surface water counterpart as it is typically of higher quality and is more dependable. In Pakistan, groundwater was the fastest-growing form of irrigation from the mid-1960s through the 1980s; the number of groundwater wells rose from some 25,000 in 1964 to nearly 360,000 in 1993 [*Postel*, 1999b]. Surface water volumes are highly dependent on seasonal precipitation events and thus can fluctuate drastically over the course of a year, making them less reliable. Groundwater is now the primary drinking water source for an estimated 1.5 billion people worldwide, and in the United States, more than 95% of the rural population depends on groundwater for their drinking water [*Sampat*, 2000].

Projected Use

Continued population growth, increasing food demands, and industrial and urban expansion are likely to compel humans to utilize an increasing share of the world's groundwater. There are a number of global studies that project total fresh water use [*L'vovich*, 1974; *Falkenmark and Lindh*, 1974b; *De Mare*, 1976; *Alcamo et al.*, 1997; *Gleick*, 1997; *Raskin et al.*, 1997; *Seckler et al.*, 1998; *Shiklomanov*, 2000], but we are unaware of any global study which specifically projects future groundwater use.

Early water projections typically relied on the extrapolation of historical and current water use trends, and assumed that the key social and economic driving forces and current management strategies that influence water use would remain the same. These "business as usual" projections often contained no variants, did not account for environmental needs, and neglected potential economic and resource limitations [*Gleick*, 2000]. The majority of these studies grossly over-estimated projected water demands as they did not account for many of the political, legal, economic, and social factors that ultimately determine how groundwater resources are managed and utilized. Actual global water withdrawals for the mid 1990s were in fact only about half of what they had been predicted to be thirty years before [*Gleick*, 2000]. Past experience has shown that it is difficult to predict what types of technological advances will arise in the future that will influence our use of groundwater.

More recent water projections have become more complex in their approach and more detailed in their temporal and spatial scales. The most recent study by *Shiklomanov* [2000] estimated that by 2025 agriculture is expected to increase its requirements for water withdrawal by 1.3 times, industry by 1.5 times, and public supply by 1.8 times. The current annual total water withdrawal in the world is about 8.4% of the renewable global freshwater resource, and by 2025, is expected to rise to 12.2% [*Shiklomanov*, 2000]. With limited surface water of sufficient quality or quantity left to utilize in many areas of the world, a significant fraction of this projected increase will likely have to come from additional ground-water withdrawals.

Projections for future groundwater use are further complicated by the uncertainty associated with global population estimates. As of mid-2000, the world population was an estimated 6.1 billion people, and by 2050 is expected to be between 7.9 and 10.9 billion [*United Nations*, 2000].

Agriculturally, many of the world's most important grainlands in China, the United States, and India are consuming groundwater at unsustainable rates [*Postel*, 1999b]. The question remains as to where farmers are going to find enough water to grow the additional food needed to feed the extra roughly 2 – 5 billion people projected to be joining the world's population by 2050 [*United Nations*, 2000]. The exact amount of additional water required will depend not only on the number of people, but also on the crop selection, crop water requirements, and irrigation efficiency. Clearly some of this water will have to come from additional groundwater development. Agricultural stresses on the ground water system will only be accentuated by continued soil degradation, salination, and excessive fertilizer and pesticide applications that already are widespread in many food producing regions.

Despite all the unknowns and uncertainties in projecting groundwater demands, it seems highly likely that humans will use a larger share of the earth's groundwater in the future to meet agricultural, industrial, environmental, and urban needs. Groundwater shortages and aquifer pollution are already creating widespread public health problems, limiting economic and agricultural development, and causing severe ecological degradation in many areas of the world [*Postel et al.*, 1996; *United Nations*, 1997]. It is now widely accepted that regions of India, China, the United States, Northern Africa, and the Middle East that heavily depend on groundwater will face groundwater crises in the coming decades because of overutilization, pollution, inadequate regulation, and/or cross boundary groundwater issues [*Biswas*, 1999]. Already in many of these areas, the rate of groundwater extraction exceeds the rate of recharge replenishment [*Postel*, 1999a].

Although the major global models predict increased water demands [*Gleick*, 1998], examining water data for the United States may provide a basis for some optimism. In the U.S., total water use and groundwater use have actually been declining in the past 20 years, a radical departure from expectations and the current and projected global trend [*Gleick*, 1998]. The total water use and per capita water use profiles mimic the groundwater profiles presented in figures 2 and 3 with groundwater consistently constituting roughly 22% of the U.S.'s annual water use. Before 1980, the U.S. experienced a 240% increase in groundwater withdrawals during a period when the U.S. population increased by 150%. Total and per capita groundwater use was on the rise, but withdrawals peaked in 1980 and since have been declining despite a growing population (Figures 2 and 3). As of 1995, [*Solley et al.*, 1998], total groundwater withdrawals had declined almost 8% and per capita groundwater use had dropped 20% since 1980 (Table 1). Whether this decrease in demand will continue remains to be seen, but with the majority of the available surface water systems currently being utilized or appropriated to environmental needs, a significant fraction of any future increases would likely have to be satisfied through further groundwater development.

Table 1. Groundwater use in the United States (in m^3/sec) [adapted from Solley et al., 1998]

Year	Public Supply	Domestic	Irrigation	Livestock	Thermo-electric	Other Uses[a]	Total Use	Per Capita Use (m^3/yr)
1960	273	83	1,314	39	40	258	2,190	388
1965	354	96	1,840	44	48	298	2,629	431
1970	412	110	1,971	48	61	350	2,979	463
1975	482	118	2,497	53	61	421	3,592	530
1980	526	145	2,629	53	70	438	3,636	505
1985	640	142	2,002	132	27	267	3,207	421
1990	662	143	2,234	118	23	301	3,478	441
1995	662	147	2,147	99	25	267	3,347	401

[a]Other Uses includes Commercial, Industrial, and Mining Applications.

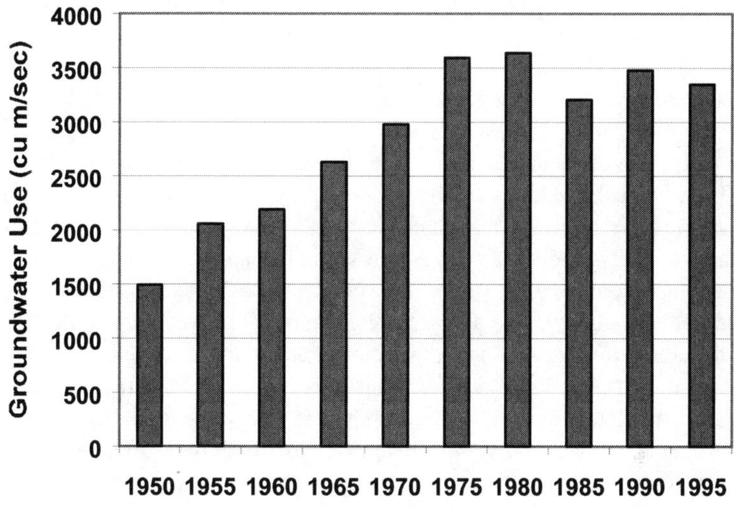

Figure 2. Groundwater use in the United States, 1950 to 1995.

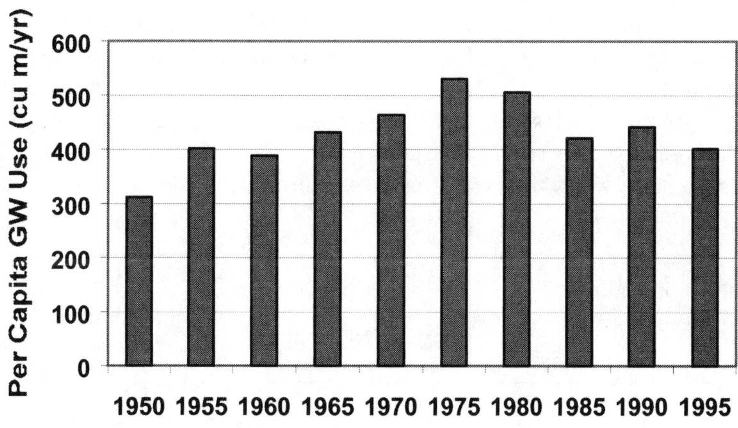

Figure 3. Per capita groundwater use in the United States, 1950 to 1995.

LONG TERM AVAILABILITY OF GROUNDWATER

Groundwater is often easily available and of high quality. The quantity of groundwater available and its suitability for human uses are dependent on local geologic and hydrologic conditions. The long term availability of groundwater is affected by overuse or other disruptions in the hydrologic cycle and by impairment of water quality. These impacts on groundwater quantity and quality are discussed in this section.

Water Quantity

Our discussion of water quantity will be centered around the notion of a water balance. The water balance for an aquifer can be expressed with three components: flows into the aquifer (both natural and human induced), flows out of the aquifer (both natural and human induced), and storage of water in the aquifer (Figure 4).

Variations in aquifer storage at the multi-year time scale can occur due to droughts or extended periods of above average precipitation. However, under natural conditions, the annual average flows in most aquifers are in approximate equilibrium with the total natural flow into the aquifer about equal to the total natural flow out of the aquifer and the storage in the aquifer remaining roughly constant. Large-scale extraction of water by human activity will change this equilibrium. Depending on the magnitude of this extraction and the aquifer-specific conditions, a new equilibrium may be achieved after some time has elapsed.

Groundwater mining and resource depletion. If the total flow of extracted water is greater than the flow into the aquifer, then the natural flow out of the aquifer will be disrupted or stopped, and the storage in the aquifer will decline. The phenomena of human induced extractions producing reductions in aquifer storage beyond those expected from natural variations is referred to as aquifer mining or overdraft. Aquifer mining is a fundamentally unsustainable practice. If groundwater is used faster than it is replenished, then the groundwater present in the aquifer will be depleted, and the ability of the aquifer to store water may eventually be diminished.

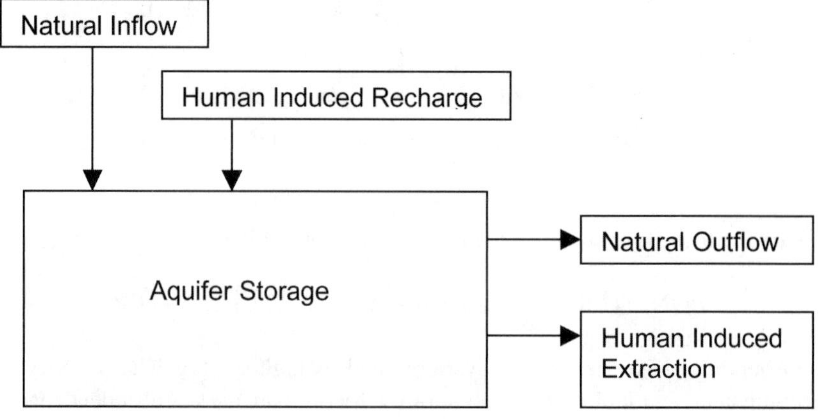

Figure 4. Schematic of flow balance in an aquifer.

The High Plains Aquifer in the United States (which includes the Ogallala formation) provides an example of aquifer mining. This aquifer extends from north central Texas to South Dakota and provides substantial irrigation water for the region. Groundwater pumping began in the late nineteenth century to support agriculture. It is estimated that before pumping began about four trillion cubic meters of water were stored in this aquifer [*Fetter*, 2001]. In addition, natural recharge to the aquifer and discharge from the aquifer to streams are both estimated to have been about 250 million cubic meters per year. At present, the net rate of storage loss is about 3.3 billion cubic meters per year. While this quantity is a small fraction of the total aquifer storage, it has already led to significant ecosystem disruption with natural discharge to streams estimated to have been reduced to about 100 million cubic meters per year. In addition, the water table has dropped as much as 30 meters in some locations [*Alley et al.*, 1999]. Large decreases in water table levels increase pumping costs and can impact the economic feasibility of farming these lands.

The High Plains example is typical of aquifer mining that is occurring around the world to support agriculture. The water table under some of the major grain-producing areas in northern China is falling at a rate of over a meter per year [*Postel*, 1999b; *National Intelligence Council*, 2000]. Water tables beneath agricultural fields in India are falling an average of 1 to 3 meters per year due to groundwater pumping [*National Intelligence Council*, 2000]. A recent World-Bank memorandum on water management in Yemen noted: "the problem of groundwater mining represents a fundamental threat to the wellbeing of the people of Yemen where in the highland plains, for example, abstraction is estimated to exceed recharge by 400 percent" [*Shah et al.*, 2000].

Groundwater mining and aquifer depletion are also associated with increased urbanization and industrialization which increase local and regional demands for water. In addition to increasing water demand, urbanization can reduce natural recharge by diverting stormwater generated from paved and roofed areas directly to streams or the sea. In South Korea's industrialized cities like Seoul and Pusan, water tables have dropped 10 to 50 meters over the last thirty years due to industrial pumping [*Shah et al.*, 2000]. In Tucson, Arizona, since the early 1950s, urban withdrawals have lowered the water table as much as 60 meters in some areas of the city [*Pagano et al.*, 2001].

Groundwater mining can also reduce an aquifer's storage capacity. Aquifer systems that contain a significant fraction of clays can be subject to subsidence when groundwater is mined. Once water is removed from these formations, subsurface pressure distributions change, and the sediments compact. Once compaction occurs, the storage capacity of the aquifer is permanently reduced [*Davis and DeWiest*, 1966]. For thick formations, this compaction can lead to significant subsidence of the land surface. Portions of Mexico City experienced as much as 7.5 meters of land surface subsidence due to heavy pumping during the period from 1930 to 1960 [*Ortega-Guerrero et al.*, 1993]. As in other regions where significant subsidence has occurred, such as Las Vegas, Nevada [*Bell*, 1997] and Tokyo, Japan [*Ministry of the Environment, Government of Japan*, 2002], Mexico City has suffered extensive damage to buildings and civil infrastructure due to differential compaction from dewatering. In settled coastal regions, the interplay of subsidence with sea level rise may cause significant flooding problems. The Romagna region of Italy along the Adriatic coast has experienced as much as 1 meter of land surface subsidence due to groundwater pumping since the 1950s. When combined with predicted sea level rises due to global climate change, this subsidence may lead to permanent submersion of significant portions of this coast [*Gambolati et al.*, 1999].

Sustainable groundwater use and ecosystem impacts. Returning to the water balance concepts depicted in Figure 4, even if the flow extracted for human use is smaller than the flow into the aquifer such that mining is not occurring, the natural equilibrium will still be disrupted. Human extraction of water will produce declines in the water table which may affect wetlands or vegetation that depend on a shallow water table. In addition, human induced extraction of water decreases the natural outflow from the aquifer such that water that would have discharged to springs, streams or elsewhere is now being intercepted and drawn for human use.

Alley et al. [1999] describe the impact of large-scale sewer and waste water treatment systems on the water balance on Long Island, New York. They estimated that the pre-development discharge of groundwater to streams, springs, and the ocean was about 1,100 million cubic meters per year. As the residential areas of the region were sewered, water that had been disposed in on-site septic recharge systems was exported to centralized waste water treatment plants and discharged directly to the sea. A recent estimate indicates that about 180 million cubic meters of waste water per year are now discharged through sewer systems directly to the sea [*Alley et al.*, 1999]. Thus, about 16% of the water that would have naturally discharged from the aquifer is diverted and artificially transported, through pumping, use, and disposal to the sea.

As the Long Island example suggests, although extracting groundwater at rates below the natural recharge rate will yield a new, sustainable equilibrium, any change in equilibrium will produce impacts on the natural system that existed before human utilization. Specifically, streams or wetlands may go dry, vegetation patterns may change, and wildlife ecosystems may be disrupted or modified. The identification of a tolerable level of change to a natural system is an ongoing policy issue at all levels.

Another threat to the existing water balance equilibrium is climate change. The inflow to the water balance for a groundwater system is related to precipitation. Climate change due to the anthropogenic discharge of greenhouse gases is expected to have significant impacts on precipitation and the hydrologic cycle. Regional responses will vary depending on numerous factors. Some locations may see increased precipitation while others will see decreased precipitation.

There is presently broad consistency between climate model projections of a decrease in groundwater recharge in central Asia, southern Africa and the Mediterranean region [*IPCC*, 2001].

Water Quality

The quality of groundwater is a function of its chemical composition, its temperature, and other physical properties that determine its suitability for an intended use. Water can be considered contaminated if it is unsuitable for a certain use (for example, human consumption), or if the quality of water has been significantly changed from a previous condition to which the local ecosystem has become accustomed. For humans and other organisms, certain contaminants have been found to be carcinogenic, disruptive to endrocine activity, toxic at high doses, and the source of a myriad of other negative health affects. For plants, high levels of contamination, such as salinity, can have deleterious impacts.

After falling as precipitation, water enters the subsurface where it is in direct contact with natural geologic and biologic materials which react with the water and consequently affect its quality. The infiltrating water may also encounter petroleum products, synthetic chemicals, or other substances emplaced in the subsurface by human activity.

Natural contamination. Water in the subsurface is in direct contact with the host geologic materials, providing the opportunity for geochemical reactions to occur between the rock and water. Calcium, for instance, which is a common constituent of groundwater, enters groundwater from the dissolution of certain types of marine sedimentary rocks. When the host materials contain toxic or hazardous materials, the groundwater can become contaminated and thus unsuitable for human consumption. For example, in Bangladesh, approximately 20 million people are at risk of arsenic poisoning as a result of concentrations of naturally occurring arsenic exceeding 50 micrograms per liter in up to one million shallow wells used for human consumption [*McArthur et al.*, 2001].

In many regions, especially near seacoasts, groundwater may be brackish or saline making it unsuitable for human consumption and for most other human uses. This problem can be exacerbated by pumping activity. In most coastal areas, under natural conditions, groundwater discharges to the ocean through seeps and springs located below sea level. If groundwater is pumped, the direction of flow may be reversed, and saline water may be drawn into the aquifer. Saltwater intrusion is a problem in numerous coastal regions of the world. For example, in Manila, groundwater overdrafting has lowered water levels in the city's aquifer fifty to eighty meters and caused the saltwater / freshwater interface to move as far as five kilometers inland, contaminating portions of the city's main drinking supply [*Sampat*, 2000].

Contamination by human activity. Groundwater contamination caused by human activity can take many forms. Factors to consider when dealing with groundwater contamination include the mechanisms by which the contaminant entered the subsurface, the concentration of the contaminant, the areal extent of the contamination, the potential for migration of the contaminant to areas of concern, and the health or ecologic impacts associated with the contaminant.

Contaminants such as petroleum products, chlorinated hydrocarbons, and heavy metals are usually associated with urbanized and industrialized settings. The typical mechanisms by which these contaminants enter the subsurface are the intentional or accidental release of highly concentrated waste streams into the environment. Once in the groundwater, concentrated plumes of contaminantion can migrate many kilometers and ultimately discharge to surface waters and/or extraction wells. The U.S. government maintains the National Priority List of U.S. sites that constitute a significant potential threat to human health or the environment; many of these sites are the result of groundwater contamination. The List is updated periodically and recently contained over 1200 sites [*Federal Register*, 2001].

Pesticides, herbicides, fertilizers, and animal wastes are usually associated with agricultural activities. These contaminants enter the groundwater system through human application to the land surface and subsequent dissolution into recharging precipitation or irrigation water. Numerous examples of groundwater contamination from agricultural activities are available. Kolpin et al. [1998] sampled a network of 303 shallow aquifer wells across 12 States of the Midwest and found relatively widespread, low-level concentrations of herbicides. In Mexico's Yucatan Peninsula, more than half of the ground water samples analyzed from wells in the region contained nitrate levels in concentrations in excess of the EPA's safe drinking water standard [*Sampat*, 2000]. In California's San Joaquin Valley, a 1988 study found the pesticide DBCP at levels ten times the EPA's current drinking water standard [*Sampat*, 2000].

Aside from the more common industrial, urban, and agricultural contaminants already mentioned, there are many other types of contaminants that may affect groundwater quality as the following examples suggest. Pharmaceutical residues present in treated municipal waste water, in septic field leachate, or in livestock production operations where anti-biotics are used may find their way into streams and groundwater systems [*Kolpin et al.*, 1999, 2002]. Airborne deposition of low-level contamination via rainfall can affect areas distant from the contaminant source as documented, for example, with herbicides by Goolsby, et al. [1997]. Biological contamination by pathogens derived from human and livestock wastes is another area of current concern and active study [*Ryan, et al.*, 1999].

POLICY AND MANAGEMENT FOR SUSTAINABLE GROUNDWATER USE

Groundwater is an integral part of the earth's ecosystem and a valuable resource for humans. It can be damaged or destroyed by improper use. The challenges faced by policy makers involve sustaining both the quality and quantity of groundwater resources while protecting ecosystems in the face of ever increasing human demands on groundwater. It is clear that groundwater should be extracted in a sustainable fashion and that significant impairment of groundwater quality should be avoided. Complications arise for the policy maker when quantification of these general goals are required or when these goals are in conflict. A variety of analysis methods and tools are available to the policy maker.

Groundwater Issues Faced by Policy Makers

Policy makers should strive to utilize groundwater sustainably. Groundwater use is considered sustainable if groundwater is extracted at a rate that does not result in groundwater mining. If a sustainable rate of groundwater extraction is maintained for an extended period of time, then a new equilibrium of the groundwater system will be established. It must be recognized however that any withdrawal of water will impact ecosystems that have come to depend on the predevelopment groundwater and surface water flow patterns. The challenges for the policy maker are to insure that groundwater use is sustainable while balancing ecosystem impacts against human needs. One approach is to select, as a policy decision, a specified level of ecosystem disruption, for example, a reduction in streamflows, and to then manage groundwater use so that this level of disruption is not exceeded. The more common approach is to extract water as demand dictates and discover the ecosystem disruptions as they occur. This approach suffers from the fact that damage to ecosystems is often irreversible.

In some settings, development demands may produce a short-term need for water that requires temporary overdrafting. This removal can still be considered sustainable if sufficient time is provided to allow water levels to recover subsequent to continued pumping. In some cases, often because of the presence of large human populations, immediate needs, and the lack of alternative sources, aquifer mining may be unavoidable. Policy makers must recognize that, under these circumstances, water use is unsustainable and alternative sources of water will eventually have to be found. In the meantime, water quality may gradually degrade, water costs may increase, and ecological impacts may become exacerbated. Furthermore, and perhaps most importantly, this valuable resource may be lost for future generations.

Policy makers should strive to maintain or improve groundwater quality. Contamination by hazardous and toxic chemicals in areas where groundwater is widely used should be particularly avoided. Extensive experience with groundwater contamination in the last several decades has shown that it is much less expensive to prevent groundwater contamination than to remediate contamination once it has occurred. Unfortunately, for some contaminants and some geologic settings, remediation is simply not practical. In these cases, the groundwater resource is lost, and alternate sources of freshwater must be found.

In some cases, the release of contaminants into the subsurface may be deemed necessary or unavoidable. Impairment of groundwater quality can be minimized if contaminants are limited to those that are known to degrade in the subsurface. It may be possible to provide adequate time for degradation to occur if contaminants are introduced far from areas of human consumption and sensitive ecological systems. It must be recognized that scientific understanding of the fate and transport of contaminants in the subsurface is limited. The degradation of contaminants in the subsurface by biological or chemical means can be incomplete and can form dangerous daughter products. Contaminants that do not degrade can migrate in unpredictable ways. The introduction of contaminants into the subsurface carries serious risks for the sustainability of groundwater resources and should be avoided whenever possible.

The issue of artificial recharge provides an example of the conflicts that can arise between the goals of sustaining groundwater quantity and preserving groundwater quality. Artificial recharge involves the introduction into the subsurface of treated wastewater and stormwater that has had some contact with human activity to artificially replenish the groundwater system. Storm water and treated waste water are common sources of water used in artificial recharge schemes, but both sources of water typically contain some trace pollutants [*Bouwer*, 1995, *Field et al.* 1998]. The subsurface does have the natural ability to filter and degrade some of these pollutants. Thus, the introduction of these waters into the subsurface may be acceptable even though pollutant levels may be higher than current standards allow as long as the disposal location is sufficiently far from sensitive ecosystems or extraction wells, subsurface processes are given enough time to effectively cleanse the water, and the contaminants contained in the water are amenable to degradation. Direct reuse of storm water and waste water may also be an option for irrigation and other processes that don't require high quality water. The challenge for the policy maker is to weigh the benefits of artificial recharge and water reuse as sources of water against the risks of releasing pollutants into the environment.

Tools for Policy Making and Management

The policy maker is involved in anticipating, assessing, and crafting solutions to water problems. Reasoned policy decisions regarding groundwater quantity and quality require strong scientific analysis and adequate data. A key ingredient of scientific analysis is hydrologic models that combine the science of fluid flow and contaminant migration with available data to produce tools for prediction and management. Ultimately, implementing management solutions requires a legal and regulatory structure to control human groundwater use. With control, pricing strategies, quotas, or other mechanisms can be used to help better manage the resource. Traditionally, groundwater has been treated as the property of the landowner whereby typical control mechanisms do not apply. The recent trend has been to develop new law that treats groundwater as a shared public resource that can be managed, thereby providing the legal structure needed to implement policy [*Grigg*, 1996].

Need for more data. Proper assessment and management of the groundwater resource requires the collection of various datasets at the appropriate temporal and spatial scales. To monitor groundwater use, inventories of active extraction wells, data on groundwater pumping rates, and water levels in wells must be collected on a regional and seasonal basis. In addition, data on the distribution of precipitation and groundwater recharge as well as the geologic and hydrologic characteristics of aquifers are necessary to determine sustainable levels of groundwater extraction.

At present, groundwater data are either nonexistent, unreliable, or irregularly collected in most parts of the world. Unlike surface water [*Global Runoff Data Center*; *Vörösmarty et al.*, 1998], there is currently no centralized global groundwater database to hold and maintain available groundwater data. Most data are collected and retained on a country wide basis. A global initiative is required to formalize the groundwater data collection process and establish a global monitoring well network if the global resource is to be properly assessed and managed. Large temporal and spatial gaps exist in the current well monitoring network. These gaps need to be filled before rational and scientifically sound decisions can be made regarding the status, use, and conservation of groundwater. Existing data sets on groundwater are simply insufficient to provide an adequate inventory of global groundwater reserves. Water quantity data are poor, but water quality data are essentially nonexistent for most parts of the world. Depending on location, problems with quality may be a larger constraint on future groundwater use than quantity limitations. This shortage of data presents challenges to policy makers at the international, national, and local levels.

Models for prediction and simulation. Models are now widely used for the analysis and management of groundwater problems. These tools are based on fundamental relationships from physics, chemistry, and biology, relevant to hydrogeology and biogeochemistry. These models are solved using widely available software. Depending on the problem addressed, models may require data on aquifer geometry, geology, hydrology, and biogeochemical properties. Models are routinely used for understanding the behavior of specific aquifers, estimating the response of an aquifer to pumping activity, predicting contaminant migration, and designing solutions to management problems [*Bear and Verruijt*, 1987; *Anderson and Woessner*, 1992]. In settings where data are inadequate for process-based models, statistical methods are available to relate hydrologic variables [*Coulibaly et al.,* 2001].

The combination of simulation models with optimization techniques produces a class of management models that are particularly useful for addressing policy issues and designing specific solutions to groundwater problems [*Willis and Yeh*, 1987; *Ahlfeld and Mulligan*, 2000] as indicated by the following selected examples. Groundwater management models have been used for control of salt water intrusion in coastal regions [*Emch and Yeh*, 1998; *Hallaji and Yazicigil*, 1996], for the design of groundwater contamination remediation systems [*Gorelick et al.*, 1993; *Tiedeman and Gorelick*, 1993] and for analysis of water supply systems [*Nishikawa*, 1998]. Barlow et al. [2003] have used management models to show that careful selection of well location and timing of pumping can limit impacts on streamflow depletion. Ahlfeld and Hill [1996] have examined the relation between remediation cost and the magnitude of regulatory requirements for groundwater contamination problems.

Use of conservation and water reuse. A promising alternative to developing new sources of groundwater is implementing intelligent water conservation and demand management programs that make better use of the groundwater that is already available [Gleick, 1998]. Water conservation has already produced significant reductions in water use in many domestic, agricultural, and industrial applications. Use of low flow toilets, more efficient washing machines, and less water intensive home landscaping can significantly reduce domestic groundwater needs. Many of these water efficient appliances and techniques are now widely used across the United States and around the world. Agricultural irrigation systems are becoming increasingly efficient. Drip irrigation has been shown to cut water use by 30 – 70% over the standard conventional irrigation procedures, and has allowed some farmers to achieve irrigation efficiencies as high as 95%.

Water reuse has also helped meet demand without having to tap additional sources. Groundwater is utilized for potable and domestic needs while lower quality surface waters and wastewater are being used to satisfy lower quality demands like irrigation. Israel currently reuses 65% of its domestic wastewater for crop production, freeing up additional freshwater for households and industries [*Postel*, 1999a]. Few reasonable estimates currently exist for water reuse at national or global levels.

CONCLUSION

Groundwater is a valuable resource for meeting human water supply needs and plays a key role in sustaining ecosystems. In many parts of the world groundwater availability has been impacted by overutilization and pollution. Increasing populations, intensive urbanization, increased agricultural activity, and climate change are likely to lead to additional pressures on groundwater use and availability in future decades.

Policy-makers must consider the impacts of water policy decisions on long-term availability and quality of groundwater. Groundwater reservoirs can provide a valuable source of high quality water; however, excessive use of groundwater can lead to non-sustainable groundwater mining. Groundwater behavior is often closely connected with surface water. Even if groundwater extractions are at a sustainable rate, the interception of water for human use will usually have an impact on natural discharge to streams and wetlands and associated ecosystems.

Establishment and implementation of groundwater related policy require sufficient data to predict the impacts of policy on long-term availability and quality. In much of the world there are inadequate data on the hydrologic characteristics of aquifers and associated watersheds. Improved data collection and better coordination of existing data are essential for the development of sound groundwater policy. With adequate data a variety of modeling techniques are available to aid in the development and implementation of sustainable groundwater use strategies.

REFERENCES

Ahlfeld, D.P. and Hill, E.H. III, 1996. The Sensitivity of Remedial Strategies to Design Criteria, *Groundwater*, 34(2), pp. 341-348.

Ahlfeld, D. P. and Mulligan, A.E., 2000. *Optimal Management of Flow in Groundwater Systems*, Academic Press, San Diego, CA.

Alcamo, J., Doll, P., Kaspar, F., and Siebert, S., 1997. Global Change and Global Scenarios of Water Use and Availability: An Application of WaterGAP 1.0. Center for Environmental Systems Research, University of Kassel, Germany.

Alley, W.M., Reilly, T.E., and Franke, O.L., 1999. Sustainability of Groundwater Resources, U.S. Geological Survey Circular 1186.

Ambroggi, R.P., 1980. Water. *Scientific American* 243(5), pp. 90 – 96, 100 – 101.

Anderson, M.P. and Woessner,W.W., 1992. *Applied Groundwater Modeling*, Academic Press, San Diego.

Barlow, P.M., Ahlfeld, D.P. and Dickerman, D.C., 2003, Conjunctive-Management Models for Sustained Yield of Stream-Aquifer Systems, *Journal of Water Resources Planning and Management*. 128(1).

Baumgartner, A. and Reichel, E., 1975. *The World Water Balance.* Vienna and Munich, R. Oldenboury Verlag. 180 p.

Bear, J. and Verruijt, A., 1987. *Modeling Groundwater Flow and Pollution*, D. Reidel Publishing Company, Dordrecht.

Bell, J.W., 1997. *Las Vegas Valley: Land Subsidence and Fissuring Due to Ground-Water Withdrawal.* Nevada Bureau of Mines and Geology.
http://geochange.er.usgs.gov/sw/impacts/hydrology/vegas_gw/

Bouwer, H., 1995. Artificial Recharge - Issues and Future, *Proceedings of the 2nd International Symposium on Artificial Recharge of Ground Water,* Orlando, FL, USA July 17-22, 1994.

Biswas, A.K., 1999. Water Crisis: Current Perceptions and Future Realities. *Water International*, 24(4), pp. 363 – 367.

Coulibaly, P., Anctil, F., Aravena, R. and Bobee, B, 2001, Artificial Neural Network Modeling of Water Table Depth Fluctuations, *Water Resources Research,* 37(4), pp.

885-896.
Davis, S. and DeWiest, R., 1966. *Hydrogeology*, John Wiley & Sons, New York.
De Mare, L., 1976. Resources, Needs, Problems: An Assessment of the World Water Situation by 2000. Institute of Technology, University of Lund, Sweden.
De Villiers, M., 2000. Water: *The Fate of Our Most Precious Resource*. Houghton Mifflin Company, Boston.
Emch, P. G. and Yeh W.W., 1998. Management Model for Conjunctive Use of Coastal Surface Water and Ground Water, *Journal of Water Resources Planning and Management*, 124(3), pp. 129-138.
Engelman, R. and LeRoy, P., 1993. Sustaining Water: Population and the Future of Renewable Water Supplies. Population Action International, 56 p.
Falkenmark, M. and Lindh, G., 1974a. Impact of Water Resources on Population. *Swedish Contribution to the UN World Population Conference*, Bucharest.
Falkenmark, M. and Lindh, G., 1974b. How Can We Cope With the Water Resources Situation by the Year 2050. *Ambio,* 3(3 –4), pp. 114 – 122.
Federal Register, 2001. *Rules and Regulations*, 66(115), June 14, 2001, pp. 32235-32242.
Fetter, C.W., 2001. *Applied Hydrogeology, Fourth Edition,* Prentice Hall, Upper Saddle River.
Field, R., M. Borst, T.P. O'Connor, M.K. Stinson, C-Y. Fan, J.M. Perdec, and D. Sullivan, 1998, "Urban Wet-Weather Flow Management: Research Directions", *ASCE J. of Water Resources Planning and Management,* 124 (3), pp. 168-180.
Gambolati, G., Teatini, P., and Tomasi, L., 1999. Coastline Regression of the Romagna Region, Italy, Due to Natural and Anthropogenic Land Subsidence and Sea Level Rise. *Water Resources Research* , 35(1).
Gardner – Outlaw, T. and Engelman, R., 1997. Sustaining Water, Easing Scarcity: A Second Update. Population Action International. 20 p.
Gleick, P., 1993. *Water in Crisis: A Guide to the World's Fresh Water Resources*. Oxford University Press, New York.
Gleick, P.H., 1997. *Water 2050: Moving Toward a Sustainable Vision for the Earth's Fresh Water*. Pacific Institute for Studies in Development, Environment, and Security. Oakland, California.
Gleick, P.H., 1998. *The World's Water 1998 - 1999*. Washington DC, USA: Island Press, 307 p.
Gleick, P.H., 2000. *The World's Water 2000 - 2001*. Washington DC, USA: Island Press, 315 p.
Global Runoff Data Center: http://www.bafg.de/html/internat/grdc/grdc.htm
Goolsby, D.A., Thurman, E.M., Pomes, M.L., Meyer, M.T., and Battaglin, W.A., 1997. Herbicides and Their Metabolites in Rainfall—Origin, Transport, and Deposition Patterns Across the Midwestern and Northeastern United States, 1990-1991, *Environmental Science & Technology,* 31(5), pp. 1325-1333.
Gorelick, S.M., Freeze, R.A., Donohue, D. and Keely, J.F., 1993. *Groundwater Contamination: Optimal Capture and Containment,* Lewis Publishers, Chelsea, MI.
Grigg, N.S., 1996. *Water Resources Management: Principles, Regulations and Cases*, McGraw-Hill, New York.
Grima, L., 2000. Water Resources Management Policies for Sustainable Development in the Beijing – Tianjin – Tangshan Region. Institute for Environmental Studies, University of Toronto. http://www.utoronto.ca/env/ies/reasia.htm

Hallaji, K. and Yazicigil, H., 1996. Optimal Management of a Coastal Aquifer in Southern Turkey, *Journal of Water Resources Planning and Management*, 122(4), pp. 233-244.

IPCC, 2001. *Climate Change 2001: Impacts, Adaptation & Vulnerability Contribution of Working Group II to the Third Assessment Report of the Intergovernmental Panel on Climate Change (IPCC)*, James J. McCarthy, Osvaldo F. Canziani, Neil A. Leary, David J. Dokken and Kasey S. White (Eds.), Cambridge University Press, UK.

Kolpin, D.W., Stamer, J.K., Goolsby, D.A., and Thurman, E.M., 1998. Herbicides in Ground Water of the Midwest—A Regional Study of Shallow Aquifers, 1991-94, U.S. Geological Survey Fact Sheet FS-076-98.

Kolpin, D. W., Weyer, P. , Meyer, M., and Thurman, M., 1999. The Occurrence of Antibiotics in Iowa Streams, *Abstracts with Programs - Geological Society of America*, 31(7), p. 453.

Kolpin, D.W., E. T. Furlong, Michael T. Meyer, E. Michael Thurman, Steven D. Zaugg, Larry B. Barber, and Herbert T. Buxton. 2002. Pharmaceuticals, Hormones, and Other Organic Wastewater Contaminants in U.S. Streams, 1999-2000: A National Reconnaissance. *Environmental Science & Technology*, 36(6), pp. 1202-1211.

Korzoun, V.I., 1978. World Water Balance and Water Resources of the Earth. UNESCO. 663 p.

L'vovich, M.I., 1974. *World Water Resources and Their Future*. 1979 English Translation, Raymond Nace. American Geophysical Union, Washington DC.

McArthur, J.M., Ravenscroft, R., Safiulla, S.and Thirlwall, M.F., 2001. Arsenic in Groundwater: Testing Pollution Mechanisms for Sedimentary Aquifers in Bangladesh, *Water Resources Research,* 37(1), pp. 109-118.

Ministry of the Environment, Government of Japan, 2002. http://www.env.go.jp/en/soe/ground.html

National Intelligence Council, 2000. Global Trends 2015: A Dialogue About the Future with Nongovernment Experts. http://www.africa2000.com/indx/trends2015.html

Nishikawa, T., 1998. Water Resource Optimization Model for Santa Barbara, California, *Journal of Water Resources Planning and Management,* 124(5), pp. 252-263.

Ortega-Guerrero, A., Cherry, J.A., Rudolph, D.L., 1993. Large-scale Aquitard Consolidation Near Mexico City, *Ground Water* , 31(5), pp. 708-718.

Pagano, T.C., Hartmann, H.C., Sorooshian, S., and Iman, B., 2001. Colorado River Basin Water Issues in a Nutshell. University of Arizona, Hydrology and Water Resources Department. http://www.ispe.arizona.edu/climas/presentations/colorado

Postel, S.L., Daily, G.C., and Ehrlich, P.R., 1996. Human Appropriation of Renewable Fresh Water. *Science*, 271, pp. 785 – 788.

Postel, S., 1999a. *Pillar of Sand: Can the Irrigation Miracle Last?* Worldwatch Institute, W.W. Norton, New York.

Postel, S., 1999b. When the World's Wells Run Dry. World Watch Magazine, September/October Issue, pp. 30 – 38.

Raskin, P., Gleick, P., Kirshen, P., Pontius, G. and Strzepek, K., 1997. Comprehensive Assessment of the Freshwater Resources of the World. Water Futures: Assessment of Long Range Patterns and Problems. Stockholm Environmental Institute, Boston, Massachusetts.

Ryan, J. N., Elimelech, M., Ard, R. A., Harvey, R. W., and Johnson, P. R., 1999. Bacteriophage PRD1 and Silica Colloid Transport and Recovery in an Iron Oxide-Coated Sand Aquifer, *Environmental Science & Technology,* 33(1), pp. 63-73.

Sampat, P., 2000. Groundwater Shock. *World Watch,* January – February 2000, pp. 10 – 22.

Seckler, D., Amarasinghe, U., Molden, D., De Silva, R., and Barker, R., 1998. World Water Demand and Supply, 1990 to 2025: Scenarios and Issues. Research Report 19, International Water Management Institute, Colombo, Sri Lanka.

Shah, T., Molden, D., Sakthivadivel, R., and Seckler, D., 2000. The Global Groundwater Situation: Overview of Opportunities and Challenges. International Water Management Institute, Colombo, Sri Lanka. http://www.cgiar.org/iwmi/pubs/wwvisn/GrWater.htm

Shiklomanov, I.A., 1993. World Fresh Water Resources. In P.H. Gleick, ed. *Water in Crisis.* Oxford University Press, pp. 13 – 25.

Shiklomanov, I.A.,1998. Assessment of Water Resources and Water Availability in the World. Report for the Comprehensive Assessment of Freshwater Resources of the World, United Nations.

Shiklomanov, I.A., 2000. Appraisal and Assessment of the World Water Resources. *Water International,* 25(1), pp. 11 – 32.

Simon, P. 1998. *Tapped Out: The Coming World Crisis in Water and What We Can Do About It.* Welcome Rain ,New York. 198 p.

Solley, W.B., Pierce, R.R., and Perlman, H.A., 1998. Estimated Use of Water in the United States in 1995. US Geological Survey Circular 1200, 71 p.

Tiedeman, C. and Gorelick, S.M., 1993. Analysis of Uncertainty in Optimal Groundwater Contaminant Capture Design, *Water Resources Research,* 29(7), pp. 2139-2153.

United Nations, 2000. World Population Prospects: The 2000 Revision, United Nations Publications, New York.

Vörösmarty, C.J., Green, P., Salisbury, J., and Lammers, R.B., 2000. Global Water Resources: Vulnerability from Climate Change and Population Growth. *Science* 289, pp. 284 – 288.

Vörösmarty, C.J., B. Fekete, and B.A. Tucker. 1998. River Discharge Database, Version 1.1 (RivDIS v1.0 supplement). Available through the Institute for the Study of Earth, Oceans, and Space / University of New Hampshire, Durham NH (USA) at http://pyramid.sr.unh.edu/csrc/hydro/

Willis, R. and Yeh, W.W., 1987. *Groundwater Systems Planning and Management,* Prentice-Hall, Englewood Cliffs, NJ.

World Resources Institute, 2000, *World Resources 2000 – 2001: People and Ecosystems – The Fraying Web of Life,* Oxford University Press. 400 p.

David P. Ahlfeld, University of Massachusetts, Environmental Engineering Graduate Program, Department of Civil & Environmental Engineering 18 Maston Hall, Box 35205, Amherst, MA 01003, ahlfeld@ecs.umass.edu.

Wes Dripps. 2421 Monroe Street, Madison, WI 53711

5

Water Development and Food Production: A Global Perspective

Mark W. Rosegrant and Ximing Cai

INTRODUCTION

Water development has been one of the most critical factors for food security in many regions of the world. The approximately 250 million hectares of irrigated area worldwide today is nearly five times the amount that existed in the beginning of the 20th century. Irrigation has not only played a major role in boosting agricultural yields and output that have made it possible to feed the world's growing population, but has also maintained stability through greater control over production and scope for crop diversification. In arid and semi-arid areas, alternatives to irrigated agriculture are rare, and water reallocation can lead to rural-urban migration and the abandonment of plots. Thus, irrigation has been vital in achieving food security and sustainable livelihoods, especially during the Green Revolution in developing countries, both locally, through increased income and improved health and nutrition, and nationally, by bridging the gap between production and demand.

The role of water development in irrigated agriculture and food security has received substantial attention in recent years (e.g., Postel, 1999). Irrigated agriculture accounts for about 72 percent of global and 90 percent of developing-country water withdrawals. Over the next 30 years, the global population is likely to increase to 7.8 billion, more than 80 percent of whom will live in developing countries, and more than 60 percent in rapidly growing urban areas. As a result, demand for food will grow at 1.3 percent per year in response to population growth and rising incomes (Cosgrove and Rijsberman, 2000). The required calorie intake and trends in diets will be translated into requirements for water to enable the necessary food production for adequate nutrition in the future.

A portion of the growing demand for water will be met through new investments in irrigation and water supply systems and through improved water management,

Water: Science, Policy, and Management
Water Resources Monograph 16

10.1029/016WM06

and some potential exists for the expansion of nontraditional sources of water supply. However, in many arid or semiarid areas—and seasonally in wetter areas—water is no longer abundant, and in many developing countries, new water supplies cost three to four times more than existing water sources (World Bank, 1993). Water availability for irrigation is also threatened in many regions by rapidly increasing nonagricultural water uses in industrial, domestic, and environmental and ecological purposes (Rosegrant and Ringler, 2000). Moreover, water pollution and groundwater mining have increased risk in irrigation water supply (Postel, 1992; Gleick, 1993). These have caused new irrigation development to slow considerably since the late 1970s. Declining expenditures are reflected in the declining growth in crop area under irrigation. Globally, the growth rate in irrigated area declined from 2.2 percent per year during 1967-82 to 1.5 percent in 1982-93. The decline was slower in developing countries, from 2.0 percent to 1.7 percent annually during the same period, but the lagged effect of declining investment in irrigation will be increasingly felt through further slowdowns in expansion of irrigated area (Rosegrant et al., 2001).

In addition, although the achievements of irrigation in ensuring food security and improving rural welfare have been impressive, past experiences also indicate environmental problems, including excessive water depletion, water quality reduction, waterlogging and salinization. Hydrological records over a long period have shown a marked reduction in the annual discharge on some of the world's major rivers. In some basins, excessive diversion of river water has led to environmental and ecological disasters for downstream areas, and pumping groundwater at unsustainable rates has contributed to the lowering of groundwater tables and to saltwater intrusion in some coastal areas. Many water quality problems have also been created or aggravated by changes in streamflows associated with water withdrawals for agriculture. Moreover, poor irrigation practices, accompanied by inadequate drainage, have often damaged soils through over-saturation and salt build-up. The United Nations Food and Agriculture Organization (FAO) estimates that 60 to 80 million hectares are affected to varying degrees by water waterlogging and salinity (FAO, 1996). These irrigation induced environmental problems can threaten entire agricultural production systems, as well as human health and natural life systems.

Water development has been, and will be, one of the most critical factors for food security in many regions of the world. Irrigation will remain as the largest single user of water, and rapidly growing municipal and industrial water demand especially in the developing countries will result in increasing water scarcity for agriculture. Whether water availability for irrigation—together with feasible production growth in rainfed areas—will provide the food needed to meet the growing demand and improve national and global food security remains a crucial and urgent question for the world. Numerous national and international studies have been conducted to explore answers to this question, and a wide range of possible responses can contribute to alleviating water scarcity, involving multiple stakeholders from government, private sector and civil society. Improving water use efficiency will be a tendency for agricultural water management in the future. Both investment and policy reforms will play a critical rule. A major challenge is to strengthen the link between science and engineering principles and relevant social, economic, institutional and legal schemes that affect water use and management for agriculture. This chapter provides a global perspective of the water and food issues, including emergent water development and management issues related to food production and exercises in integrated water and food modeling and analysis at the global scale. Based on some global water and food modeling exercises, we discuss implications for water development regarding food security.

EMERGENT SCIENCE AND POLICY ISSUES IN WATER DEVELOPMENT FOR AGRICULTURE

Expansion of Irrigation

The development of irrigation and water supplies has become increasingly expensive. In India and Indonesia, for example, the real costs of new irrigation have more than doubled since the late 1960s and early 1970s. Costs have also increased by more than 50 percent in the Philippines, 40 percent in Thailand, and more than 300 percent in Sri Lanka (Rosegrant and Svendsen, 1993). Irrigation has absorbed over half of all agricultural investment in China, Pakistan and Indonesia, and it has accounted for 30 percent of all public investment in India. In addition, once established, irrigation projects become some of the most heavily subsidized economic activities in the world, both directly and indirectly. In the mid-1980s, average subsidies to irrigation in six Asian countries covered an average of 90 percent or more of total operating and maintenance costs. During the 1990s, subsides have declined somewhat as most countries worldwide have officially adopted the stated goal of full recovery of operation and maintenance (O&M) costs. No country in either the developed or the developing world has fully eliminated subsidies, however, progress on this front has been uneven, with Chile and South Africa implementing particularly innovative programs. Despite these ongoing reforms, O&M cost recovery remains dismal in most major irrigators. In Pakistan, for example, the gap between O&M expenditures and recoveries in the Punjab region was 62 percent in 1994-1995, increasing to 74 percent in 1995-1996, while the Sindh region performed even worse with the gap as high as 88 percent in 1995-1996 (Dinar, et al., 1998).

Heightened national and international concerns over the broad environmental and human effects of large irrigation projects will make it very difficult to proceed with many new projects, particularly when combined with evidence of relatively low returns to irrigation investment. Evidence from India (Evenson, et al., 1999; Fan et al., 1999) and China (Fan and Zhang, 2000) shows very low rates of return to irrigation investment when compared with other public investment priorities.

In developing countries more generally, a number of problems have beset large irrigation systems, including low water use efficiencies at the field level; low cropping intensities, yields, and irrigated areas; significant environmental impacts; and escalating costs (World Commission on Dams, WCD, 2000). World Bank analysis indicates that the supply of funds to agricultural water development has declined in recent years, with the sum of investment from Development Assistance Committee members and the World Bank falling to a low of $1.0 billion in 1995, although investment subsequently increased to $1.4 billion by 1997, indicating a slight turnaround. Along with changing funding levels, money in this sector is increasingly being directed toward reforming existing irrigation infrastructure, including policies, institutions and the overall development approach; enhancing the physical productivity of existing projects through rehabilitation; and focusing dollars for new construction on small-scale projects (WCD, 2000).

Small-scale irrigation is often considered to be more cost-effective than large-scale irrigation. However, a recent review of the World Bank's experience with irrigation shows that irrigation projects display significant economies of scale (i.e. the rates of return to large projects have been higher than returns to small-scale projects) (Jones, 1995). These estimates are incomplete in that they do not incorporate the full range of negative externalities generated by irrigation projects and do not account for the economic, environmental, and social consequences of projects not being developed, but they do indicate that a shift away from large irrigation projects to smaller projects could have mixed results. While small-scale irrigation projects can have considerable advantages over large-scale projects in terms of their overall sustainability, the bureaucratic mode of implementation has effectively eliminated these potential advantages in many cases (Rosegrant, 1995; Rosegrant and Perez, 1997). Both large and small systems often share a number of common characteristics: high capital costs per hectare and per farmer; bureaucratic, costly, and inefficient management; low technical efficiency; low settler incomes; and zero or negative returns (Adams, 1990). On balance, selective development of new irrigation must still play a role in future water resource development, but it is difficult to imagine expansion of irrigated area much in excess of the current situation.

Dams for Irrigation

According to the International Commission on Large Dams (ICOLD, 1998), 48 percent of multipurpose dams store water for irrigation. Irrigation is the main reason for dam construction in South Africa, Australia, India, China, Egypt, Cyprus, Syria, Tunisia, and Zimbabwe, to name but a few countries. While dams built for irrigation are numerically important, they are usually smaller than hydropower dams. Therefore, a debate over large dams and their associated environmental and social impacts is concerned less with irrigation dams than with hydropower dams per se. However, irrigation is increasingly mentioned as the main purpose of new dams, a trend possibly reflecting growing food and water needs in many developing countries (WCD 2000). The International Commission on Irrigation and Drainage (ICID) concludes that the need for food security, especially the urgent need to improve the standard and quality of the poorer strata of their societies calls for urgent steps to build water storage facilities in an environmentally sound and sustainable way. Keller et al. (2000) argue that aquifers, small and large reservoirs all have an indispensable role in water storage and comparative advantages under specific conditions. Where possible, substantial gains can be achieved by combining technologies—large reservoirs, small reservoirs, links between them and groundwater storage—in an integrated system.

Recycling and Reuse of Water for Irrigation

Additional water can be conserved on the demand side through recycling (reuse in the same home or factory) and wastewater reuse (collection, treatment and redistribution of used water to other locations). In developed countries, pollution control laws and incentive pricing ('polluter-pays' principle) have been primary motivators for industrial water recycling. In the U.S.A., for example, total industrial water use fell 36 percent while industrial output increased nearly fourfold between 1950 and 1990 (Postel, 1992). Similar conservation efforts have also begun in the urban areas of some water-scarce developing countries. In Beijing, China, for example, the rate of water recycling increased from 61 percent in 1980 to 72 percent in 1985; and between 1977 and 1991, total industrial water use declined steadily while output increased by 44 percent in real terms (Nickum, 1994).

The rate of expansion of wastewater reuse will depend on both quality and public acceptance of the final product. Israel serves as a good example of the potential for wastewater reuse in agricultural production. Israel undertakes the largest wastewater reuse effort in the world: 65 percent of its treated sewage is used for crop production, accounting for 30 percent of the nation's agricultural water supply, with a possible increase to 80 percent by 2025, thus freeing up significant freshwater for nonagricultural uses (Postel, 1999). Worldwide, however, treated municipal wastewater irrigates only about 500,000 ha of cropland, accounting for only two-tenths of 1 percent of the world's irrigated area. Moreover, given the relatively high costs of wastewater treatment and transport to agricultural areas, it is likely that wastewater can make up an important share of agricultural water supply only in arid regions characterized by prohibitive new water supply costs and a dearth of other, cheaper conservation techniques. On the other hand, agricultural areas lie upstream of major urban concentrations and agricultural return flows carry agricultural chemical and fertilizer residues, and concentrated salts from soil or the incoming irrigation water. The return flows make up a portion of a river's discharge as they pass the urban area, affecting the quality of water withdrawn for municipal use and that available for in-stream uses, including natural ecosystems.

Groundwater Development and Management

Groundwater, as a distributed source of stored, often high quality water, plays an increasingly important role in water supply for food production. In many arid and semi-arid zones, groundwater is the primary source used to meet both domestic and agricultural needs. In other regions, groundwater serves as a buffer to overcome dry spells or periods of drought. In many agricultural areas, groundwater is used as a supplement to surface irrigation, increasing efficiency, reliability and flexibility of water supplies and providing greater security. The surface and groundwaters are interdependent and need to be considered together as a renewable water resource, which should be used conjunctively. Influent and effluent streams, infiltration and interflow, recharge due to riverflow and re-emergence of streams in river low stages, are manifestations of such interdependence. However, over-development of groundwater as a resource for food production is becoming a serious concern. In many areas, like India, China, the United States, North Africa and the Arabian Peninsula, aquifers have been over-exploited and water tables are falling. Water quality in some aquifers is also declining due to salinization or fertilizer and pesticide residues leaching into groundwater from both irrigated and rainfed land (Postel, 1999). It may take years for such pollution to become problematic and when it shows up it may be too late to resolve.

Most countries can do a better job of sustainably developing their groundwater resources. The massive expansion of private-sector tubewell irrigation in India, Pakistan, and Bangladesh is a highly most successful example of private-sector irrigation development in the developing world. Private tubewells have grown most rapidly in areas characterized by reasonably good roads, research and extension systems, accessible credit, and relatively inexpensive electric or diesel energy. Private tubewells have largely developed in and around the command areas of large surface-irrigation systems, since tubewells can access aquifers recharged by deep percolation losses from the surface systems.

Inter-Basin Water Transfers

Plans for inter-basin water transfers were heralded in the 1960s and 1970s as the best solution to acute water shortages in adjacent basins or sub-basins, particularly in arid and semiarid regions and those regions anticipating significant transfers of water from agricultural to urban and industrial users. However, most of the large-scale proposals never materialized due to huge capital costs; the scope for less expensive alternative sources of new water or water savings; and increasing concerns over negative economic, environmental, and social impacts in the exporting basin. Potential negative impacts in the exporting region include the elimination of future development opportunities, social disruption, and irreparable environmental damage. On the other hand, micro-level basin transfers over short distances have proven to be viable options in many regions, although the potential economic and social costs of these transfers for the area of origin must still be taken into account. All water transfers involve tradeoffs; while the purchase of water rights by the city of Los Angeles from the Owens Valley of Eastern California had a devastating impact on the Valley from which it never recovered (U.S. Office of Technology Assessment, 1993), the Imperial Irrigation District reaped large profits from selling water to the Metropolitan Water District (Postel, 1999).

Non-Conventional Sources of Water

Several non-conventional sources of water have occasionally been developed for special purposes. These sources comprise water that is generally considered to have reached a sink, or become unfit for use as fresh water, such as seawater, incompletely treated urban wastewater, or brackish water. Increasingly in water scarce areas, urban wastewater is reused in agriculture after treatment. However, the amount of water available is relatively small compared to agricultural demands, and quality concerns persist. Brackish and moderately saline water can be used for aquaculture, or to produce salt tolerant crops. These waters can be desalinated and made fit for any use, though the cost is still high. Desalination is now sometimes employed to produce drinking water where other sources are unavailable and where society can afford it. However, new reverse osmosis technology is bringing the cost down, and if electricity costs fall as predicted, desalinated water may become a feasible source of supply for producing high value agricultural crops using precision technology. It is, however, unlikely to become a common source of agricultural water (Gleick, 1993).

Rainfed Agriculture and Water Harvesting

Rainfed agriculture is still the most important crop production system. In rainfed agriculture, rainfall, and in certain cases capillary rise, are the only mechanisms for replenishing soil moisture and no artificial supplies are provided. Rainfed agriculture is practiced under a wide variety of conditions. Under conditions of reliable rainfall, rainfed agriculture has advantages of not requiring costly irrigation infrastructure and leaving flowing or stored water in rivers, lakes and aquifers where it is available to support ecosystems and for other human use. Given adequate land holding sizes, such rainfed agriculture can still provide reasonable livelihoods for farmers. Where rainfall is not reliable, however, drawbacks exist. Under these less favorable conditions, improved farming practices, like land leveling, tie ridging, and others that increase water intake after rainfall, can improve the retention of water and nutrients in the soil and increase yields. Practices that reduce non-beneficial evaporation, such as mulching, can have similar effects. Better agricultural practices combined with the result of bio-technical advances (drought resistant varieties) and modern technology can contribute considerably to the improvement of rainfed production systems.

Rainfall harvesting is the capture, diversion, and storage of rainwater for plant irrigation and other uses, and can be an effective water conservation tool, especially in arid and semi-arid regions. Water harvesting can provide farmers with improved water availability, increased soil fertility, higher crop production in some local and regional ecosystems, and broader environmental benefits through reduced soil erosion. Although improved water harvesting is often considered in connection with traditional agriculture, it also has potential in highly developed agriculture. Advanced tillage practices can also increase the share of rainfall that goes to infiltration and evapotranspiration. Contour plowing, which is typically a soil-preserving technique, should also act to detain and infiltrate a higher share of the precipitation. Precision leveling can also lead to greater relative infiltration, and therefore a higher percentage of effective rainfall. Rainfall harvest has had a significant impact on food supplies and incomes in limited regions (Rosegrant et al., 2002).

Water-Saving Technology

Availability of appropriate technology can be an essential component for generating water savings, if improved demand management introduces incentives for water conservation. As the value of water increases—or subsidies that make water artificially cheap are removed—the use of more advanced technologies such as drip irrigation utilizing low-cost plastic pipes, low-pressure sprinklers, and system level computerized control systems (used widely in developed countries) could have important results for developing countries.

Evaluation of the impacts of these technologies must take into account the difference discussed above between consumptive use of water and water withdrawals or applications. All of these advanced technologies can significantly reduce the amount of water applied to a field, but, to the extent that the saved water simply reduces the amount of drainage water that is reused, the actual water savings will be lower than the apparent efficiency gains (Seckler, 1996). Nevertheless, if the scarcity value of water is high enough, appropriate use of new technologies appears to offer both real water savings and real economic gains to farmers.

Technological opportunities also exist at the irrigation system level. For example, in North Africa, modern irrigation systems using hydraulically operated diversion and measuring devices were developed as early as the late 1940s, and were employed in irrigation schemes constructed in the 1950s. Modern schemes in this region deliver water on demand to individual farmers, allowing water users to be charged according to the volume of water delivered, thereby encouraging conservation and efficient use of water. Some of these irrigation techniques have been transferred to the Middle East, and in pilot projects to other developing countries (World Bank, 1993). Continued increases in the value of water could make these capital-intensive irrigation distribution systems more widely feasible in other regions of the world.

In addition to these high tech options, small-scale appropriate irrigation technology such as treadle pumps appears to have an increasing role in promoting efficient water management in some developing countries. A treadle pump can be a *suction* pump, which is a foot-operated device that uses bamboo, PVC, or flexible pipe for suction to pump water from a shallow aquifers of 1-2 meters; or alternatively a *pressure* pump that operates in similar fashion, but that can be used to lift water from slightly deeper sources, in excess of 4 meters. Suction treadle pumps have spread to over 600,000 ha in Bangladesh. The potential for greater adoption of treadle pumps in South Asia has been estimated to be huge depending upon extraordinary marketing expansion, maintenance of quality control in manufacturing, and the progress in reducing subsidies on diesel fuel in the region.

Technology holds considerable promise for future water savings—and for benefits to farmers in specific agroclimatic niches—but there is no simple solution for solving water scarcity. Both advanced technologies and simple small-scale technologies have covered only a small proportion of irrigated area, and future expansion is highly dependent on improved incentives to induce farmers to invest in water-saving technologies. Unless water prices—or scarcity values of water generated in water markets—provide correct signals about the value or water, the uptake of water-saving technology will remain limited.

Water Demand Management

New investments in irrigation and water supply systems, and some expansion of nontraditional sources will supply a portion of growing future water demand. In many regions, however, neither of these sources will be sufficient to meet rapidly growing nonagricultural demands for water or to mitigate the effects of water transfers out of agriculture to rapidly growing municipal and industrial water demands. A large share of the water necessary to meet new demands will have to come from existing uses through comprehensive water policy reform providing the necessary incentives to conserve water in municipal, industrial, and agricultural uses. Such reform will not be easy: Long-standing practices and cultural and religious beliefs historically have treated water as a free good, and entrenched interests benefit from the existing system of subsidies and administered allocations of water (Rosegrant, 1995).

The task of demand management is to generate physical and economic water savings by increasing crop output per unit of evaporative water loss (generating more crop per drop), by improving water utilization before it reaches salt sinks (increasing water use efficiency in the river basin), and reducing water pollution. At the farm level, more output per unit of water can be generated by using less water on a given crop, by shifting of water applications to more water-efficient crops, by changing the crop mix to crops that have a higher value per unit of water utilized, and by adopting more efficient irrigation technology. The potential for reducing water scarcity and enhancing food production by improving water management is basin-specific. Although, as shown above by the powerful impact of changes in river basin efficiency on food production and prices, improved demand management that increases water use efficiency offers considerable scope for improving water availability for agriculture.

The most significant reforms on the demand side involve changing the institutional and legal environment for water consumption to provide correct signals—including environmental externalities—regarding the real scarcity value of water and empower water users to make their own consumptive use decisions. The innovative institutional and policy reforms required for efficient water management require a action from the public sector, market, and civil society, and will vary depending on the location, level of institutional and economic development, and degree of water scarcity. For agricultural water management, key issues in reform include the role of user-managed irrigation and the impacts of water rights, pricing and markets. These are briefly discussed below.

User-Managed Irrigation. In many countries, poor performance of centralized administrative management, together with fiscal pressures from mounting O&M costs, has provided a major stimulus for transferring management responsibility for both irrigation and domestic water-supply systems from agencies to user groups. As a result, there has been a strong trend toward decentralization of irrigation management in many developing countries during the last decade. The level of devolution of systems ranges from participation of farmers in irrigation management to full irrigation management transfer. The former concept refers to increasing farmer responsibility and authority in irrigation management, and the latter typically involves a shift of management responsibility from a centralized government irrigation agency to a financially-autonomous local-level non-profit organization, which is either controlled by the irrigation water users or in which water users have a substantial voice (Svendsen et al., 1997). Irrigation management transfer can have positive effects for farmers, including improved irrigation service and maintenance, a sense of ownership of resources that provides incentives to improve operational performance, increased (financial) accountability and transparency and greater accessibility to irrigation system personnel, and reduced conflicts among users.

Irrigation districts, groundwater districts, co-operatives, irrigator associations, village-based organizations, and more informally constituted user groups all represent viable user-based institutions. Studies of farmer-managed irrigation systems have also shown a wide diversity of rules for within-system allocation: By timed rotation of water depth and land area or shares of the flow. Property rights represent a critical factor affecting the viability of water management institutions. The cohesive force of property is important in many aspects of water management, but is especially critical for allocation. Secure property rights—including ownership of the actual irrigation facilities and/or water rights—render water assets fungible and facilitate knowledge sharing, greatly easing and expanding the decision set presented to users and their associations. Property rights ultimately form the basis for relationships among irrigators and become the social basis for collective action by irrigators in performing various irrigation tasks. so-called "traditional" rights, usually an amalgam of official assurances and local indigenous methods of recognizing ownership, are appropriate in some cases, but are usually not recognized by formal legal system. The seeming egalitarianism of these more informal rights also frequently masks their reinforcement of existing inequalities (Rukuni, 1997). Moreover, water user groups also tend to be more effective if they build upon existing social capital or patterns of cooperation and are homogeneous in background and assets (though heterogeneity is a manageable impediment) (Meinzen-Dick et al., 1994).

Water Rights, Pricing and Markets. Irrigation water in nearly the entire developing world has been essentially free-of-charge historically, providing farmers with little incentive to economize on its use. The virtually free provision of water to farmers has led to three major environmental problems in intensified areas: salinity build-up, waterlogging and excessive reliance on monoculture rice. Increasing water use efficiency through opportunity cost pricing or market valuation of water would have substantial environmental benefits and would not adversely affect yields; however, few developing countries have employed this powerful means for improving the sustainability of the resource base.

In principle, markets in tradable water rights may have considerable advantages over other allocation mechanisms, particularly in terms of efficiency. These advantages include (1) the empowerment of water users by requiring their consent to any water reallocation and ensuring compensation for any water transferred; (2) the encouragement of investments in water-saving technologies by ensuring security of water rights to users; (3) the provision of incentives for users to consider the full opportunity cost of water when making water use decisions, including its value in alternative uses; (4) the provision of incentives to water users to account for the external costs imposed by their water use, thus reducing the pressure to degrade resources.

Despite these potential benefits, the unique physical, technological and economic characteristics of water resources pose special problems for establishing tradable water rights regimes. Tradable water rights or water-use rights are costly to establish and allocate; impose costs on certain economic actors; and require public enforcement. The fundamental importance of water to farm production and income also raises serious equity concerns when major shifts in water allocation are considered. Furthermore, the lack of adequate information regarding water supply and demand will also impede development of effective markets in developing countries. These complexities render the establishment of markets in tradable water rights a longer-term solution in much of the developing world, with more extensive implementation for groundwater and surface water irrigation systems concentrated, at least initially, around major urban areas. However, innovative approaches that combine the benefits of water markets and user management can form initial forays into the realm of market-type incentives (Easter et al., 1998).

INTEGRATED WATER AND FOOD MODELING AT THE GLOBAL SCALE

In the past decade, numerous international and national efforts have been undertaken to assess the science and policy issues discussed above, and agricultural water development has been given a high priority. Most recently, the *World Water Vision* (WWV) provided a comprehensive study on the status of global and regional water supply and demand, the existing problems and development potentials (Cosgrove and Rijsberman, 2000). However, in terms of methodology, the previous studies either took a physically based approach that did not reflect properly market forces and economic considerations, or took a mainly economic oriented approach, and none of them were able to capture adequately the complex interactions between physical, engineering, and economic linkages and feedback mechanisms in water resources management. The WWV provided pieces of information for the hydrologic, agronomic, and economic components, and it is a challenging and promising task to combine these pieces into an integrated modeling framework, particularly for analyzing the relationships between water availability, water infrastructure development, water management polices, and food production, demand, and trade at both the global and regional scales. In the rest of this chapter, we will review some integrated water and food models at the global scale that have been used as a tool to derive implications for water development in agriculture.

Models

With increased competition for scarce water and land resources, a main concern is providing additional food necessary to feed growing populations. In this context, three agencies—the International Water Management Institute (IWMI), the United Nations organization of Food and Agriculture (FAO), and the International Food Policy Research Institute (IFPRI)—have developed integrated water and food models/approaches to address these issues. As an example of many such models that exist, these models are briefly reviewed in the next section, which largely follows World Bank (2001).

IWMI developed the "Policy Interactive Dialogue Model (PODIUM) (IWMI, 2001). The model projects water demand and supply for irrigated agriculture covering about 100 countries. The model converts projected food production, which is based on assumptions concerning population growth, daily calorie intake and composition of diets, and conversion ratios from feed stuff to animal products, into water demand. Water demand, actual water diversions, and the available renewable water resources in the base year (1995) are calibrated to the actual data. For food production, the model projects cereal production based on the expected yields and cultivated area, under both irrigated and rain-fed conditions. Finally, the results are analyzed by grouping the countries by the degree of water scarcity, assuming several criteria for classification of degree of water scarcity. The model is presented in Excel spreadsheet format and the results obtained from the individual country model are grouped to provide a global scenario of water demand and supply by 2025.

The FAO Study "Agriculture: Toward 2015 and 2030" (FAO, 2000) projects the likely expansion in irrigated area and total water demand by 2015 and 2030. A simple soil water balance model is used to calculate actual evapotranspiration and surface runoff. The water balance is assumed to be a function of precipitation, reference evapotranspiration, and soil moisture storage properties. The computation of water balances was carried out by grid-cells of 10 by 10 km, using a GIS database. The water balance is calculated in monthly time steps and consists of annual values by grid-cell for the actual evapotranspiration and runoff. Water withdrawals by 2015 and 2030 are calculated and calibrated per country with data on water resources for each country using the FAO AQUASTAT information system. Projection of food demand-supply balance is based on the FAO World Food Model for GOL (Grains Oilseeds Livestock) commodities. The model also incorporates the food prices and the influence of income growth on the food demand projections. For food production, the Global Agro-ecological Zone (GAEZ) database and FAO/UNESCO Digital Soil Map are used to identify potential areas for rainfed and irrigated agriculture. The actual crop area and yield for the base year takes the average of a 1995-97 database available from FAOSTAT and the irrigated/rainfed area and yield are estimated based on diverse sources of data and knowledge and judgment contributed by the different specialists in many countries. All databases listed above can be accessed from FAO's web site (www.fao.org).

The IFPRI model (Rosegrant and Cai, 2001) attempts to project and analyze how water availability and demand would evolve over the next three decades (from a base year of 1995), taking into account the availability and variability in water resources, water supply infrastructure, and irrigation and non-agricultural water demands, as well as the impact of alternative water policies and investments on water supply and demand. The study uses an extended version of the International Model for Policy Analysis of Agricultural Commodities and Trade (IMPACT) model for the projection and analysis of food supply and demand, and the Water Simulation Model (WSM) model for projections of water supply and demand. In the IMPACT-WATER model, the world is divided into 69 spatial units. The spatial units include individual countries and regions, and China,

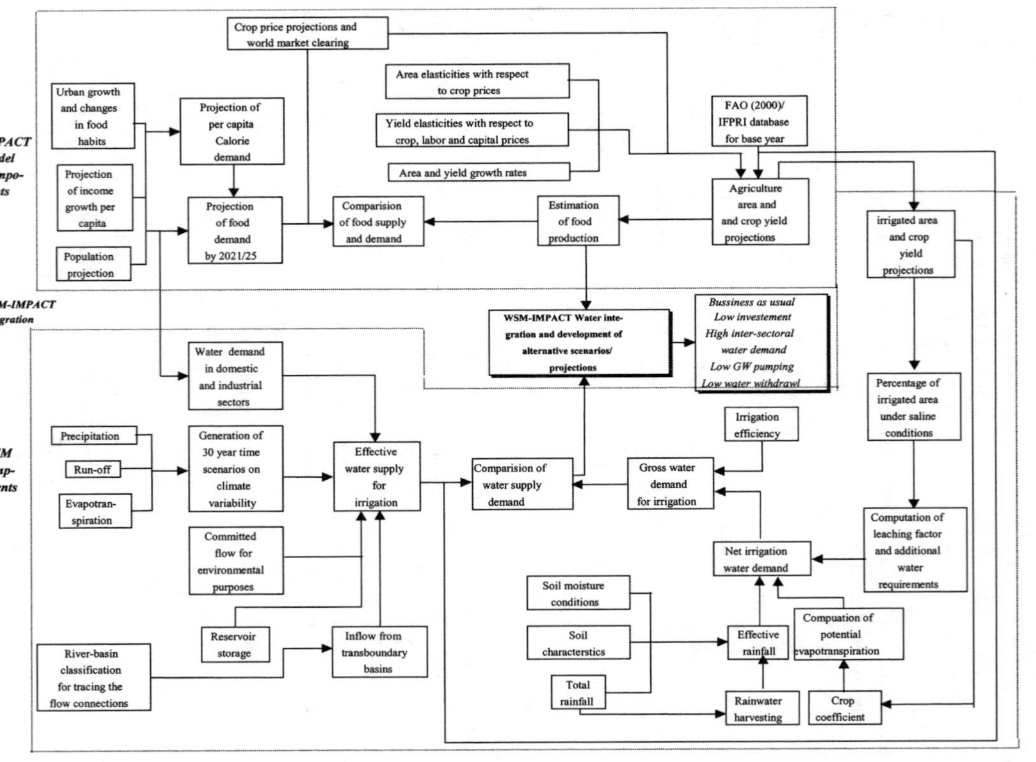

Figure 1. WSM-IMPACT Water Model Components and Integration for Estimation of Water Balance and Projections of Alternative Scenarios.

India, and the United States, which together produce about 60 percent of cereals globally, are disaggregated into major river basins. Water supply and demand, and food production are assessed at the river basin scale, and food production is summed to the national level, where food demand and trade are modeled.

The model also assesses irrigation water demand under several alternative scenarios and considers changes in irrigated area and cropping patterns, water use efficiency, rainfall harvest technology, and water allocation among sectors, as the main drivers of policy shifts. Though global climate change could also significantly affect future irrigation water supply and demand, this aspect is not directly incorporated into the current modeling framework.

The model structure and components are shown in Figure 1. In the WSM module, water available for food production is simulated as a function of precipitation, runoff, water supply infrastructure, and socioeconomic and environmental policies. Crop water demand and water supply for irrigation are simulated, taking into account the year-by-year hydrologic fluctuations, irrigation development, growth of industry and domestic water uses, environmental and other flow requirements (committed flow), and water supply and use infrastructure.

IMPACT provides a consistent framework to examine the effects of different food policies, rates of investment in agricultural research in crop productivity growth, and income and population growth on long-term food demand and supply balances and food security (Rosegrant et al., 2001). Separate crop-wise area and yield functions for rainfed and irrigated crops are implemented in the model. These functions include water availability as a variable, through which IMPACT is connected with the global water simulation model (WSM).

The combined water-food modeling framework provides a wide range of opportunity for analysis of water availability and food security at basin, country and global scales. Many policy-related water variables are involved in this modeling framework, including potential irrigated area and cropping patterns, maximum allowed water withdrawal due to infrastructure capacity and environmental constraints, water use efficiency, water storage and inter-basin transfer facility, rainfall harvest technology, allocation of agricultural and non-agricultural uses, and allocation of instream and offstream uses. For the sake of exploring alternative futures, investment and management reform can influence the future paths of these variables, which influence food security at both national and global scales.

Modeling Results—Implications for Water Development

In terms of technology, the three models reviewed above have taken advantage of advances in hydrological science, agronomic science, and system modeling technology, as well as numerous international and national efforts in global and regional water resources and agriculture assessments. The common perspectives reflected in the projections of the three models include:

1. Irrigation is and will remain, the largest single user of water. But com pared with domestic and industrial water demand, the irrigation sector would be the most slowly growing water sector.
2.. Irrigation efficiency needs to be improved, and demand management will play an important role in improving irrigation efficiency in water scarce regions.
3. Considering the storage loss due to sedimentation, the minimum new storage required might be to replenish the storage loss.
4. Considering the overdraft of groundwater in some countries and regions, the total amount of additional water storage and conveyance might be required.
5. Alternatives exist for inter-basin transfers as being considered in India and Northern China.

However, the models are based on different assumptions and all admittedly provide only a partial picture of the projections of irrigation water demand and supply by 2025 and 2030. For demonstration purposes, this chapter provides results from various scenarios of the IMPACT-WATER model, including a baseline and several alternative scenarios, as an example for integrated water and food modeling and analysis at the global scale.

The model projections are briefly discussed in the following paragraphs, inclduing a baseline scenario and two alternative ones. The baseline scenario is the starting point for this analysis. Total non-irrigation water consumption in the world is projected to increase from 370 cubic kilometers (km^3) in 1995 to 620 km^3 in 2025, an increase of 68 percent. The largest increase of about 85 percent is projected for developing countries. Moreover, instream and environmental water demand is included as committed flow that is unavailable for other uses, and ranges from 10 percent to 50 percent of the runoff depending on runoff availability and relative demands for instream uses in different basins. The global potential irrigation water demand is assessed as 1758 km^3 in 1995 and projected to be 1992 km^3 in 2021/25, an increase of 12.0 percent. The developing world is projected to have much higher growth in potential irrigation water demand than the developed world over this period, with potential consumptive demand in the developing world rising from 1445 km^3 in 1995 to 1673 km^3 in 2021–2025, a 13.6 percent increase.

Moderate increases are projected for water withdrawal capacity, reservoir storage, and water management efficiency, based on estimates of current investment plans and the pace of water management reform. Water demand can be defined and measured in terms of withdrawals and actual consumptive uses. The potential demand or consumptive use for irrigation water is defined as the irrigation water requirement to meet the full evapotranspirative demand of all crops included in the model over the full potential irrigated area. Potential demand is thus the demand for irrigation water in the absence of any water supply constraints. Actual irrigation consumptive use is the realized water demand, given the limitations of water supply for irrigation. Total global water withdrawals are projected to increase by 23 percent between 1995 and 2025, from 3906 km^3 (groundwater pumping 817 km^3) in 1995 to 4794 km^3 (groundwater pumping 922 km^3) in 2025. Reservoir storage for irrigation is expected to increase by 621 km^3 (18 percent) over the next 25 years. The worldwide average basin efficiency increases from 0.56 in 1995 to 0.61 in 2025.

Global consumptive use of water is expected to increase by 16 percent, from 1800 km^3 in 1995 to 2085 km^3 in 2025. Assuming non-irrigation water demand will be given first priority, water available for irrigation water consumption will only increase by 3.9 percent, from 1430 km^3 to 1485 km^3, which is considerably lower than the 12 percent increase in potential irrigation demand. Therefore, it is critical to note that irrigation water demand will be increasingly supply-constrained, with a declining fraction of potential demand met over time. The situation is especially serious in developing countries, where potential demand increases by 13.6 percent, and the supply increases by only 4.4 percent (therefore the increase in actual consumptive use of irrigation water).

This tightening constraint is shown by the Irrigation Water Supply Reliability index (IWSR), which is defined as the ratio of water supply available for irrigation over potential demand for irrigation water. For developing countries, the IWSR declines from 0.79 in 1995 to 0.71 in 2025. Relatively dry basins that face rapid growth in domestic and industrial sectors, experience slow improvement in river basin efficiency or rapid expansion in potential irrigated area show even greater declines in water supply reliability. The developed countries as a whole show a sharp contrast to the developing world. Irrigation water supply in the developed world is projected to increase by 7.0 km^3, while the corresponding demand decreases by 5.0 km^3. As a result, after initially declining from 0.86 in 1995 to 0.84 in 2010, the IWSR improves to 0.89 in 2025 as domestic and industrial demand growth slows in later years (and actually declines in the United States and Europe) and efficiency in agricultural water use improves. The divergence between trends in the developing and developed countries indicates that agricultural water shortages will become worse in the former even as they improve in the latter, providing a major impetus for the expansion in virtual water transfers through agricultural trade.

Given the tighter baseline water situation described above, would policy or behavioral changes in water availability and demand have a major impact on future food availability and prices? Several alternative scenarios are briefly discussed in the following paragraphs. A Low Infrastructure Investment (LINV) scenario attempts to examine the effect of low investment on global and regional food production. Further, it assumes that any improvements in these drivers due to already committed investments and water management reforms will be offset by faster depreciation of existing infrastructure and an increased rate of reservoir siltation. The major assumptions involved in LINV include: 1) *Reductions in the growth rates of reservoir storage for irrigation and water supply*: The net increase of global reservoir storage for irrigation and water supply only increases by 325 km^3 between 1995 and 2025 under the LINV scenario compared to an increase of 621 km^3 under the baseline. 2) *Reductions in water use efficiency*: Under this scenario, the global average basin irrigation efficiency is assumed to increase to 0.57, compared to 0.62 under the baseline. 3) *Smaller increase in irrigated area*: The increase of irrigated harvested cereal area from 1995 to 2025 is 5.0 million hectares, compared to 23.9 million hectares under the baseline scenario.

The results show that the net increase in water withdrawal between 1995 and 2025 is only 345 km^3 under the LINV scenario compared to 859 km^3 under the baseline. The irrigated cereal production will be reduced by 125 million metric tons (mt) by 2025. The irrigated cereal production is expected to decline by 6 percent for the developed countries and 12 percent for the developing countries. This would further result in a 16 million ton increase in food imports for the developing countries.

Many regions, including northern India, northern China, some countries in West Asia and North Africa (WANA), and the western U.S. have experienced significant groundwater depletion due to pumping in excess of groundwater discharge. A Lower Groundwater Pumping (LGW) scenario attempts to examine the effects of potential limits to groundwater pumping, especially in those countries/regions that are unsustainably using their water, and phases out groundwater overdraft over the next 25 years through a reduction in the ratio of annual groundwater pumping to recharge at the basin or country level. Under the LGW scenario, groundwater pumping in these countries/regions is assumed to decline by 163 km^3 compared to 1995 levels (including a reduction by 11 km^3 in the U.S., 30 km^3 in China, 69 km^3 in India, 29 km^3 in WANA and 24 km^3 in other countries). The projected increase in pumping for areas with more plentiful groundwater resources remains almost the same as under the baseline scenario, however, the total global groundwater pumping in 2025 falls to 753 km^3, representing a decline from the 1995 value of 817 km^3 and from the baseline 2025 value of 922 km^3. Under this scenario, irrigated cereal production is expected to decline by 4 percent and cereal imports to developing countries are expected to increase by 7 percent (16 million tons) by 2025. However, the biggest impacts under the LGW scenario are concentrated in the basins that currently experience large overdrafts. Compared with the baseline scenario, LGW results in a significant reduction in cereal production, leading to an increase of cereal imports of 7.0 million mt in India, 6.5 million mt in China, 1.7 million mt in WANA and 0.8 million mt in other developing countries, in 2021–2025.

Simulations of improvements in basin irrigation efficiency targeted within the overdrafting basins show that to compensate for the loss in production and consumption from reduced groundwater pumping, water sector policies would be required to focus beyond the most affected basins. Although improvements in irrigation efficiency in the specific overdrafting basins could in theory compensate for these declines in groundwater use, the required efficiency improvements would be huge and likely unattainable. In the Indus basin, an improvement in basin efficiency from 0.59 in 1995 to 0.76 in 2025 would be required to generate enough cereal production to compensate for the reduction in groundwater overdraft. In the Yellow River basin in China, the basin efficiency would be required to improve from 0.62 in 1995 to 0.82 in 2025.

Alternatively, could increased rainfed cereal production within the overdrafting countries compensate for the irrigated production decline due to reduced groundwater pumping? A scenario combining elimination of groundwater overdraft with higher rainfed agriculture development addresses this question. The reductions in irrigated production due to reduced groundwater pumping can be offset by an increase in rainfed area and yield within the same regions, but the required increase in rainfed cereal yields would be very large. In 2025, average rainfed cereal yields would need to be 13 percent or 0.6 metric tons per hectare higher than baseline projections in China, 20 percent or 0.3 metric tons per hectare in India, and 0.3 metric tons per hectare in WANA. In addition, rainfed cereal area would need to increase by 0.6 million hectares in China, 0.8 million hectares in India, and 0.10 million hectares in WANA. Greater expansion in rainfed area is itself environmentally damaging, however, requiring encroachment on fragile lands. Moreover, the yield increase would require substantial additional investments in agricultural research and management for rainfed areas and it is not clear that these yield increases are achievable in rainfed areas, even with increased investments.

CONCLUSIONS

Irrigation will remain the single largest user of water in the next thirty to fifty years, while rapidly growing municipal and industrial water demand (particularly in the developing countries) will result in increasing water scarcity for agriculture. Climate change may further positively or negatively impact water supply through changes in rainfall and water requirements such as evapotranspiration, but during the next three decades these impacts are likely to be much smaller than demands for water due to population increases and improved living standards. A wide range of possible responses could contribute to alleviating water scarcity, involving multiple stakeholders from government, the private sector and civil society.

Expanding water supply capacities, including storage and withdrawal capacities, seems to be still necessary, especially in developing countries. However, due to the scarcity of water sources and increasing environmental concerns, increased food production will mainly depend on producing more food from existing water and land resources. A critical issue is therefore to look for ways of improving water use efficiency, using both physical and non-physical means. Fortunately, there is plenty of scope to improve water use efficiency in both irrigated and rainfed systems, as their current efficiencies are so low.

Scenario results from IMPACT-WATER show that investments and policy reforms in water and irrigation management are necessary to improve water use efficiency. The baseline scenario shows that water for irrigation will become increasingly scarce in developing countries. Changes in investments and policies have major impacts on water availability for agriculture and on food production. If investments and projected rates of water use efficiency grow more slowly than under the baseline scenario, food production will decline sharply compared to the baseline. These results indicate the importance of careful country- and basin-level analysis of the necessary investments and water management reforms that would most cost-effectively boost the efficiency of agricultural water use and sustain crop yield and output growth to meet rising food demands in the face of rapidly increasing non-agricultural demands for water. Further development of modeling of the linkages between water and food would permit more disaggregated assessment of the allocation of investment across alternative projects and programs, and the balance between infrastructure investment and investment in water management reform.

Therefore, there is a major challenge to link science and engineering principles with relevant social, economic, institutional and legal schemes that affect water use and management for agriculture. More direct outreach initiatives (e.g. via the UNESCO/WMO HELP program) are also needed to strengthen the links among science, engineering, and the implementation of investments and policy reforms in order to sustain food security through efficient water resources development and management.

REFERENCES

Adams, W.M, How beautiful is small? Scale, control, and success in Kenyan irrigation, *World Development*, 18(10), 1309-1323, 1990.

Cosgrove, W. J., and F. Rijsberman, *World Water Vision: Making Water Everybody`s Business*: World Water Council & World Water Vision & Earthscan, London, 2000.

Dinar, A., T.K. Balakrishnan, and J. Wambia, Political Economy and Political Risks of Institutional Reform in the Water Sector, World Bank Working Paper 1987, World Bank, Washington, D.C., 1998.

Easter, K. W., M.W. Rosegrant, and A. Dinar, The Future of Water Markets: A Realistic Perspective, in *Markets for water: Potential and Performance*, edited by K. Easter, M.W. Rosegrant, and A. Dinar, Kluwer Academic Publishers, Boston, 1998.

Evenson, R.E., C.E. Pray, M.W. Rosegrant, Agricultural research and productivity growth in India. International Food Policy Research Institute, Washington, D.C., 1999.

Fan, S., L. Zhang, and X. Zhang, Growth & Poverty in Rural China: the Role of Public Investments. Environment Discussion Paper No. 66. International Food Policy Research Institute, Washington, D.C., 2000.

Fan, S., P. B. R. Hazell, and S. Thorat, Linkages between government spending, growth, and poverty in rural India, Washington, D.C.: International Food Policy Research

Institute, 1999.

FAO (Food and Agriculture Organization of the United Nations), Food Production: the Critical Role of Water, World Food Summit, Rome, Italy, 1996.

FAO 2000, Agriculture: Towards 2015 and 2030: Technical Interim Report, FAO, Rome (available from FAO web-page).

Gleick, P.H. ed. *Water in crisis: A Guide to the World's Water Resources*. New York: Oxford University Press, 1993.

ICOLD (International Commission on Large Dams), *World register of dams*. ICOLD, Paris, 1998.

IWMI (International Water Management Institute), PODIUM – the Policy Dialogue Model, http://www.cgiar.org/iwmi/tools/podium.htm, 2000.

Jones, W., *The World Bank and Irrigation*, A World Bank Operations Evaluation Study, Washington, DC, 1995.

Keller A., R. Sakthivadivel and D. Seckler, *Water, Scarcity and the Role of Storage in Development*, Research Report 39, International Water Management Institute, Colombo, Sri Lanka, 2000.

Meinzen-Dick, R.S., M. Mendoza, L. Sadoulet, G. Abiad-Shields, and A. Subramanian Sustainable water use associations: lessons from a literature review, Paper presented at the World Bank Water Resources Seminar, Lansdowne, Virginia, 1994 December 13-15, & Chantilly, Virginia, 1995 December 11-13, 1994.

Nickum, J.E., Beijing's maturing socialist water economy, in *Metropolitan water use conflicts in Asia and the Pacific*, edited by J.E. Nickum and K.W. Easter, East-West Center and United Nations Centre for Regional Development, Westview Press, Oxford, 1994.

Postel, S., *Last oasis: Facing water scarcity, The Worldwatch Environmental Alert Series*, W. W. Norten, New York, 1992.

Postel, S., *Pillar of Sand—Can the irrigation miracle last?* W.W. Norton, New York, 1999.

Rosegrant, M. W. and M. Svendsen. Asian Food Production in the 1990s: Irrigation Investment and Management Policy, *Food Policy* 18(February): 13-32, 1993.

Rosegrant, M. W., Water Transfers in California: Potentials and Constraints. *Water International* 20: 72-78 (June), 1995.

Rosegrant, M.W. and Perez, N.D., Water resources development in Africa: A review and synthesis of issues, potentials, and strategies for the future. Environment and Production Technology Division Paper No. 28, International Food Policy Research Institute, Washington, D.C, 1997.

Rosegrant, M.W. and C. Ringler, Impact on Food Security and Rural Development of Reallocating Water from Agriculture, Environment and Production Technology Division Discussion Paper No. 47, International Food Policy Research Institute, Washington, D.C., 1999.

Rosegrant, M., X. Cai, S. Cline, and N. Nakagawa, The Role of Rainfed Agriculture in the Future of Global Food Production, Environment and Production Technology Division Discussion Paper No. 90, International Food Policy Research Institute, Washington, D.C., 2002.

Rosegrant, M. W. and X. Cai, Modeling Water Availability and Food Security: A Global Perspective: The IMPACT-WATER Model, Working paper, International Food Policy Research Insitute, Washington, D.C., 2001.

Rosegrant M.W., M.S. Paisner, S. Meijer, and J.Witcover, Global Food Projections to

2020: Emerging Trends and Alternative Futures. International Food Policy Research Institute, Washington D.C., 2001.

Rukuni, M., Creating an enabling environment for the uptake of low-cost irrigation equipment by small-scale farmers. Irrigation technology transfer in support of food security. Proceedings of a Subregional Workshop, Harare, Zimbabwe, 14-17 April 1997, Land and Water Development Div., FAO, Rome, 1997.

Seckler, D., The new era of water resources management: from "dry" to "wet" water savings, Research Report 1, International Irrigation Management Institute, Colombo, Sri Lanka, 1996.

Svendsen, M., and M.W. Rosegrant, Irrigation development in Southeast Asia beyond 2000: Will the future be like the past? *Water International* (19)1: 25-35, 1994.

Svendsen, M., J. Trava, and S.H. Johnson III, Participatory Irrigation Management: Benefits and Second Generation Problems, Lessons from an International Workshop held at Centro Internacional de Agricultura Tropical (CIAT) Cali, Colombia, February 9-15, 1997.

U.S. Office of Technology Assessment, Preparing for an uncertain climate, Government Printing Office. OTA-O-567, Washington, D.C. <http://enso.unl.edu/ ndmc/ mitigate/ policy/ota>, 1993.

World Bank, Water resources management, A World Bank policy paper, World Bank, Washington, D.C., 1993.

World Bank, *Prospects of Irrigated Agriculture: Whether the Irrigated Area and Irrigation Water Must Increase to Meet Food Needs of the Future.* Water for Food and Rural Development Team, Rural Development Department, The World Bank, Washington, D.C., 2001.

World Commission on Dams (WCD), Dams and Development: A New Framework for Decision-Making, Report of World Commission on Dams (http://www.dams.org), 2000.

Mark Rosengrant and Ximing Cai, International Food Policy Research Institute, Environmental and Production TEchnology Division, 2003 K. Street, NW, Washington, DC 20006-1002; m.rosengrant@cgiar.org; x.cai@cgiar.org

Introduction To Section 2

This section introduces us to the policy sector and the various issues that characterize discussions in that sector. Policy is a very broad term covering legislation, regulation, negotiated agreements, standards and guidelines, incentives, assessments and evaluations, and economic and other instruments. Here we consider he role of law in the management of water in the United States, particularly as it relates to water quality and sustainability issues. Policies guide the implementation of projects and programs that benefit water management. Adaptive management is a desired approach for dealing with water issues because the solutions to many problems are inextricably linked to level of infrastructure development in a basin. Although physical resource limitations create demands for new policy initiatives, economic research and public surveys can often point the way to support such new initiatives. Discussions here, for example, demonstrate the benefits of managing surface and ground water conjointly.

The international aspects of water management introduce added complexity to policy issues particularly where developments and long-term planning are involved. Cross-border issues can be exacerbated by substantive inequalities that may exist on opposite sides of a border. These inequalities relate to the level of service expected and the level of data, technology, and support for people to engage in a science-policy dialogue. While this section does not provide a comprehensive review of all of the possible policy options, it does describe a number of policy approaches in which science can make a substantial contribution.

Water: Science, Policy, and Management
Water Resources Monograph 16

10.1029/016WM07

6

Water Policy for the West: Who Makes It? What Is It? Will It Lead Us to a Sustainable Future?

Denise D. Fort

INTRODUCTION

Sustainability is the great challenge of the western United States, because of the aridity of much of the West and the demands of its growing population. Sustainability is intended to bring about the long-term survival of societies, in part by protecting the natural resources that permit their existence. Although it would seem to be a self-evident goal for a society, it is not the principle under which water is managed. Rather, water management is the domain of a myriad of federal, state, tribal, and local governments, each with disparate mandates and motivations. The mandates of these agencies largely were shaped when resource use was the dominant paradigm, and the politics of water reflected the interests of water owners.

Managing for sustainability not only requires integration of agencies and interests, but also of disciplines. The physical dimension of water shortages is undeniable, but it is social and institutional choices that lead to water crises. The relationship between scientists and policymakers can be an uneasy one because the solutions that present themselves to scientists are often dismissed by politicians. This has resulted in a frustrating disconnection between what could be achieved toward reaching sustainability and the likelihood that inertia and political stalemate will result in greater societal injury.

The management of water in the West largely reflects historic attitudes toward natural resources. Water scholars from law, economics, political science, geography, and the natural sciences have pointed to dysfunctions in water management. Periodically the U.S. engages in a revisiting of national water policy, with the last such effort undertaken by the U.S. National Water Commission in 1973. The most recent reassessment of western water policy was focused on the western United States: the Western Water Policy Review Advisory Commission (here-

Water: Science, Policy, and Management
Water Resources Monograph 16

10.1029/016WM08

inafter the Commission). This commission was established by an act of Congress, and was comprised of citizens and ex officio federal officials[1]. The report of the Commission, *Water in the West: Challenge for the Next Century* [*WWPRAC*, 1998], details the growing population in the West, the deterioration of ecological systems and the extinction of native fishes, development pressures on agricultural lands, and the voracious thirst of western cities.

Western water management raises issues that are common to many areas of resource management, as our society's increasing scientific competence repeatedly collides with policies that reflect special interests and a preference for the status quo. When we look at environmentally injurious decisions that were made in the past we often excuse them because of the ignorance of ecological understanding in those eras. But water policy is unfolding in our own time, and this forces us to ask whether natural resource decision making in the United States will escape its longstanding pull toward the "lords of yesterday."[2] This paper sketches the challenges that achieving sustainability present to the West. It then reviews the underlying political dynamics that have shaped resource policies. Finally, it examines some of the policy prescriptions that might bring about a change toward sustainability.

THE WESTERN WATER POLICY REVIEW ADVISORY COMMISSION

Congress authorized the Commission in 1992 to review federal water policies and to make recommendations about future directions of these policies. It stands in a long tradition of reports about the nation's water. In fact, the legislation establishing it was modeled on that which established the National Water Commission. Measured by direct policy change these reports have had a mixed record, but Congress' continued convening of outside experts demonstrates the urgency it attaches to securing appraisals of the Nation's water situation.

---

[1] I served as chair of the Commission. The report of the Commission is available through the National Technical Information Service and at www.den.doi.gov/wwprac. There are several discussions of the work of the Commission, including Denise Fort, "The Western Water Policy Review Advisory Commission: Watershed and Basin Management Receive the Attention of a New Generation," Journal of the American Water Resources Association (1999) and "The Western Water Policy Review Advisory Commission: Another Look at Western Water," 37 Natural Resources Journal 909 (1997).

[2] This phrase was coined by Professor Charles Wilkinson in *Crossing the Next Meridian: Land, Water and the Future of the West*, Island Press, 1992. It refers to the foundation of the old West: the Mining Law of 1872, the public rangelands and forestlands, dams and the prior appropriation doctrine. The dichotomy in the West can also be presented as one between the "Old West" of ranching, logging, mining, and other resource based activities, and the "New West" of tourism, technology, financial services, and other economic activities that are not unique to the West.

The Commission identified population growth as the primary challenge to the sustainability of the West's water. The West is growing at a rate that is disproportionate to the remainder of the nation, as Americans and immigrants move to Sunbelt states. The difficulty of supporting large populations and preserving natural systems in such an arid region is intuitively obvious, but public policies rest on the assumption that any natural impediments to growth can be overcome by technology.

The crisis in western water takes several forms. At its center are the competing demands of a growing number of people for a fixed quantity of water. Increasing water for one use will affect other users; the era of free lunches in western water is over. The crisis of conflicting demands is visible anywhere in the West. For example, the California Department of Water Resources estimates that, because of population pressure, California will face water shortfalls of between 4 and 9 million acre feet per year by 2020. [*WSWC*, 1997, Appendix 1]. There is contemporary evidence of what the future holds for the most populous state in the union; increasing urban and environmental improvement demands have already resulted in some agricultural users receiving only 50% of their contracted supply. [Id].

The crisis in supply for human uses will be "solved," in the sense that municipal and industrial needs will be met, through transfers of waters from other uses, water conservation, and other approaches that now being utilized across the West. These solutions do not come without costs, although some of the innovative approaches discussed below can result in outcomes with positive attributes for all parties.

Transferring water from agricultural to municipal uses can change the communities that were based on agricultural economies. [National Research Council, 1992, but see *Tarlock*, 1999]. The loss of agriculturally based communities is a source of contention in the western United States. The mobile nature of many businesses has led to scenic rural valleys sprouting "trophy homes," "ranchettes," and traditional subdivisions. [*Riebsame*, 1997(a)] Water transfers typically are not driving these changes, however, but rather consolidation of farming operations, economic pressures, and the development pressures that the population boom is putting on formerly isolated areas. [*Riebsame*, 1997(b)] Nonetheless, the sale of water rights from agriculture to other uses is a powerful symbol of the death of the traditional culture.

It is possible to mitigate the costs associated with transfers by conservation improvements that keep some lands productive while transferring conserved waters, and through payments to third parties in the community. Negotiations among affected interests can ensure that all parties gain something out of settlements. Profligate water practices by other user communities also can be addressed with conservation measures, slowing down the need for water transfers or new projects.

The ecological crisis resulting from population growth and increased pressure on rivers is seen in the loss of river ecosystems and of species that are water dependent. Other changes have quietly occurred, such as the drying up of springs due to groundwater pumping, and the loss of flows in intermittent streams. In fact, because of the lack of regulatory interest in these ecosystems[3], the ecological changes that water development has brought have not been measured systematically.

Hydrologic modifications, such as the construction of dams and the channelization of rivers, also have affected river ecosystems. [See e.g. *Minckley*, 1997]. There are few proposals for the construction of new large-scale dams to meet population and agricultural needs. Instead, agencies are pursuing off-stream reservoirs, paid for by state, not federal funds, water reuse and recycling projects, additional groundwater development, and an intensified effort to get the maximum benefit out of all available water supplies.

IS THERE A FEDERAL WATER POLICY FOR THE WEST?

There is no single federal water policy: no statute that comprehends all water management purposes, nor any single agency that has a mandate that encompasses all of the federal duties with respect to western water. Instead, there are a myriad of agencies with water related responsibilities. [See *WWPRAC,* 1998; see also *Neuman*, 2001] Of course, federal agencies are only part of the picture: state, regional, tribal, municipal, and other private entities also manage water.

The federal role in western water is a complex one and has been many years in the making. In parts of the West federal agencies have a minor role, and state or other agencies are the most visible governmental entities involved in river management. This perspective would be familiar in many Eastern states, where the federal presence is relatively low key. But, in many Western river basins, the federal presence is far more dominant, and the institutional structures for water management are strongly oriented toward federal control. The management of the lower Colorado River is perhaps the best illustration of this: the Secretary of the Interior actually administers the river, with authority over issues that are of critical importance to the affected states. The role of the Corps of Engineers and the Bonneville Power Authority in the Columbia River basin is another example of dominant federal control. In both instances, control of physical infrastructure and water resources has made Washington, D.C., not state capitols, the seat of power for these water decisions.

---

3 Regulatory interests protect these ecosystems in limited instances, for example, when the Endangered Species Act applies. Wetlands protection programs occasionally apply, but there are many limitations on their applicability because water withdrawal, per se, is not regulated by section 404 of the Clean Water Act. Similarly, the Clean Water Act applies to discharges, not withdrawals of water. There is no federal law that protects riparian and aquatic ecosystems as such.

The Bureau of Reclamation built the infrastructure for agricultural development in many western states, and Bureau projects provide 30.9% of the West's water. [*WWPRAC*, 1998, at Figure 2-3] In many river basins the Bureau is the dominant actor, because it controls the reservoirs and irrigation works of the river. The Corps of Engineers began its operations at the time of the Nation's birth, but found a role for itself in the developing West, providing storage for hydro facilities, as well as flood control devices and thinly disguised storage projects. (This paper is limited to freshwater issues; the Corps is also active in the West in harbor maintenance, and other coastal responsibilities.) Competition among these agencies for projects is legend in the West, and these conflicts spill into the halls of the Congress.

Environmental programs added a new role for federal agencies. Federal laws passed in the early 1970s established minimum standards for air, discharges to water, and later, management of hazardous waste. The Clean Water Act created a national program to control the discharge of pollutants, including the requirement that states create stream standards and a permitting system for dischargers. The architecture of the regulatory program is one of national effluent limitations for categories of industries, linked to state-created water quality standards. The link is a system of permits for discharges to rivers and lakes, called the National Pollutant Discharge and Elimination System (NPDES) permits. The NPDES system typically is operated by states, although EPA regional offices issue permits in some states. Tension between the federal government and states over the requirements of the Act mostly has abated, but control of nonpoint source pollutants remains contentious. Nonpoint sources are exempt from the federal requirement to have a NPDES permit. These sources are usually agricultural operations, which tend to be deferred to by state regulators. Environmental groups used a provision of the Clean Water Act, the Total Maximum Daily Loading program, to attempt to force states to deal with nonpoint pollution. Mandatory regulation of nonpoint sources is yet to be achieved under the Act.

The protection of endangered species has proven to be the most difficult goal of all to manage, because it often requires that a property owner leave her property undisturbed, rather than merely restrict the discharge of pollutants. [*Sax*, 1993]. Some portion of every major river system in the West contains a species of endangered fish. Both agencies now spend substantial amounts on programs to protect species or habitats. [WWPRAC, 1998] The federal presence in river management remains highly controversial despite the emphasis that agencies have put on collaborative processes in recent years. The Klamath Basin imbroglio, with conflicts among salmon, irrigation, and tribal rights, illustrates the centrality of the federal government to water management in the West.

The federal role as a manager of water for economic purposes and its role as a protector of environmental values need not be in conflict, but it must be admitted that the reconciliation of these mandates occurs, if at all, on a case by case basis. The Bureau of Reclamation and the Corps of Engineers have powerful supporters in their traditional constituents. These constituents typically have little to gain from the insertion of new values in federal policy. Furthermore, conservative interests vocally oppose an expanded federal role in western water. The new mission of environmental protection draws resources from the traditional constituents and may conflict with their interests.

Another important dimension of the federal role is as trustee for tribes. Some tribes have interests in water, or in water management, through treaty rights. The Supreme Court found that others have water rights that were created when tribal reservations were established. [See *Winters v. U.S.*, 207 U.S. 564 (1908)]. In many instances, these rights have not been quantified through judicial proceedings, leaving some future reallocation of water, or monies, likely.

Finally, the federal government has potential control over water through its ownership of federal land in the West. The federal government reserved water rights when it created special status for some types of land, such as national parks. [*U.S. v. New Mexico*, 438 U.S. 696 (1978)]. The assertion of federal rights is extremely unpopular among state officials. The Forest Service is considering withdrawing a policy that protects streams on federal lands because of concerns raised by municipal diverters. The Secretary of the Interior recommended that the Department of Justice not appeal an Idaho Supreme Court decision denying instream rights to a federal wildlife reserve. [*Barker*, 2001]

There is no entity in Washington, D.C., other than the President, with the legal authority to integrate all of the federal water agencies. President Reagan allowed the Water Resources Council, which had been established for this purpose (and which still exists in the statutes), to quietly go defunct in 1981. Its passing was mourned by few and there is little momentum for its resurrection.

Despite the lack of a formal coordinating entity, on specific projects one can find coordination among water agencies, but it is ad hoc and dependent on the individuals and administration involved. This coordination may occur at various levels of government, from the most informal exchange of information among low-level field staff to meetings of subcabinet officials. This occurred in the California Bay-Delta, where water quality, endangered species, water demand, and other conflicting water interests were addressed by coordinated federal agencies. (State agencies were also coordinated and involved by the State of California.) The result was CALFED, a joint enterprise that involved federal and state agency officials in working toward a variety of water-related goals, including restoration of aquatic environments, water supply, water management for agricultural and municipal uses. [CALFED's official website can be viewed at http://calfed.ca.gov, see also *Wright*, 2001; *Rieke*, 1996.]

The Congress is as much of an actor in creating and implementing federal water policies as are the federal agencies. Budgets, substantive lawmaking, and oversight for the various water agencies are divided among different Congressional committees. [*WWPRAC*, 1998, Appendix C]. Because of the powers these roles give to members of Congress, it is unlikely that members will vote to divest themselves of control over these agencies.

The quest to rationalize water management within the federal government is likely to be quixotic. Indeed, conflict among agencies can be healthy for the public interest. Each agency represents the ascendancy of a constituency. Congressional power reinforces interest groups' claims over a particular agency's mission. One cannot capture the appropriate goals of water policy under a single principle; the conflict among principles represents the very real conflicts among the many dimensions of water. On the other hand, the federal government could function better, by elevating water decision making to a higher level, by requiring coordination among agencies, and by increasing the visibility and opportunities for public comment.

STATE WATER MANAGEMENT

States, not the federal government, aggressively assert control over water within their boundaries. There are many exceptions to this rule, but in the main, states set policies and administer the allocation of water within a state. States face the challenge of adjudicating complex water rights claims, where thousands of competing uses need to be determined in accordance with state laws. Claims that a user is infringing on another also will be brought to a state agency for resolution. State courts play a role in the resolution of disputes as well. As water grows in importance greater expectations for a proactive management role will be put on state agencies, although the agencies often lack the statutory powers to comprehensively manage water needs.

State legislatures have been slow to incorporate the changing demands on western water and the changing constituencies for these new uses into their statutory schemes. The state agencies with jurisdiction over water generally reflect the historic dominance of agricultural interests. Municipal water interests have established roles for themselves in the agencies and legislatures, which is understandable given the relative numbers of people affected. In debates before Congress over environmental issues, the state agencies are likely to represent the interests of irrigators. As western states grow more diverse and new voices are heard in water debates, some states are beginning to incorporate these perspectives in water management, with less antipathy to environmental protection and greater cooperation in endangered species recovery.

WILL STATE AND FEDERAL POLICIES LEAD US TO A SUSTAINABLE FUTURE?

When it is time to make a prediction about the likelihood of our achieving sustainability in our water policies, one reveals oneself to be either a Cassandra or a Dr. Pangloss. One can read our environmental history as the portrait of a people who defeated nature and triumphed over the limitations that the natural world appears to impose, or as a story that is not yet finished, where the limitations of the Earth will be vividly revealed to all doubters. This debate accompanied the release of *Beyond the Limits*[4] and *A Moment on Earth*[5], and the reporting of the famous wager between Paul Erlich and Julian Simon over the future price of certain minerals. [See *Julian Simon, 65, Optimistic Economist Dies*, NY Times, Feb. 12, 1998, at B 11, col.1] The most pronounced difference in perspective might be found in how one views the importance of ecosystems; if one looks at the ability of humans to build impressive economies of ever increasing wealth then the limitations of the natural world seem puny, but if one looks at the havoc wreaked on ecosystems by the industrial age, then the damage seems high. Indeed, the ability of humans to limit pollution, and to protect human health from environmental harm has been proven with respect to many pollutants in the last few decades. We do have better water quality in rivers than we did thirty years ago and air pollution is reduced. There is a broad chasm between the perspective that finds evidence of abundant human progress and that which sees escalating ecological loss.

It is a precept of the sustainability movement that there is no necessary conflict between economic development and ecological sustainability. (See, for example, the President's Council on Sustainable Development (1996): "Economic growth, environmental protection, and social equity are linked. We need to develop integrated policies to achieve these national goals.") In many parts of the less developed world, the effect of ecosystem deterioration on humans is unquestioned, because of the obvious links between ecosystems and human needs. The crisis in the Aral Sea is both a human and ecological crisis, as are the effects of the construction of the Three Gorges Dam, and the degradation of watersheds. In these instances, the water crisis has led to developments in which the linkage between the sustainability of human and natural systems is fully evident. China, for example, acknowledges that the environmental consequences of over harvesting timber in mountainous zones includes massive human and economic costs downstream, because the degraded watersheds no longer protect against downstream flooding. In the western U.S., by contrast, the consequences on humans of environmentally adverse projects are not so stark. From the perspective of a more affluent society, able to utilize natural resources on a global basis, the ecological sustainability of a particular river or lake appears to be a matter of choice, not a necessity to preserve human life. Thus affluent nations can adapt to the loss of an offshore fishery by paying to import fish products from remote oceans.

---

[4] Meadows, D. H., Meadows, D. L., & Randers, J., *Beyond the Limits: Confronting Global Collapse, Envisioning a Sustainable Future*, Chelsea Green, Post Mills, Vt., 1992.

[5] Easterbrook, Gregg, *A Moment on Earth: The Coming Age of Environmental Optimism*, Viking, New York, 1995.

The sustainability of western water systems can also be viewed from these two perspectives. In one view, cities with booming populations and strong economies have grown up in regions with almost no natural supplies of water. Presumably the valuation of economic life in these regions reflects whatever importance the market attaches to water shortage; one senses that the financial markets are confident that water will be provided for the life of their financing obligations. Southern California's ingenuity in meeting its water needs is justly famous among popular (the movie "Chinatown") and academic sources. [*Ostrom*, 1990]. The population of the lands beyond the 100th meridian exceeds any prediction that could have been made a century ago, but water still flows from the newcomers' taps. [*Maddock*, 1995]

The other perspective would ask whether the manner in which this growth has been pursued is sustainable. The reliance of the West on mined groundwater is by definition unsustainable because this groundwater is being exhausted. Furthermore, global climate warming is expected to change many factors that are related to water availability and demand, such as higher surface temperatures, altered seasonal precipitation, extreme weather events, etc. [*McCarthy*, 2001]. Additionally, population increases will require hard choices and volatile disputes as water is sought out for new users.

Most importantly from the environmental perspective is that river and other aquatic ecosystems have been severely affected by a century of dam construction, irrigation diversions, channelization, pollution, introduction of nonnative species, and other actions. The loss of the native fishes of the West, and the endangered status of fishes and other species that are dependent on the aquatic environment all point to a crisis of sustainability in the natural world.

The debate over whether the United States should protect and restore its western waters is ultimately unnecessary, in that there is no reason one must choose between a sustainable ecosystem and human uses of water. Regardless of the necessity of ecosystem protection, the American people strongly support the Endangered Species Act. It has been amply demonstrated that ecosystem protection is not at odds with development, and indeed, can enhance it. The most significant development in western water has been the movement to restore balance to water management, permitting healthy economies to co-exist with healthy ecosystems.

WHAT ARE THE INSTITUTIONAL FACTORS THAT MILITATE AGAINST SUSTAINABILITY?

Sustainability is not the basis of our natural resources policies, nor of our economy at large. Our institutions value the exploitation of resources; the protection of natural systems is only a minor addition to a massive economic system that is indifferent to ecosystem protection. But counterbalancing factors are beginning to give more weight to ecosystems (see Section VI) as a means of improving water management.

The Value of Water

The primary threat to sustainability can be expressed in economic terms; we do not value the use of water commensurate with the full range of effects that should be included to guide social choices. A recent Congressional news release makes the point: Senator Pete Domenici of New Mexico announced that he was seeking a federal appropriation to provide desalinized water for a group of constituents. They would pay nothing for the water. [*Water Supply Security Act of* 2001, S.B. 1309, 107th Congress]. This type of federal subsidy was the rule for many years of western development of agriculture. To extend it to residential users is the logical extension in federal subsidies for settlement of the West. As groundwater supplies are mined, and municipalities are unable to find relatively inexpensive water for growing populations, federal subsidies will be sought by the next generation of settlers. The problem is not federal largesse, but the well-known effect of subsidies: poor choices are made by water users when there is no information about the true costs.

What one pays for water in the West depends on where one lives and who controls water rights and infrastructure. Those who own water rights may pump groundwater for free, except for the costs of power and well construction. Members of irrigation districts will pay an assessment for water. Customers of municipal water systems will pay a fee that is many times higher than that paid by irrigation districts. [*MacDonnell*, 1999].

The future of water subsidies is uncertain. Here Congress is the entity with motive and opportunity, as constituents realize that federal funds may make up for regional planning disasters. The line between "pork" and "a national investment" is admittedly dim. For example, some of the current water related expenditures include federal operation of Bureau of Reclamation and Corps of Engineers dams. There is a movement to turn some of these facilities over to local entities, thereby reducing the federal presence. Multiple constituencies use these facilities, and turning them over to a single interest might be contrary to a broader public interest. The Corps provides flood control and some navigation projects in western states. Flood control, such as building levees, need not be a federal concern (it was not historically), but the generous federal cost-share makes it popular with state and local officials. Fish and wildlife, and river restoration projects are a significant part of the Bureau's budget and a growing part of the Corps budget. This is an area of expenditure where some conservatives have questioned the growing federal role.

As agriculture loses its political clout, we would expect to see urban areas of the West demanding additional subsidies for their water supply needs. The potential demand from this constituency is unlimited. These projects may include off-stream storage, groundwater storage, treatment and injection of groundwater, cleanup of contaminated groundwater, etc. There are no agreed upon principles about the federal role in funding these ventures, so that it is difficult to predict whether projects of this sort will become new roles for the Bureau and the Corps. However, every time that a federal subsidy is provided the comparison with alternative means of addressing demands is distorted. Water conservation or water transfers may be less expensive than new water supplies, but are unlikely to appeal to members of Congress.

Organizational Dysfunctions

Water management repeatedly has been criticized for overlaps and conflicts among agencies. The conceptual division between administration of water allocation and the administration of programs to protect water quality confuses the physical relationship between these aspects of water. Water allocation was a precondition for development of the western states. Water quality, on the other hand, largely was unregulated until the Congress passed its comprehensive water quality act in 1972. Most states have separate agencies to deal with each of these programs and coordination may be poor between the agencies. Adherence to the disparate statutory frameworks may discourage states from creating programs that link the two objectives.

The federal Clean Water Act contains a provision that the Act is not intended to affect state water allocation powers. [See 33 U.S.C. 1251(b) (West 2001)]. The EPA does not interpret the Act to require states to protect instreamflows. In a related context, one lower court recently held that the withdrawal of water, which would result in the concentration of pollutants, was not subject to the Clean Water Act's permitting requirement, because no addition of a pollutant had occurred. [*Colorado Wild, Inc. v. U.S. Forest Service*, 122 F.Supp. 2d 1190 (D. Colo. 2000)]. In the hierarchy of water management decisions, states have given primacy to water allocation decisions. Water quality is protected only if the permitted withdrawals have left sufficient water to protect. The ecological values of water in the stream are not generally protected, unless an endangered species is present or state law requires a minimum streamflow.

The State of Washington is anomalous in including water quantity as part of water quality, which is typically viewed as a chemical property of water. Washington took the unusual measure of applying minimum flow requirements to a proposed hydrogeneration facility, acting through its delegated Clean Water Act authority. The U.S. Supreme Court upheld the flow requirements as part of the state's water quality standards. Although the standards were state-created, they were enforceable under the federal Clean Water Act. The Court explained why water quality standards might appropriately contain flow standards: "...In many cases, water quantity is closely related to water quality; a sufficient lowering of the water quantity in a body of water could destroy all of its designated uses, be it for drinking water, recreation, navigation or, as here, as a fishery." [*P.U.D. No. 1. of Jefferson County v. Washington Dep't of Ecology*, 511 U.S. 700, 719 (1994)].

A second conceptual divide affects water management, the division that is made by water management agencies between surface and groundwater. Surface and groundwater often are hydrologically connected. Even when they are not hydrologically connected, users may switch from one source of supply to another. In California's Central Valley, agricultural users have increased their withdrawals from groundwater when surface water supplies are restricted. States have been slower to assert jurisdiction over groundwater, with some asserting control only in certain basins, others leaving control to substate units, and some imposing no effective controls. The Commission recommended the conjunctive management of water resources. To do so would result in a more sustainable water policies, because water would be managed as an integrated resource, regardless of its source.

The Prior Appropriation Doctrine

Even casual students of western water policy are aware of the prior appropriation doctrine. While its history is complex, it resembles John Locke's notion that the use of property was the means by which ownership rights in nature were created [see *Locke, J.* (1960)]; through the removal of water from a stream and the addition of human labor, a property right was created. The exigency behind the development of this doctrine was to enable miners and irrigators to perfect rights, in turn creating the certainty necessary for those economies to develop. Arguably, the prior appropriation doctrine is not, in and of itself, an impediment to protection of streams. In most states, water codes have been modified to give some legal protection to water left in a stream although western states have conditioned these rights in a variety of ways (for example, so-called "instreamflow" rights usually must be held by governmental entities).

Most of the waters of the West are fully appropriated (if not over appropriated) and have been for some time. At the time that rights were perfected, the concept of protecting instreamflows was unknown. Thus, the prior appropriation doctrine, by creating property rights in water, has contributed to water being held by those whose predecessors claimed the rights perhaps a century or more ago. The economic burden rests on those who would restore water to rivers, not those who would withdraw those flows.

Prohibited "Takings"

The Fifth Amendment to the U.S. Constitution prohibits the government from taking private property without compensating the owner. This principle has become a formidable impediment to governmental regulation, the product of a conservative judiciary faced with a set of cases contending that environmental regulations have "taken" private property without compensation.

When the government constructs a building on private property, it is clear that the property owner should be compensated. But, when the government prohibits buildings over a certain number of stories from being built on that property, or requires compliance with a building code before a building can be occupied, the private property owner does not have a right to compensation. [See e.g. *Penn Cent. Transp. Co. v. City of New York*, 438 U.S. 104 (1978)]. The question raised in water related regulation is the degree to which regulation of these rights amounts to a takings.

Water scholars have long maintained that water is unique; that ownership of rights to do some things with water does not mean that one is free of obligations to the greater public. [*Sax,* 1993; *Thompson,* 1998]. In fact, the context in which water rights are exercised is very different from the context surrounding the exercise of other types of property rights. In water short years, water may be rationed. Water quality laws may protect water owners, as well as moderating their conduct as dischargers. The ability to transfer water is likely to be highly regulated by state government, if it is permitted at all. Water held under federal contracts is subject to federal laws and water users may not have a right to contract renewal.

The most likely context in which a Fifth Amendment takings claim would arise is when a diverter is unable to use water because of a conflict with the needs of an endangered species. Because of the desire of the U.S. Fish & Wildlife Service to avoid this sort of conflict, and the compromises reached in most instances, there are very few cases considering what restriction of property interests in water constitutes a compensable taking. However, in a recent case the Court of Claims (a federal court that has jurisdiction over claims against the United States for monetary claims) held that farmers who were unable to use leased water because of an endangered species should be compensated for the value of the lost water. [*Tulare Lake Basin Water Storage Dist. v. U.S.*, 49 Fed. Cl. 313 (2001)].

Federalism

Another jurisprudential trend threatens federal environmental regulation and thus the protection of sustainability. The authority of the Congress to address a range of social and other dysfunctions is based on the Constitutional grant of power to regulate interstate commerce. A series of cases testing this grant in the context of Civil Rights legislation and environmental regulation led to the belief that most regulated activities have some connection with interstate commerce and that Congressional authority requires no particular showing of that connection. [See e.g., *Hodel v. Indiana*, 452 U.S. 314 (1981); *Heart of Atlanta Motel, Inc. v. U.S.*, 379 U.S. 241 (1964)].

Recently, the U.S. Supreme Court revived a doctrine that had little viability during the expansionist era of the Warren Court and the activist Congress of that era. In a series of decisions the Court has indicated that Congressional power to legislate can no longer be taken for granted. The Court has yet to squarely face the issue in the environmental arena, but there is little question where the current majority is heading. In a recent case the Court determined that the Clean Water Act did not regulate the destruction of wetlands when the wetlands in question were not connected to any body of navigable water. [*Solid Waste Agency of Northern Cook County v. U.S. Army Corps of Engineers,* 531 U.S. 159 (2001)]. The Court held that the term "navigable waters," which had been taken to be coterminous with the Congress's powers under the Commerce Clause, should instead be limited to waters that were navigable or connected to navigable waters. The Court opined, however, that serious constitutional questions would be raised had the Congress intended to extend regulation beyond navigable waters or waters connected to navigable waters. This dictum stands as a warning of the scrutiny that the Court will apply to environmental regulation that is based on the Commerce Clause. [See, *Mank,* (2002)]

This case and others reconsidering national power under the Commerce Clause have significant implications for long term ecological protection because they make federal regulation less likely to be extended to wetlands that are isolated from navigable waters. They are particularly significant for the western states, where some of the most valuable wetlands are isolated playa lakes and springs.

WHAT DOES SUSTAINABILITY REQUIRE OF WATER POLICY?

Sustainability will require reform of many of the precepts of water policy. A reform movement can already be seen in many basins across the West and public policies that would further sustainability can be gleaned from these examples. First, sustainability requires broad participation in water resource decision making and a recognition of the public interest in how water decisions are made. Second, sustainability requires that the ecological systems that are dependent on water be protected, and, when feasible, restored. Third, the economic signals sent about water use should reflect its value to society. Fourth, society's equitable interests in providing access to water as a basic human right and the claims to water of Native American tribes must be incorporated into water policy.

Public Participation

A pivotal change in water resource management has been the involvement of additional stakeholders in water decision making. Because of the assignment of water decisions to the arena of private property rights, many water decisions have been treated as if they were bilateral, limited to two private owners. The public's role, as represented by a state agency, was comparable to that of a recording clerk; that is, noting changes in deed ownership, and perhaps deciding disputes among conflicting property owners. But, while water ownership might be relatively straightforward, interests in rivers are not. People are interested in rivers because they live near them, fish or swim in them, or merely enjoy their beauty. They may enjoy using reservoirs for recreation, or may have treaty rights to fish that are impeded by a dam. Competing paradigms collide, as broader public interests meet the world of private property rights. Environmental interests meet property-based water rights at right angles.

For example, water quality standards are set by public bodies with many opportunities for public review. The implementation of the Endangered Species Act allows public participation in the decision to list species, and in other steps leading to the recovery of a species. Insofar as these sorts of interests are involved, a private (or public) water rights holder is not able to unilaterally exercise water rights and often must participate in multiparty proceedings. Tribal governments also may assert interests in water related decisions concerning watershed management or water basin management, because these processes may affect unadjudicated water rights, treaty rights, or interests in tribal land and water.

The U.S. has experimented with different approaches to natural resource management. Scientific based management was the policy approach of the Forest Service for many years [Raymond, 1999], but the movement to allow citizen participation has toppled any single area of expertise as appropriately having hegemony over water. Stakeholder participation in decision making is incorporated in most federal statutory schemes. Water management is also the focus of litigation, where water holders, agencies, and citizens may all seek relief. These multiparty cases can spin off into non-adjudicatory negotiations, or these negotiated processes can be the result of the threat of litigation.

It is particularly critical that water resource management be acknowledged to contain broad public interests because the trend in U.S. water is likely to be a greater private role and a decreasing public role. As the demand for new uses of water grows, water transfers will inexorably play a larger and larger role. The commodification of water spells market opportunities, and corporations are racing to fill the need. Unlike a federal agency, a private entity is not subject to the review of Congressional appropriations processes, the environmental review of the National Environmental Protection Act, nor other administrative processes designed to bring public and environmental concerns into agency decision making. Public interest groups have protested the transfer of municipal water functions and of federal facilities. Indeed, the broader movement against globalization has embraced the cause of keeping public water supplies in public ownership. Regardless of ownership, it is imperative that the public interest in water management decisions be recognized. A recent report by the Pacific Institute for Studies in Development, Environment, and Security, "The New Economy of Water: The Risks and Benefits of Globalization and Privatization of Fresh Water" published in 2002, presents a considered discussion of these issues and measures to ameliorate the effects of privatization.

On a smaller scale, watershed management captures the public's interest in rivers and their health. The greatest emphasis in watershed management has been the protection of watersheds from non-point-source pollution, motivated in part by the lack of a statutory requirements and the need to create a different sort of pressure on polluters. Watershed councils often are voluntary, and include both agency officials and citizens. Because of the citizen interest that they can command they have been hailed as an effective means of restoring degraded watersheds. The success of the watershed movement demonstrates the benefits of public participation in furthering sustainability.

Protection and Restoration of Ecological Systems

The Commission called for a national program to restore damaged river ecosystems. There is no law mandating the protection of ecological functions, nor is there a law mandating the restoration of ecosystems, although the Endangered Species Act can act to require them both in certain circumstances.

The United States is at a critical point in its public policies concerning the value of river ecosystems. The scientific support for river restoration was voiced in the National Research Council's *Restoration of Aquatic Ecosystems*[6], which explained the importance of these systems and called for a national program for their restoration. The nation has already begun restoration of important systems, most notably that of the Florida Everglades, which is the subject of an $8 billion effort under a joint state and federal program. States and the federal government have already spent billions of dollars on the California Bay Delta and the Columbia River. Smaller scale efforts are underway across the West, from the Platte River, to the lower Colorado River, to the Verde River in Arizona.

The widespread support for restoration of river ecosystems contains a paradox: the legal mandate behind restoration is often the Endangered Species Act, which is widely denounced by western water managers and irrigation interests. Significant federal funding for habitat maintenance and restoration before species become endangered can reduce the political pressure emanating from the application of the ESA. Restoration is often popular with a variety of water interests, despite the increased federal presence that federal funding often implies.

---

6 National Research Council, *Restoration of Aquatic Ecosystems: Science, Technology, and Public Policy*. National Academy Press. Washington, D.C., 1992.

Value of Water

Water has a special nature in the West, neither purely a commodity, nor purely a public resource. But decisions that are made about water are too often insulated from the value of water to society because of subsidies and inappropriate pricing. Groundwater mining provides one example. Groundwater mining refers to the practice of consuming groundwater at a rate that is higher than the rate of recharge. In common usage, it refers to aquifers that are recharged very slowly, perhaps only over millennia. For many aquifers in the arid West, there is essentially no recharge, so that any use of groundwater constitutes mining. Some have suggested that groundwater mining be banned entirely, but that suggestion is not politically feasible.

Under the laws of many western states, groundwater pumping may be unregulated, or subject to appropriation under state law. While it may be owned by the state, it can be used by individuals who satisfy the requirements of state law, without any further payment for the water. An alternative to prohibiting groundwater mining would be to ensure that the uses that are made of the groundwater represent the value to society of the depleted resource. Under the prior appropriation doctrine individuals may have a right to utilize groundwater but that right is subject to societal controls. An individual with mineral rights may be subject to severance taxes, for example, on the mining of those minerals. Similarly, a state could impose a tax on the mining of groundwater, capturing the value to society of the lost water and ensuring that the use to which the water is put is economically meaningful.

Equity

Any discussion of pricing should be accompanied by a discussion of the protection of equitable interests in water. In the western U.S., there are still communities that lack basic potable water supplies and wastewater treatment. In particular, there are many tribal lands where communities are remote and water scarce. If one concedes that there is a basic human right to sufficient potable water for domestic needs, then that right should not be subject to onerous charges. The same is true of water pricing for municipal uses, where lessons from utility charges can be applied to provide adequate water supplies to the poor.

Tribal interests in water were poorly protected by the federal government for much of the 20th century, and federally funded development of water proceeded without regard to tribal water rights. More recently, the federal commitment to protection of tribal interests has strengthened and tribes have retained their own counsel to assist in vindication of their rights. Although some claims are resolved in courts, settlements that are ratified and funded by the Congress are a preferred route. Nonetheless, undetermined tribal water claims are a potential claim against substantial volumes of water in many river basins. Ultimately, a sustainable water regime requires justice toward those whose water is being used by others and clarity about the ownership of water rights. In practical terms, this means that claims need to be resolved and funding from Congress procured to adjust burdens among users.

CONCLUSION

Sustainability in western water requires nothing less than a reorientation of a consumption-based regime to one in which a longer-term view is taken of what best serves human development. Cultural and institutional change will be required to move away from the short-term perspective that now dominates water management toward one focused on sustainability. The scientific literacy of our society assumes increasing importance in the transition to a sustainable future. Citizens need to be able to critically evaluate arguments from biologists, economists, hydrologists, and a myriad of other disciplines in understanding the policy dilemmas that we confront.

REFERENCES

Barker, R. and Bremner, F., *Wildlife Refuge Water-Rights Case Could Affect Other Federal Claims*, The Idaho Statesman, Page 1, Wednesday, September 5, 2001.

Easterbrook, Gregg, *A Moment on Earth: The Coming Age of Environmental Optimism*, Viking, New York, 1995.

Locke, J., *Two Treatises of Government*, (P. Laslett rev. ed. 1960) (1st ed. London 1690).

MacDonnell, Lawrence J., *From Reclamation to Sustainability: Water, Agriculture, and the Environment in the American West*, University Press of Colorado, 1999

Maddock, T. S., and Hines, W. G., *Meeting Future Public Water Supply Needs: A Southwest Perspective*, 31 Water Resources Bulletin 2, 1995.

Mank, Bradford C., *Protecting Intrastate Threatened Species: Does the Endangered Species Act Encroach on Traditional State Authority and Exceed the Outer Limits of the Commerce Clause*, 36 GA.L. Rev. 723 (2002).

McCarthy, J. and Canziani, O. (Eds.), *Climate Change 2001: Impacts, Adaptation, and Vulnerability: Contribution of Working Group II to the Third Assessment Report of the Intergovernmental Panel on Climate Change*, Cambridge University Press, 2001.

Meadows, D. H., Meadows, D. L., and Randers, J., *Beyond the Limits: Confronting Global Collapse, Envisioning a Sustainable Future*, Chelsea Green, Post Mills, Vt., 1992.

Minckley, W. L., *Sustainability of Western Native Fish Resources*, in *Aquatic Ecosystems Symposium*, National Technical Information Service, Springfield, VA, 1997.

National Research Council, *Restoration of Aquatic Ecosystems: Science, Technology, and Public Policy*. National Academy Press. Washington, D.C., 1992.

National Research Council, *Water Transfers in the West: Efficiency, Equity, and the Environment*, National Academy Press, Washington, D.C., 1992.

Neuman, Janet, *Federal Water Policy: An Idea Whose Time Will (Finally) Come*, 20 Va. Env. L. Rev. 107, 2001.

New York Times Obituaries, *Julian Simon, 65, Optimistic Economist Dies*, N.Y. Times, B

11, col.1, Feb. 12, 1998.

Ostrom, Elinor, *Governing the Commons: The Evolution of Institutions for Collective Action*, Cambridge University Press, New York, 1990.

Raymond, L. and Fairfax, S. K., *Fragmentation of Public Domain Law and Policy: An Alternative to the "Shift-to-Retention" Thesis*, 39 Nat. Res. J. 649, 1999.

Riebsame, W. E., et al. (Ed.). *Atlas of the New West.* W. W. Norton & Company, New York. 1997(a).

Reibsame, W. E., *Western Land Use Trends and Policy: Implications for Water Resources*, National Technical Information Service, Springfield, VA, 1997(b).

Rieke, Elizabeth A., *The Bay Delta Accord: A Stride Toward Sustainability*, 67 U. Colo. L. Rev. 341, 1996.

Sax, Joseph L., *Property Rights And The Economy Of Nature: Understanding Lucas V. South Carolina Coastal Council*, 45 Stan. L. R. 1433, 1993.

Tarlock, A. Dan, *Can Cowboys Become Indians? Protecting Western Communities As Endangered Cultural Remnants*, 31 Ariz. St. L. J. 539, 1999.

Thompson, Barton H., *Water Law as a Pragmatic Exercise: Professor Joseph Sax's Water Scholarship*, 25 Ecology L. Quarterly 363,1998.

Western States Water Council (WSWC), *Water in the West Today: A State's Perspective*, National Technical Information Service, Springfield, VA, 1997.

Western Water Policy Review Advisory Commission (WWPRAC), *Water in the West: Challenge for the Next Century,* National Technical Information Service, Springfield, VA, 1998.

Wilkinson, Charles F., *Crossing the Next Meridian : Land, Water, and the Future of the West*, Island Press, Washington, D.C., 1992.

Wright, Patrick, *Fixing The Delta: The CALFED Bay-Delta Program and Water Policy Under The Davis Administration*, 31 Golden Gate U. L. Rev. 331, 2001.

Denise D. Fort, University of New Mexico, School of Law, Albuquerque, NM 87131; fortde@libra.unm.edu.

7

Moving Borders from the Periphery to the Center: River Basins, Political Boundaries, and Water Management Policy

Robert G. Varady and Barbara J. Morehouse[1]

WHY TRANSBOUNDARY BASINS ARE SIGNIFICANT

Boundaries are a fact of life. We owe much of our understanding of the world, and of biophysical as well as societal interactions, to the order imposed by borders and boundaries. Yet, at the same time, we recognize "borderlands" as important areas that transcend political boundaries. These zones of interaction provide a place for utilizing things held in common (such as shared water resources) as well as reinforcing differences (for example, differing regulatory structures). Borders sometimes serve as legal or political barriers to scientific data gathering and collaboration. Of equal or even greater importance, borders often impede the rational application of scientific knowledge to the problems it is meant to solve. As Aaron Wolf and others have noted, the problems of managing water and other resources are exacerbated when those resources cross political boundaries and jurisdictions [*Wolf, et al.*, 1999; *Udall and Varady,* 1994]. While this reasoning has been persuasive, the persistence of political and administrative practices that treat border areas as peripheries has prevented transboundary management issues from receiving the attention they deserve—an observation that is especially meaningful in the case of river basins. The distinctiveness of the transboundary condition should be of particular interest to scientists because of the contrast between the disparities engendered by political borders and the continuity found in natural processes.

[1]Respectively, deputy director and research professor at the Udall Center for Studies in Public Policy; and deputy director of the Institute for the Study of Planet Earth, and adjunct assistant professor of Geography–all at The University of Arizona.

Water: Science, Policy, and Management
Water Resources Monograph 16

10.1029/016WM09

By and large, managers, decisionmakers, and the public-at-large accept that transboundary basins are distinct from watersheds that do not cross borders, and allow that the transboundary cases may require special policies and procedures. But because borders are so easily imagined as being "somewhere else" and "somewhere distant," transboundary basins tend to be treated as exceptions to the norm. Table 1 demonstrates conclusively that such a perception is far from correct. This listing of the world's 35 largest river basins shows that all but six (i.e., 83 percent) extend through more than one nation. On average, each of those 35 major basins flows through 4.3 countries.[2] The figures are even more convincing when basins cross significantly disparate domestic jurisdictions; in that case, 100 percent can be considered to be transboundary. In other words, far from being rare and interestingly peculiar cases, river basins that require management approaches tailored to their multinational or multijurisdictional status represent the overwhelming norm.

BORDERS, BOUNDARIES, AND RESOURCE-MANAGEMENT POLICY

Physiographic boundaries exist, such as rivers, but more typically political and jurisdictional boundaries ignore biophysical patterns and functions. Boundaries define many kinds of geographies in the world, including, for example, countries, provinces, states, administrative districts, indigenous territories, military cantonments, counties, census tracts, irrigation districts, voting precincts, protected areas, marketing or trade zones, and school districts.

Within boundaries, agencies and jurisdictional authorities at various levels pursue their own, often conflicting, missions. Their activities often affect each other. Nowhere is this truer than in natural resource management. Here, geopolitical boundaries slice across geographical areas where natural and human processes and interactions converge to produce very particular landscapes, as well as resource-use opportunities and pressures. Differences in laws, rules, practices, and values on either side of the boundary often increase the pressures on both the natural and human systems.

Disparity and Complexity

A different kind of boundary-drawing occurs in science: here, boundaries serve to define that which is within the scope of data collection, modeling, and analysis from that which is not. Boundaries allow the scientist to specify initial conditions, conduct a controlled analysis of identified variables, stimulate perturbations to those variables and/or their interactions, and measure the results [*Morehouse*, 2000]. Such data collection and analysis may benefit from the rationality imposed by boundaries. However, ordering

[2]Roger Pulwarty (2001) offers a similar analysis of transboundary river basins to stress their prevalence and importance.

Table 1. Major River Basins and Number of Countries

RIVER	CONTINENT	SIZE ($10^3 km^2$)	NO. CTRIES	INCL. DEVEL. CTRY.[a]	INCL. ARID TERRIT.[b]
Amazon	S. Amer.	5,866	8	✓	X
Congo	Africa	3,699	11	✓	X
Mississippi	N. Amer.	3,226	2	X	✓
Nile	Africa	3,038	12	✓	✓
La Plata/ Parana	S. Amer.	2,967	5	✓	✓
Ob	Asia	2,735	3	✓	✓
Yenisey	Asia	2,498	2	✓	✓
Lena[c]	Asia	2,430	1	---	---
Niger	Africa	2,118	11	✓	✓
Yangtze[c]	Asia	1,950	1	---	---
Amur	Asia	1,884	4	✓	✓
Mackenzie[c]	N. Amer.	1,800	1	---	---
Ganges	Asia	1,665	6	✓	✓
Volga	Europe/Asia	1,554	3	✓	✓
Zambezi	Africa	1,388	9	✓	✓
Hsi	Asia	1,350	2	✓	X
Nelson	N. Amer.	1,109	2	X	X
Indus	Asia	1,086	4	✓	✓
Murray[c]	Australia	1,070	1	---	---
St. Lawrence	N. Amer.	1,055	2	X	X
Orinoco	S. Amer.	959	2	✓	✓
Tarim	Asia	950	7	✓	✓
Orange	Africa	948	4	✓	✓
Tocantins[c]	S. Amer.	900	1	---	---
Yukon	N. Amer.	830	2	X	X
Juba-Shibeli	Africa	805	3	✓	✓
Tigris-Euphrates	Asia	794	6	✓	✓
Mekong	Asia	780	6	✓	X
Danube	Europe	780	17	X	X
Huang He[c]	Asia	745	1	---	---
Okavango	Africa	709	4	✓	✓
Columbia	N. Amer.	668	2	X	✓
Colorado	N. Amer.	651	2	✓	✓
Dnieper	Europe	496	3	X	X
Senegal	Africa	437	4	✓	✓
SUMMARY			Avg: 4.3	Tot.: 22	Tot.: 20

Sources: Wolf, et al. (1999); World Bank (2002); UNESCO (1977)
Notes:
[a] Transboundary basin includes at least one developing country.
[b] Transboundary basin includes arid or semiarid territory.
[c] Not a transboundary basin.

✓ = Yes
X = No

social and natural processes into tidy, bounded units is likely to generate as many problems as it solves, for border areas where things tend to blur can seldom be eliminated.

These often hazily defined areas are the interface where differentiated systems, processes, cultures, or social structures come together, where ordered assumptions are most apt to break down. The problem may be examined from several perspectives.

First, "borders separate problems and solutions." Because borders are typically distant from the centers where policies are conceived and applied, potential solutions are likewise formulated in locations far removed from where the problems are felt. Second, "borders create perverse economic opportunities." Globalization erodes incentives for conservation, sustainable growth, and equitable taxation. This process is most pronounced at the geographical extremities of political jurisdictions, where authority is weakest and cooperative management most difficult to achieve. Third, "borders aggravate perceived inequalities." In most countries, law, administration, and infrastructure financing are the province of national governments, resulting in a disregard for the specifics of local conditions. This disregard, in turn, has the effect of perpetuating inequitable practices that manifest in border regions. Finally, "borders obstruct grassroots problem-solving." In general, bottom-up, community-based, stakeholder-driven efforts to manage water and other resources tend to be inhibited by more powerful, centralized decisionmaking forces—a common inhibition in border regions. In those areas, which are distant from the loci of national influence, citizens may feel more disenfranchised than in centrally located communities.[3]

A symposium on transboundary challenges associated with climate and water management in the Americas, held in the summer of 2000, highlighted an array of boundary issues that affect the management of shared water resources.[4] The issues identified ranged from high variability in the quality, quantity, and accessibility of scientific data to the nature of laws and other institutional arrangements that inhibit development and implementation of policies that allow for rational, cooperative management of shared resources.

The symposium also highlighted the need to harmonize data across scientific boundaries, such as the boundaries of grid cells used in climate models. In these types of cases, what may appear to be an externality to a particular cell might in fact have relevant effects when complex inter-cell dynamics are taken into account. For example, in the case of the U.S. Southwest, due to complex terrain and related high variability across the area, the boundaries of a grid cell may fail to recognize important snowpack or other hydroclimatological influences originating in a nearby grid cell on conditions within the cell in question. While the concept may be abstract, the implications of such boundary-related scientific assumptions are critical when applying such principles to problems affecting human and natural systems. Nowhere is this more

---

[3]These characteristics, based on an example in the U.S.-Mexico border, are described in a 1994 essay in *Environment* (*Ingram, Milich, and Varady*, 1994).

[4]Symposium on "Climate, Water and Transboundary Challenges in the Americas," University of California-Santa Barbara, Santa Barbara, California, July 16-19, 2000.

clearly seen than in the development of finer-scale models that can be coupled to coarser, larger-scale models. Such modeling requires addressing uncertainties and instabilities occurring at these kinds of "edges" in order to produce valid, usable information to apply to specific locales in the finer-scale model.

Differing National Priorities

Transboundary water-resources management brings into sharp relief differences in the goals and strategies of the parties involved. At the national level, priorities on one side of a border may favor improving urban water delivery, while on the other they may promote environmental protection. If the upstream state pursues protection, the downstream state could benefit by gaining cleaner water or greater instream flows. However, if protection is the downstream state's priority while its upstream neighbor suffers from inadequate supply and treatment infrastructure, the downstream state is unlikely to achieve its priority. This common type of disparity impedes cooperative transboundary management.

In the San Pedro River basin that straddles the U.S.-Mexico border, grassroots organizations have arisen to address just this type of problem. The river, a tributary of the Colorado River, originates in Mexico, outside the copper-mining town of Cananea, Sonora. It contributes to the community's urban and mining water needs and sustains agricultural activities downstream, before crossing into Arizona. North of the border, while the river continues to support agriculture and urban development, a stretch, the San Pedro Riparian National Conservation Area, is protected for its nationally and internationally recognized value as bird habitat for both resident and migratory species, and for its recreational opportunities and its importance as one of the few functional riparian areas remaining in Arizona. Local residents strongly support the environmental values of the area and continue to work for its preservation. On each side of the border, a diverse collection of stakeholders vies for influence over allocation, habitat protection, and water-quality concerns. Since the late 1990s these often-competing interests have organized cooperative watershed associations: the U.S.-based Upper San Pedro Partnership and the Mexico-based Asociación Regional Ambiental de Sonora y Arizona (ARASA). Separately and jointly, these organizations are attempting to overcome divergent national priorities and some of the disadvantages of being marginalized along the border [*Varady, Moote, and Merideth* 2000; *Varady and Browning,* in press].

Regardless of how robust civil-society institutions may be, severe drought or flooding usually exposes underlying institutional barriers to effective cooperation. Strong increases in natural resource demands on either side of a border–from population growth, industrial/commercial growth, or some combination of these factors—often confront contradictions embedded in notions of sovereignty, local control, and other such institutional arrangements. Differences in levels of infrastructure development also serve as barriers. Across their border, for example, residents in the United States and Mexico have access to substantially different levels of water-supply and treatment infrastructure [*Ingram, et al.,* 1995]. Pronounced industrialization and urbanization in Mexico strain the capacity of local systems to keep up with growing demand as ever more residents pour into the cities. In many of the poorest *colonias* (unplanned settlements), water is delivered by truck, not through pipes. The cost of water is very high, and its quality is generally poor, especially among the poor. By contrast, on the U.S. side of the border water-supply infrastructure has accommodated population growth–with the notable exception of some very poor colonias in southern Texas and New Mexico [*Lemos, et al.,* 2002]. As a result, water remains relatively inexpensive and of higher quality.

In the U.S.-Mexico region, there has been some progress in addressing such imbalances. As one example, paired crossborder communities such as El Paso-Juárez, Ambos Nogales, Yuma-San Luís Río Colorado, and San Ysidro-Tijuana have developed dialogues aimed at solving shared water-resource management problems. Yet much remains to be done there and elsewhere in the world. Models such as the above one, which rely on public participation in decisionmaking, show signs of effectiveness. But they may be difficult to adapt to *developing-country* transboundary river basins (three-quarters of the largest such basins, according to Table 1), where political traditions are generally inhospitable to public involvement in policymaking. For obvious reasons of resource paucity and data scarcity and inconsistency, developing countries are particularly vulnerable to water-management problems [*Vörösmarty, et al.,* 2001]. Extreme climatic conditions–drought, heat, and flood–strain already-inadequate systems. The impacts of such stresses generally spill across borders, affecting neighboring systems and services [*Morehouse, et al.*, 2000].

In areas where water scarcity is a consistent or recurrent issue (Table 1 above shows that two-thirds of transboundary basins include some arid or semiarid territories), allocation decisions are influenced most heavily by policies articulating a hierarchy of use values. Meeting basic human survival needs tends to be paramount, and where legal provisions so state, agricultural water use may be deferred in favor of addressing human and municipal needs. Yet rules like these typically have validity only within individual countries, states, or other jurisdictional units—not across political boundaries. Obviously, climate and its impacts do not stop at jurisdictional lines [*Morehouse,* 2000]. How to negotiate science and policy interactions across borders, particularly international ones, remains a significant challenge to both scientists and policymakers. This question will be addressed following a brief discussion of key institutional forms, below.

Diverse Transboundary Institutions

The institutional mechanisms associated with decisionmaking and implementation can differ substantially. Nearly always, where local control over water supplies meets a nationalized system at an international boundary, cooperation is difficult [*Kliot, et al.*, 2001]. Local interests are likely to resist national-level assertions of authority, but managers of those interests lack power to negotiate internationally–a manifestation of sovereignty that nation-states strongly protect.

Typically, negotiations take the form of international treaties among two or more sovereign nations. In the realm of transboundary water management, examples include the Mexico-United States Treaty of 1944 (for the Colorado and Rio Grande/Río Bravo Rivers), Danube Declaration, Indus Water Treaty, and Plata Basin Treaty. In some instances deriving their authority from such treaties, but more frequently arising from separate multilateral agreements or conventions, nations that possess territories within large river basins form multinational commissions. Prominent examples are the Mekong River Commission, Niger Basin Authority, Organization for the Development of the Senegal River, and Rhine Commission. Sometimes, these institutions are created though exogenous actions, as with the Action Plan for the Environmentally Sound Management of the Common Zambezi River System, an instrument facilitated by the United Nations Environment Programme. In some instances, permanent institutions such as the [U.S.-Mexico] International Boundary and Water Commission, have established protocols allowing the participating nations to agree to specific, ad-hoc actions–for example, through officially approved amendments to existing treaties.

Nearly all of the world's river-basin management instruments are rooted in an earlier age, when they were molded principally by geostrategic considerations. Not until the last two to three decades have such other concerns as pollution, human health, habitat protection, recreation needs, ecological costs of engineering structures, land-tenure rights, and relationship between water and climatic processes become accepted elements of river-basin management. Not surprisingly, the organizations created to manage transboundary rivers reflected the times of their origins. As instruments of diplomacy, they were part of nations' foreign-policy infrastructure and exhibited the features and tendencies of those establishments: they were typically top-down, secretive and tightly possessive of information, highly dependent on engineering/earth-moving solutions to water problems, and inclined to regulatory approaches. Until the 1970s, and since then only in selected places, nongovernmental stakeholders and the public-at-large were virtually excluded from helping to determine policies that affected livelihoods, local environmental conditions, and public health in the regions within the province of these transnational commissions. And when the "needs, constraints, and practices" of local users are ignored, watershed management often fails [*Johnson, et al.*, 2001b].

Only since the 1970s, and only in selected countries, have transboundary organizations shown signs of becoming more democratic, incorporating such features as openness, transparency, and public participation into their procedures. Such changes have been most evident along the U.S.-Mexico border and in western Europe [*Milich and Varady*, 1998, 1999]. Elsewhere, especially in some former communist nations of eastern Europe, nongovernmental organizations in are showing interest in adopting and adapting similar processes.[5] But even in areas where processes have been reformed and integrated approaches adopted, cooperation among nations remains mainly crisis-driven, thus inhibiting steady, long-term collaboration [*Van der Zaag and Savenije*, 2000].

SCIENCE AND TECHNOLOGY, AND DECISIONMAKING

A recent paper on transboundary river-management institutions identified their greatest vulnerability: they operate at the intersection of two largely incompatible world views, hydrology and politics [*Kliot, et al.,* 2001]. A large body of literature on the integration of science and technology into decision-making has attempted to understand this incompatibility [e.g., *Baldwin*, 2000; *Gibbons,* 2000; *Gibbons, et al.,* 1994; *Jasanoff and Wynne*, 1998; *Latour and Wolgar*, 1979; *Weiss*, 1978]. But application of models and concepts in empirical studies, particularly to transboundary settings, is not yet fully developed. This deficiency is very apparent in the use of scientific and technological tools for transboundary river-basin management. Not infrequently, disparities in data collection, analysis, archiving, and dissemination retard the use of information for transboundary regions.

As importantly, the degree and quality of collaboration between scientists, policymakers, and stakeholders varies from country to country. Whereas western-style democracies tend to emphasize openness and transparency, many societies restrict access and participation. Even in otherwise open societies, certain limitations may be imposed. For example, in Croatia and Bulgaria the authorities often restrict dissemination of climate information they believe could be misused (e.g., for setting forest fires). Elsewhere, water-supply information can be classified as sensitive, particularly where rights to shared water resources remain unresolved. Or even more commonly, technological developments such a state-of-the-art dams and delivery systems are typically considered internal matters, even when they adversely affect downstream users. Almost always, insufficient institutionalization of international water law obstructs resolution of such conflicts.

These kinds of challenges are exacerbated when nations assert their sovereignty to justify control of the amount, nature, and destination of information domestically and across borders. They are further worsened by national policies that affect where and how scientific results are kept, and by laws that curtail international collaboration and communication among scientists, decisionmakers, and other interested parties.

---

[5]One such effort, facilitated by Z. C. Kovács and P. Balogh, is taking place in Hungary, in the Tisza basin. International Water Conference, "Globalization and Water Management: The Changing Value of Water" (7 Aug. 2001, Dundee, Scotland).

Several international examples illustrate the difficulties of integrated, international management of shared watershed. In the case of the Tigris-Euphrates, the countries sharing this basin are at peace with each other, but competing interests, mutual suspicions, and exercise of raw power assure that politics trumps science and technology.[6] At the extreme, elsewhere in the Middle East, current conflict between Israel and Palestine poses an even more immediate and seemingly intractable challenge for cross-border cooperation on water issues. During calmer times, the 1995 Israeli-Palestinian Interim Agreement on the West Bank and Gaza Strip articulated Israel's recognition of Palestinian water rights in the West Bank and Gaza Strip. Both parties also recognized the need to develop additional water, assure sustainability, and adjust utilization to changing hydrological and climatic conditions [National Research Council, 1999]. The agreement, one of many such technical accords between the two entities, showed that scientists and technocrats can play useful roles in overcoming political differences. But when two nations are in a state of war, it is impossible to envision productive crossborder collaboration.

Transboundary scientific research can be problematic even between countries that are at peace and have similar levels of development and scientific expertise. Sometimes, arrangements to share data and scientific knowledge rely more on established scientific forums (publications, conferences, exchanges), supplemented by informal or personal networks than on institutional arrangements. Even where formal institutions exist, such as the International Joint Commission established by the United States and Canada to harmonize water management along the U.S.-Canadian border, collaboration on transboundary issues may be restricted to those narrowly defined by the agency as being within its mission.

Even more complex are the scientific and technological issues associated with river basins lying in more than two countries. The Plata River (Río de la Plata), which drains lands in five South American countries, has been the subject of international management efforts since the late 1960s [*Tucci and Clark,* 1999]. In 1969 the five nations signed the Plata Basin Treaty, which envisioned integrated development and identified joint objectives, but failed to include specific actions or mechanisms to implement the goals. In the early 1980s, within this framework, the member countries created the Hydrologic Warning Operations Center for the Plata Basin. While far from solving the basin's many problems, this warning system institutionalized the application of scientific and technological expertise to managing risk. To the extent that the member countries have the capacity and will to provide the data needed to develop forecasts for the basin, the Center offers a focal point for anticipating and planning for emergencies. However, with its narrow mission, it is not empowered to carry out much-needed transboundary collaboration on land and resource use in the Basin.

[6] Turkey controls the headwaters, but the basin includes five other nations, of which Iraq and Syria depend most heavily on the supply.

STRESSES, SURPRISES, SENSITIVITY, AND ADAPTATION

Issues associated with transboundary water resources frequently remain inchoate, until stresses become too great to ignore [*Van der Zaag and Savenije*, 2000]. Such stresses may include increases in demand where supplies are already limited, degradation of water quality to levels below that required for intended use, political tensions manifested in conflicts over sharing water resources (whether or not the basic contest is over water, per se), technological changes that modify the nature and quantity of flows across a border, or climatic changes that alter water availability and flow regimes. Surprises are stresses that arise unexpectedly, such as flash floods. While plans may be devised to address surprises, the possibility always exists that an event will exceed an anticipated level or that other unanticipated factors may confound planned action. Linking predictions with policymaking and decisions is, in the best of circumstances, fraught with obstacles [*Pielke, Jr., and Byerly, Jr.*, 2000]. In contexts involving political and jurisdictional boundaries, the barriers may multiply exponentially.

Crossborder collaborations that assess the sensitivity of human and natural systems to stresses and surprises, including climate variability and change [*Morehouse et al.,* 2002], are crucial to development of contingency plans. Such analyses help establish current baseline conditions, which then permit evaluation of potential impacts under different sets of conditions [*Morehouse et al.,* 2000]. Vulnerability assessments that seek to identify who is vulnerable to external stresses, how, and under what conditions they are vulnerable, are also valuable management tools for contexts of uncertainty and multiple stresses. The impacts of climate variability and change on transboundary water resources, for example, are likely to be unevenly experienced both within and across a border [*Liverman, et al.,* 1997]. As Liverman (2001) has emphasized, the most vulnerable are often those having the least flexibility and fewest resources with which to respond. Knowing where resilience and flexibility are least, and thus where adaptation capacity is most limited, allows for holistic appraisal of what sorts of planning are needed and where resources are likely to be most needed.

In border contexts, sensitivity analyses and vulnerability assessments carried out strictly within political and jurisdictional boundaries may result in crucial sensitivities and vulnerabilities being overlooked. Policies and actions that are based on narrow definitions of the bounds of analysis may exacerbate cross-border tensions. Examples of this sort of outcome may be found in conflict arising from downstream flooding of the Ganges-Brahmaputra river system in India and Bangladesh, generated by heavy upstream snowpack accumulation in Nepal and Tibet; and in tensions associated with potential implications of decreased snowpack in the mountains of British Columbia, foreseen in climate-change models, for summer water availability in the U.S. Northwest [*Miles, et al.,* 2000].

CONCLUSIONS

Key Issues Complicating Transboundary River Basin Management

There is no question that managing water resources in crossborder contexts is extraordinarily complex. Much of the complexity added by the transnational dimension is attributable to human agency. Questions relating to sovereignty and jurisdictional autonomy, economic forces, and the applications of technological advances are the chief complicating factors.

Sovereignty and jurisdictional autonomy. The decisionmaking process typically is multidimensional, nuanced, and difficult to fathom. Further, it varies enormously from one nation to another, even when those nations are neighbors. Institutional and organizational frameworks that apply at international, national, regional, and local scales require differing levels of autonomy and authority to make decisions. Thus, local communities sharing a boundary and common problems may face insuperable barriers to cooperative management of water resources because of national policies that view any such cross-border interaction as the purview of the state.

A useful illustration of this tension between domestic priorities and a desire to transcend international borders is the potential impact on river-basin management of climate variability and change. Issuance of official hydrologic and climate forecasts is typically restricted to national-level entities, and the forecasts themselves are solely for areas within a nation's territories. To obtain a border-area forecast, users must reconcile or harmonize forecasts issued by each country, or turn to less official ones (though not necessarily less accurate or skillful) like those issued by research centers such as Columbia University's International Research Institute. In some cases, stakeholder-relevant forecasts may not be available at all, or even when they are, the latitude to take action based on those forecasts may be limited in different ways–by international tensions, for instance; such conflicts may prevent action in border zones.

Economic drivers. The formulation and enforcement of water policy is strongly influenced by economic factors. One of the most common such factors is the dominance of one or more sectors in a border area (e.g., agriculture, mining, or industrial activity). For example, upstream diversion of water for agricultural irrigation may significantly diminish water availability in the neighboring downstream country or jurisdictional unit. In such cases, tensions may quickly mount on both sides of the border. As contests over the waters of the Rio Grande/Rio Bravo and Colorado River have demonstrated, prolonged and sometimes acrimonious negotiation is required to resolve the issues.

Interbasin water transfers are another controversial, economically driven form of river management. Such projects can raise potent issues for the affected region–even when they are within a single nation, and much more so when they are transnational. They may degrade ecosystems in one or both watersheds, or diminish the vitality of entire communities (as when diversion of water and elimination of agriculture dependent on that water causes a domino-like fall of commercial and retail activities that generate local revenues and jobs).

A third economically propelled phenomenon is the rapid increase in urbanization in international borders areas. Massive urban growth may result from many trends. For example, intensive, spontaneous migration may swell populations over short time periods. Likewise, borderland urban growth may be an outcome of intentional policy, such as stimulating economic development through establishment of *maquiladoras* (foreign-owned manufacturing plants south of the U.S.-Mexican border) operations in border communities. In any event, the resulting urban agglomerations constitute an on-the-ground *fait-accompli* that challenges the ability of each neighboring country to supply sufficient resources to all sectors. In such situations, the government may provide potable water to meet residential needs while at the same time delivering water to competing users such as commercial and industrial concerns.

Applications of technological advances. Basic research and technological innovation, which can be the leavening for improved management [*Vörösmarty, et al.,* 2001], are complicated by transboundary dimensions. Attempts to apply the fruits of science to basin management often encounter serious obstacles, ranging from data-access limitations to governmental policies that serve to inhibit interactions among researchers, managers, and stakeholders from different countries or even from different jurisdictions.

The San Juan River Basin Project, covering parts of Costa Rica and Nicaragua, typifies the challenges to internationally conducted river-basin research [*Rucks,* 1999]. The region is characterized by high poverty rates and rising population; it also features important geological reserves and conservation areas. The San Juan is considered essential to serving future development in the semiarid Pacific Slope areas of the two countries. Efforts to improve water quality and reduce erosion and sedimentation have had high priority with both governments, as has the development of a basin-wide information system containing data on regional and local hydrology and climate. Similarly, the two countries have agreed to develop an effective bilateral, basin-wide planning process, strengthen public and private institutions and organizations, and develop environmental education programs. The project represents an ambitious undertaking that requires substantial investment of financial and human resources, as well as of goodwill. But underlying problems have slowed collaborative attempts at integrated management. These have included inadequate planning and administration capacity, weak institutional and monetary support, and insufficient stakeholder participation in decision making.

Like the Río de la Plata and San Juan projects, the North American Monsoon Experiment (NAME) is an attempt at transboundary, water-related science.

NAME would sponsor the installation of observation systems in the monsoon region of Mexico and possibly the southwestern United States. The program aims to develop spatially and temporally rich datasets to improve scientific understanding of the dynamics driving the monsoon and its spatial and temporal variability. NAME is designed to satisfy several needs of stakeholders in each country: good summer half-year precipitation forecasts that are issued during the previous winter or spring; better data for input to regional and global-scale climate models; and improved understanding of interactions between the monsoon and other climate processes over seasonal, annual, decadal, and centennial timescales. The ideal outcome of NAME would be better transboundary hydrologic and climate forecasts; the extent to which such forecasts will include both countries on maps and other graphics remains to be seen and will surely be subject to the forces slowing binational environmental cooperation more generally. Still, NAME is likely to benefit from the high degree of existing transboundary interaction and interlinkage within the shared borderlands and between the United States and Mexico. Forecasts spanning the Canada-U.S. border may be strongly justified for similar reasons.

Looking Ahead

Certain trends in science, combined with the dynamics of globalization and innovations in communications, provoke speculation about what the future of transboundary water management will look like. In the realm of water management, the past decade has seen a growing emphasis on funding society-relevant and policy-relevant science, supporting regionally focused and catchment-based integrated-assessment efforts, and producing "usable" science. Key examples of this approach are such relatively new programs as the HELP (Hydrology for Life, the Environment and Policy; described elsewhere in this volume) Initiative, the Dialogue on Water and Climate, and UNESCO's World Water Assessment Programme for Development, Capacity-building and the Environment [HELP 2001; *Kabat, et al.,* 2002; UNESCO 2001].

These progressive trends in the epistemology of water management, if sustained, hold promise for reducing sensitivities and vulnerabilities to multiple stressors and promoting adaptive capacity among decisionmakers and ordinary citizens. Yet realizing the potential benefits of these and similar initiatives requires explicit commitment to expend time and resources to grapple with complicated border and boundary issues of all kinds. And, as the International Panel on Climate Change has discovered, while science is essential, it is insufficient to address society-environment problems. There is no alternative to wading into the messy world of social process–including the effects of drawing, justifying, and enforcing boundaries.

Meanwhile, globalization processes–including the restructuring and relocation of manufacturing operations, growth of border cities and associated water demand, and reallocations of capital–intensify the need to translate scientific advances into improved transboundary water-resources management. Added to all this is the rapid diffusion of communications technologies that permit instant and global dissemination of information of all kinds–scientific, political, economic, and environmental. As governments are discovering, attempts to exert sovereign rights over information dissolve in the borderless flows emanating from Internet, satellite, and other communications sources. Further, in an era of global trading of financial instruments and commodities of all kinds, many have argued that because under-valuation of water is the largest contributor to shortages, international trading in water and water futures is highly likely to occur [*Johnson, et al.*, 2001a]. Serious questions have been raised about what such commoditization of water will mean for equitable access around the world.

These and related changes necessitate the construction and acceptance of water-resource institutions that can respond to all the forces at play. Finally, most significantly, in view of the prevalence and "centrality" of transboundary river-basins, it is clear that these institutions will have to "be" transboundary in both design and outlook.

Acknowledgments The authors are grateful for assistance from Robert Merideth, Leah Stauber, and Allison Davis at the Udall Center. Some of the ideas are extensions of previous work, cowritten with Helen Ingram, Lenard Milich, Vera Pavlakovich-Kochi, and Doris Wastl-Walter. Additionally, W. James Shuttleworth, Michael Bonell, and James Wallace, formulators of the HELP initiative; and Aaron Wolf and Roger Pulwarty have, perhaps unknowingly, contributed influential insights. Finally, we acknowledge the Ford Foundation, the Morris K. Udall Foundation, the National Oceanic and Atmospheric Administration, and the Science and Technology Center for the Sustainability of Semi-Arid Hydrology and Riparian Areas (SAHRA) for their past and present support for investigations on river-basin management.

REFERENCES AND SELECTED BIBLIOGRAPHY[7]

Baldwin, S. 2000. Interactive social science in practice: new approaches to the produc tion of knowledge and their implications. *Science and Public Policy* 27, 3:183-194.

Blatter, J., and H. Ingram (eds.). 2001. *Reflections on Water: New Approaches to Transboundary Conflicts and Cooperation*. Cambridge: MIT Press. 356 pp.

Gibbons, M. 2000. Mode 2 society and the emergence of context-sensitive science. *Science and Public Policy* 27, 3:159-163.

Gibbons, M., C. Limoges, H. Nowotny, S. Schartzman, P. Scott, and M. Trow. 1994. *The New Production of Knowledge*. London: Sage Publications. 179 pp.

HELP (Hydrology for the Environment, Life and Policy) Task Force. 2001. *The Design*

[7] All but four of the titles (90%) are cited in the text. The remaining are useful, closely related works.

and Implementation of the HELP Initiative. Paris: UNESCO International Hydrological Programme (IHP)-V Technical Documents in Hydrology No. 44. 67 pp.

Ingram, H., N. K. Laney, and D. M. Gillilan.1995. *Divided Waters: Bridging the U.S.-Mexico Border*. Tucson: University of Arizona Press. 262 pp.

Ingram, H., L. Milich, and R. G. Varady. 1994. Managing transboundary resources: lessons from Ambos Nogales. *Environment* 36, 4:6-9, 28-38.

Jasanoff, S. and B. Wynne. 1998. Science and decisionmaking. In S. Rayner and E. L. Malone (eds.), *Human Choice and Climate Change: The Societal Framework, Vol. 1*, pp. 1-88.

Johnson, N., C. Revenga, and J. Echeverria. 2001a. Managing water for people and nature. *Science* 292:1071-1072.

Johnson, N., H. Munk Ravnborg, O. Westermann, and K. Probst. 2001b. User participation in watershed management and research. *Water Policy* 3:507-520.

Kabat, P., et al. (c. 40 contributing authors). 2002. *Dialogue on Water and Climate: First "White" (Position) Paper*. Dialogue on Water and Climate: Delft and Wageningen. 113 pp.

Kliot, N., D. Shmueli, and U. Shamir. 2001. Institutions for management of transboundary water resources: their nature, characteristics and shortcomings. *Water Policy* 3:229-255.

Latour, B., and S. Wolgar.1979. *Laboratory Life*. Beverly Hills: Sage. 272 pp.

Lemos, M. C., D. Austin, R. Merideth, and R. G. Varady. 2002. Public-Private Partnerships as Catalysts for Community-based Water Infrastructure Development: The Border WaterWorks Program in Texas and New Mexico Colonias. *Environment and Planning C: Government and Policy* 20:281-295.

Liverman, D. M. 2001. Vulnerability to drought and climate change in Mexico. In (eds.) J. X. Kasperson and R. E. Kasperson, *Global Environmental Risk*, New York: Earthscan, United Nations University Press, pp. 343-352.

Liverman, D. M., C. Conde , and V. Magaña. 1997. *Climate Variability and Transboundary Freshwater Resources in North America: U.S.-Mexico Border Case Study*. Montreal: Commission for Environmental Cooperation. 37 pp.

Miles, E. L., A. K. Snover, A. Hamlet, B. Callahan, and D. Fluharty. 2000. Pacific Northwest Regional Assessment: The impacts of climate variability and climate change on the water resources of the Columbia River Basin. *Journal of the American Water Resources Association* 36:399-420.

Milich, L. and R. G. Varady. 1999. Openness, sustainability, and public participation: new designs for transboundary river-basin institutions. *Journal of Environment and Development* 8, 3:258-306.

Milich, L., and R. G. Varady. 1998. Managing transboundary resources: lessons from river-basin accords. *Environment* 40, 8:10-15, 35-41.

Morehouse, B. J. In Press. Theoretical approaches to border spaces and identities. In Pavlakovich-Kochi, V., B. J. Morehouse, and D. Wastl-Walter (eds.), *Challenged Borderlands: Transcending Political and Cultural Boundaries*. Ashgate Press, 2003.

Morehouse, B. J. Boundaries in Climate Science-Water Resource Discourse. 2000. Paper presented at the Symposium on Climate, Water, and Transboundary Challenges in the Americas, University of California – Santa Barbara, Santa Barbara, CA, 16-19 July.

Morehouse, B. J. 1995. A functional approach to boundaries in the context of environmental issues. *Journal of Borderlands Studies* 10, 2:53-73.

Morehouse, B. J., R. H. Carter, and P. Tschakert. 2002. Sensitivity of urban water resources in Phoenix, Tucson, and Sierra Vista, Arizona, to severe drought. *Climate Research* 21: 283-297.

Morehouse, B. J., R. H. Carter, and T. W. Sprouse. 2000. The implications of sustained drought for transboundary water management in Nogales, Arizona, and Nogales, Sonora. *Natural Resources Journal* 40: 783-817.

National Research Council. 1999. *Water for the Future: The West Bank and Gaza Strip, Israel, and Jordan.* Report prepared by the U.S. National Academy of Sciences, Royal Scientific Society of Jordan, Israel Academy of Sciences and Humanities, Palestine Academy for Science and Technology. Washington, DC: National Academy Press. 244 pp.

Pielke, Jr., R. A., and R. Byerly, Jr. 2000. *Prediction: Science, Decision Making and the Future of Nature*. Washington, DC: Island Press. 400 pp.

Pulwarty, R. S. 2001. Transboundary river flow changes. In *Handbook of Climate, Weather, and Water*. Ed. by T. Potter. McGraw-Hill. Pp.1-20.

Rucks, J. 1999. Manejo integrado de recursos hidricos en cuencas transfronterizas: el caso de la cuenca del Río San Juan Costa Rica-Nicaragua. *Proceedings of the Third Inter-American Dialogue on Water Management: "Facing the Water Crisis in the 21st Century."* Organization of American States. Panama City, Panama. http://www.iwrn.net/D3_Proceedings.pdf. Accessed 10 May 2002.

Symposium on "Climate, Water and Transboundary Challenges in the Americas," University of California-Santa Barbara, Santa Barbara, California, July 16-19, 2000.

Tucci, C. E. M., and R. T. Clarke. 1999. Environmental issues in the Plata Basin. *Proceedings of the Third Inter-American Dialogue on Water Management: "Facing the Water Crisis in the 21st Century."* Organization of American States. Panama City, Panama. http://www.iwrn.net/D3_Proceedings.pdf. Accessed 10 May 2002.

Udall, Stewart N., and R. G. Varady. 1994. Environmental conflict and the world's new international borders. *Transboundary Resources Report* 7, 3:5-6.

UNESCO. 2001. *Fitting the Pieces Together: The World Water Assessment Programme (WWAP) for Development, Capacity-building and the Environment.* Paris: UNESCO. 15 pp.

UNESCO. 1977. *World Distribution of Arid Regions*. Map Scale: 1/25,000,000. Paris: UNESCO.

Van der Zaag, P., and H. H. G. Savenije. 2000. Towards improved management of shared river basins: lessons from the Maseru conference. *Water Policy* 2:47-63.

Varady, R. G., and A. Browning-Aiken. In Press. The birth of a Mexican watershed council in the San Pedro basin in Sonora. In *Planeación y Cooperación Transfronteriza en la Frontera México-Estados Unidos (Transboundary Planning and Cooperation in the U.S.-Mexico Border Region)*, ed. by C. Fuentes and S. PeZa. C. Juárez: El Colegio de la Frontere Norte.

Varady, R. G., M. A. Moote, and R. Merideth. 2000. Water allocation options for the Upper San Pedro Basin: assessing the social and institutional landscape. *Natural Resources Journal* 40, 2:223-235.

Vörösmarty, C., P. Green, J. Salisbury, and R. Lammers. 2001. Global water resources: vulnerability from climate change and population growth. *Eos* 82:54.

Weiss, C. 1978. Improving the linkage between social research and public policy. In L. E. Lynn (ed.), *Knowledge and Policy: The Uncertain Connection*. Washington, DC: National Academy of Sciences, pp. 23-81.

Wolf, A. T. 1998. Conflict and cooperation along international waterways. *Water Policy* 1:251-265.

Wolf, A. T., J. A. Natharius, J. J. Danielson, B. S. Ward, and J. K. Pender. 1999. International river basins of the world. *International Journal of Water Resources Development* 15, 4: 387-427.

World Bank. 2002. *World Development Report 2003: Sustainable Development in a Dynamic World—Transforming Institutions, Growth, and Quality of Life*. Washington, DC: The World Bank. 231 pp.

Robert G. Vardy, Udall Center for Studies in Public Policy, The University of Arizona; 803 East First St., Tucson, AZ 85719; rvarady@email.araizona.edu

Barbara J. Morehouse, Institute for the Study of Planet Earth, University of Arizona, 715 N. Park Ave, Tucson, AZ 85721

8

Economics of Conjunctive Use of Groundwater and Surface Water

Eric G. Reichard and Robert S. Raucher

INTRODUCTION

Although commonly treated as separate resources, ground water and surface-water interact in most areas (Winter et al., 1998). The National Research Council (1997) defines conjunctive use as "any integrated plan that capitalizes on the combination of surface and groundwater resources to achieve a greater beneficial use than if the interaction were ignored." Assessment of the viability of a conjunctive use program requires both economic and hydrologic analyses. This chapter describes approaches that have been employed to quantify the economic costs and benefits of conjunctive use and presents some specific examples in California.

Figure 1 provides a schematic representation of the potential components of conjunctive use programs in both urban and agricultural/rural settings. The connection between surface water and groundwater can be through infiltration within a stream channel, spreading in an infiltration pond, injection and extraction in wells, and indirectly through the use of delivered surface water in lieu of pumpage. Surface-water sources include local runoff, imported water, or recycled (and highly treated) sewage effluent. The scale of the project can be intra-basin (the source and use of all water is within the basin) or inter-basin (the source and(or) use of at least some of the water is outside the basin). Regardless of the specific combination of components that are present in a particular conjunctive use program (details of three existing programs are presented at the end of the chapter), thorough economic assessment requires explicit quantification of benefits and costs, determination of the baseline and timing, and rigorous incorporation of the hydrology.

Water: Science, Policy, and Management
Water Resources Monograph 16

10.1029/016WM10

BENEFITS OF CONJUNCTIVE USE

Economic benefits of conjunctive use have been identified by economists and hydrologists (Todd, 1965; Reichard and Bredehoeft, 1984; Tsur, 1990; Tsur and Graham-Tomasi, 1991; National Research Council, 1997; Donovan and others, 2002). Several of the types of benefits mentioned in the literature include:

- insurance (buffer, or option) value
- storage value (use of groundwater basin as reservoir)
- conveyance value of groundwater basin
- treatment value of groundwater basin
- reduced pumping lifts
- subsidence control
- seawater-intrusion control
- "nonuse" benefits

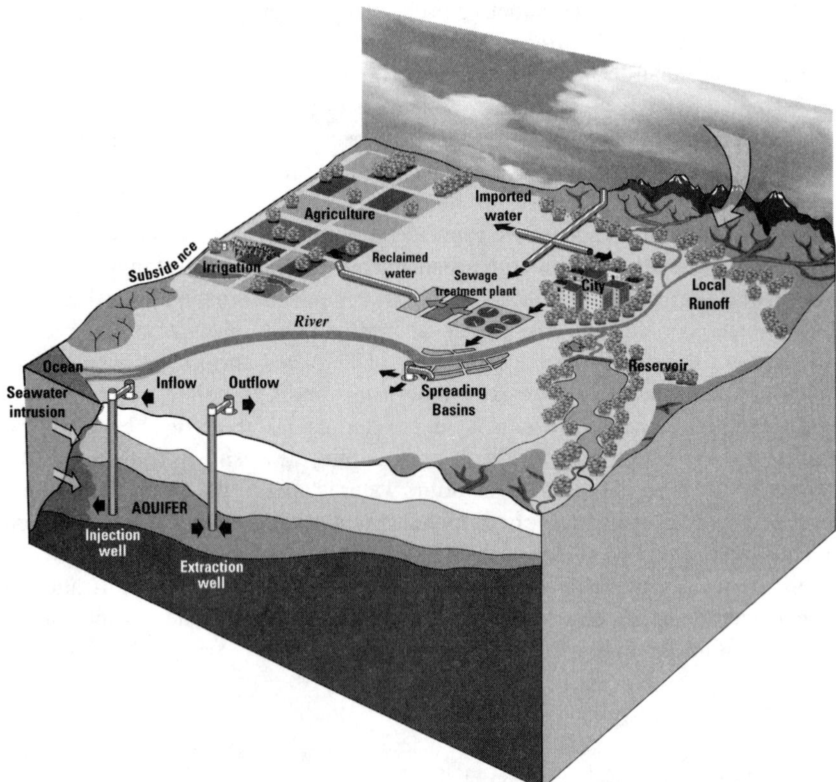

Figure 1. Schematic of components of conjunctive use.

These various terms and categories for the potential benefits of conjunctive use are not necessarily comprehensive or mutually exclusive. In addition, economists often use terms that may appear inconsistent with or duplicative of the semantics applied by other professions or even other economists. Thus, the key point for practitioners is to ensure that they properly identify all the relevant benefits and costs of a conjunctive use program, and include them insofar as feasible when evaluating the relevant policy options for conjunctive use.

For example, the term "buffer value" has been used to describe the gains realized by the use of groundwater to "mitigate undesired fluctuations in the supply of surface water" (Tsur and Tomasi, 1991). This is the same concept that natural resource economists have typically labeled as "option value" (e.g., Weisbrod, 1964; Fisher and Raucher, 1984). Option value reflects the value people place on the option to have access to (and use of) a resource—above and beyond current use value (expected consumer or producer surplus)—when there is uncertainty about the future supply of (or demand for) a resource.

In the context of conjunctive use, groundwater resources often provide a back-up source of water (e.g., for irrigation) in years when surface supplies cannot meet demands (e.g., in drought years when surface supplies are below normal, or in years when demands may be above average, or both). In effect, this type of "option value" or "buffer benefit" is also reflective of "storage value" since the groundwater stock reflects past surface water (including rainfall) that is held in reserve until years in which the water needs exceed surface supplies. Therefore, analysts should be aware that often the same element of economic value may have different labels. The key for practitioners is to make sure that all the legitimate and important benefits of conjunctive use are identified and included where feasible, so that there is neither a double counting of benefits nor the omission of potentially important benefits.

Regardless of how the various types of benefits are defined and characterized, the core point is that groundwater has a total economic value (TEV) that exceeds its current value in extractive uses. As noted by the National Research Council (1997), TEV is the sum of the extractive value and the *in situ* value. Extractive-use values typically are more apparent and generally can be estimated directly according to the value that the water generates for users (e.g., the value of the marginal product of water in agricultural irrigation). In contrast, *in situ* values can be more obscure—conceptually and empirically—and it is these values that often constitute the rationale for conjunctive use programs. Therefore it is important to recognize these types of *in situ* values that apply to conjunctive use.

Most of the *in situ* values of groundwater systems managed through conjunctive use programs are ultimately related to human uses of the water resources. For example, pumping and recharge schemes that are designed to limit seawater intrusion are intended to preserve freshwater aquifers for various current (and(or) future) human uses such as potable water supply, irrigation, and so forth. As such, the economic value of these aspects of conjunctive use generally can be estimated according to the value of the water in those enhanced or preserved applications (e.g., the users' willingness to pay for the additional potable or irrigation water made possible by the conjunctive use program).

Likewise, conjunctive use in the form of groundwater recharge that reduces subsidence can be valued in terms of reduced pumping costs at wells that extract water from the aquifer, the avoided costs of infrastructure repairs that would have been necessary (or necessary sooner) because of subsidence, and so forth. In similar fashion, where conjunctive use includes water-quality improvements as the water passes through the underground system, then such benefits may be reflected in reduced treatment costs at the point and time of extraction. Several of these use-related types of *in situ* benefits will be discussed in the case-studies offered later in this chapter.

Nonuse benefits associated with conjunctive use may be the most difficult to grasp conceptually and express in monetary terms. Nonuse values are generally characterized in the economics literature as consisting of existence and bequest values (e.g., Fisher and Raucher, 1984). Existence value refers to the intrinsic value individuals place on maintaining the availability and quality of a resource, and refers to values above and beyond any current or anticipated use. Bequest value is the willingness to pay for passing on the resource—in suitable quantities and quality—to future generations (National Research Council, 1997). These nonuse benefits are also sometimes referred to using terms such as stewardship, preservation, and inter-generational equity values.

Difficulties in quantifying nonuse benefits are discussed by many authors. As a practical matter, nonuse benefits may not be estimable for most practitioners looking to develop benefit-cost analyses for conjunctive use, but there is evidence that the benefits exist and are positive in many circumstances (e.g., McClelland et al, 1992). Therefore, these nonuse values should at least be mentioned as a qualitative benefit if monetized values cannot be derived.

COSTS OF CONJUNCTIVE USE

Conjunctive use programs also have costs that need to be properly identified and estimated for use in a benefit-cost context. Many of the costs are fairly direct and can be estimated based on observed (or projected) out-of-pocket expenses. For example, an analyst can readily derive the cost of developing and using wells to inject water into (and (or) extract water from) a groundwater source.

In addition to direct, out-of-pocket implementation costs, there may be many other indirect, less obvious costs of conjunctive use programs. Such indirect costs may be difficult to identify and (or) estimate in monetary terms. Nonetheless, in some instances such nonfinancial or indirect costs can be significant. Where this may be the case, such costs need to be considered in the benefit-cost framework.

Some of the potential costs of conjunctive use programs in which surface water is stored in underground systems may include (Logan and Allen-King, 2001): (1) water-quality degradation, where contaminants present in either the introduced water, native groundwater, or unsaturated zone affect water quality and(or) increase treatment requirements of the water when it is extracted for use; (2) decreased aquifer storage capacity (or increased subsidence) that may develop where chemical or physical reactions between recharged water and existing groundwater and aquifer materials decrease the porosity of the system; and (3) impacts on riparian-zone ecosystems due to withdrawing water from streams, or reducing return flows to those streams. In many instances, the degree to which these costs arise and should be counted depends on what actions would be taken absent the conjunctive use program. Therefore, an essential element in a benefit-cost analysis of conjunctive use is establishing the appropriate baseline.

DEFINING THE APPROPRIATE BASELINE

In any benefit-cost analysis, a key issue is defining the appropriate baseline. In a conjunctive use context, the baseline should be defined according to what would likely happen absent the conjunctive use program. In other words, analysts should compare the outcomes in a framework that compares water-resource uses and impacts between a "with" and "without" conjunctive use context.

For example, using streamflows in wet months or years as a source of aquifer recharge may impose some costs in terms of reduced instream flows to support ecologic systems, recreational services, and so forth during those wet periods. However, if the stored water is then used to supplement surface flows in dry periods, then the conjunctive use may actually increase the benefits associated with instream flows during such critical dry spells. In other words, the loss of instream flow values in wet periods is often likely to be more than offset by the benefits of enabling greater instream flows during dry periods. The point here is that absent the conjunctive use program, the instream flow impacts are likely to be more severe than they are under the conjunctive use regime.

Defining the suitable baseline helps ensure that the benefits and costs are properly identified and attributed to the program in a suitable manner. For example, if conjunctive use is intended to provide backup water in dry periods, then the benefit-cost analysis should be anchored according to what would likely happen if that added reserve were not available in dry periods. Absent the conjunctive use program's supply of reserve water, the water users would have to either forgo some water uses (e.g., reduce irrigation, or endure municipal water use restrictions) or obtain water from an alternative source. If water uses are foregone, then there are costs (e.g., reduced agricultural yields) that are avoided due to having the reserve water available, and these costs avoided due to the conjunctive use program should be counted as benefits of the program. Likewise, if the forgone conjunctive use water is replaced by an alternative supply, then the costs of obtaining that alternative supply should be counted as benefits of the conjunctive use program. The costs of obtaining alternative supplies in a dry period can be significant (assuming water is available at all); hence the benefits of the conjunctive use program would likely be significant.

In many cases, it may not be entirely clear what would happen "without" the conjunctive use program. In such cases, the analysis should be developed using multiple plausible scenarios. An example of this is provided in the case study of a conjunctive use program in the Santa Clara Valley of California discussed below.

FRAMEWORK FOR TIMING ISSUES IN COST-BENEFIT ANALYSIS OF CONJUNCTIVE USE

A complete economic analysis of a conjunctive use program must compare all benefits with the costs of the program. Not only must these benefits and costs be anchored by baseline scenarios that reflect the benefits and costs of alternative water-use or supply options, but the analysis must also account for the timing of the benefits and costs.

The costs of conjunctive use are typically incurred at the outset of the program, whereas the benefits may not be fully evident for several years into the future. This raises a classic issue of discounting future benefits and costs so that the overall project can be evaluated on a net present value basis. Resource economics provides a framework for completing such an analysis. Pioneering work by Burt (Burt, 1964, 1966, 1967; Provencher and Burt, 1993), Gisser (Gisser and Mercado, 1972; Gisser and Sanchez, 1980), Cummings (Cummings and Winkelman, 1970; Cummings and McFarland, 1974), and others presented analytic solutions that maximized the discounted net benefits of groundwater use. This work recognized the fact that groundwater is a common resource, extracted by multiple users. Some of the analyses addressed the linkage between groundwater and surface water. Policy tools for achieving economically optimal water use include water-rights structures, market mechanisms, taxes, subsidies, and quotas.

The net discounted benefits of a conjunctive use program can be expressed as:

$$\text{Net benefits} = \sum_{t=1}^{T} \frac{(B_t - C_t)}{(1+r)^t} \qquad (1)$$

where: B_t is the benefit of a conjunctive use program in period t, C_t is the cost of conjunctive use program in period t, T is the planning period, and r is the discount rate.

COST-EFFECTIVENESS ANALYSIS

The preceding discussion has focused on benefit-cost analysis, which is a common and powerful program-evaluation technique applied to water resource issues. A related although somewhat simpler tool for program evaluation that can be used in lieu of a benefit-cost approach is called "cost-effectiveness analysis."

In cost-effectiveness analysis, the evaluation is based on the cost of achieving a given objective, and that objective is typically measured in physical (as opposed to monetary) terms. For example, a cost-effectiveness analysis may be structured to examine options for assuring a given level of water supply (e.g., acre-feet per year), in which case the analyst simply characterizes each option according to the dollar cost per acre-foot delivered. This type of analysis allows policy makers to evaluate options based on which is the most efficient (least cost) approach for delivering the desired supply of water.

Cost-effectiveness analysis has several advantages and limitations relative to benefit-cost analysis. Among the key advantages are that cost-effectiveness analysis is typically much simpler to develop and apply than benefit-cost analysis. In benefit-cost analysis, the analyst must identify, quantify, and then monetize every important type of outcome. In contrast, a cost-effectiveness study simply identifies the key objective, limits the evaluation to more readily observed and measured units of quantification (e.g., acre-feet), and divides these single physical output measures into the dollar cost of the option.

However, the simplicity of cost-effectiveness is counter-balanced by the limitations in interpreting the findings. Cost-effectiveness analysis can help identify the least-cost option for attaining an objective, but there is no assurance that the agency or society is better off with the least cost option versus taking no action at all. Only benefit-cost analysis can provide an indication of whether a program is actually efficient in terms of net positive gains to social welfare (i.e., whether benefits outweigh costs). In addition, benefit-cost analysis is a more flexible and suitable tool when there are multiple objectives (e.g., different types of benefits and(or) costs) that are relevant to a conjunctive use program.

APPROACHES FOR INCORPORATING HYDROLOLGY

Hydrologic assessment of a conjunctive use program is necessary to determine the physical and chemical functioning of the planned joint operation of surface-water and groundwater systems. For most programs, this requires development of simulation models. To determine the economic benefits and costs of a program requires the additional step of correlating hydrologic variables (e.g., predicted ground-water levels, contaminant concentrations) with economic variables (e.g., cost savings from reduced pumping lifts, costs of water treatment). This hydrologic-economic analysis may be done in a comparative framework where discrete alternatives are compared with a baseline. This is the perspective used in the case studies presented later, where specific economic aspects of existing conjunctive use programs are presented. Hydrologic-economic analysis also may be done in an optimization framework wherein economic objectives are maximized or minimized subject to constraints. Relevant methodologies developed for addressing the operation and management of surface-water reservoirs and for simulation-optimization of groundwater systems are discussed below.

Operation and Management of Surface-Water Reservoirs

In assessing the economics of conjunctive use programs, it is helpful to take advantage of the extensive body of work on the operation of surface-water reservoirs (Yeh, 1985). Although the specific hydrologic components (see fig. 1) and geohydrological complexities of conjunctive use programs may differ, consideration of the analogy between management of a groundwater reservoir and the operation and management of surface-water reservoirs can provide a useful initial framework for analysis. As illustrated in Table 1, there are many parallels between constraints and objectives for managing surface reservoirs and groundwater.

Table 1. Comparison of objectives and constraints in surface-reservoir management and groundwater management.

Surface reservoir	Groundwater
Maintain minimum reservoir levels for recreational needs.	Maintain protective water levels: control (limit) negative outcomes (e.g., migration of contaminants, land subsidence).
Maintain reservoir capacity for flood control and capture of inflow for future water supply.	Maintain available storage to capture future water for recharge.
Release sufficient water for water supply.	Pump sufficient water for water supply.
Minimize "wasted" spillage of water from reservoir that is not providing benefits.	Minimize outflow of groundwater outside the basin.
Maintain downstream flow so as to preserve ecological resources.	Maintain spring flows so as to preserve ecological resources.
Maintain structural limits on maximum storage levels in reservoir.	Maintain upper limits on groundwater levels to prevent high water level hazards such as water logging and liquefaction.
Specify stochastic and deterministic components to inflow to reservoir.	Specify stochastic (e.g., local runoff) and deterministic (e.g., purchased supplies of imported or reclaimed water) sources of water for artificial recharge.
Balance navigation, power generation, other uses of reservoir-stored water within a multi-objective social welfare context.	Balance ecologic discharges (e.g., to minimum stream and spring flows), future water needs, and other uses of groundwater within multi-objective context.

Philbrick and Kitanidas (1998) developed a conjunctive use optimization model that expanded on previous frameworks for reservoir operation and management. Their model included both surface and subsurface storage as state variables and reservoir release, pumpage, and groundwater recharge as decision variables. They explicitly considered the costs of pumping, recharge, and shortages.

Simulation-Optimization of Groundwater Systems

Bredehoeft and Young (1970) were the first to link economics with detailed geohydrologic models that incorporate the spatial heterogeneity of groundwater systems. In a series of papers, they used an approach that iterated between an agricultural production model and a groundwater simulation model to address economic questions, including the conjunctive use benefits of decreased revenue variance. Gorelick (1983), Yeh (1992), Wagner (1995), and Ahlfeld and Mulligan (2000) describe the computational strategies for linking groundwater models with optimization methods. Groundwater simulation-optimization studies explicitly addressing aspects of the economics of conjunctive use include Young and Bredehoeft (1972), Bredehoeft and Young (1983), Chaudry et al. (1974), Danskin and Gorelick (1985), Matsukawa et al. (1992), and Nishikawa (1998).

CONJUNCTIVE USE CASE STUDIES

Information on three existing conjunctive use programs in California is presented below. Locations of the three areas are shown in figure 2. The economic attributes of these programs, which have been operating for decades, provide useful information for the planning of new programs. Although no new benefit-cost analyses are conducted for these examples, components of their respective costs and benefits are discussed.

Santa Clara Valley

In the Santa Clara Valley in the San Francisco Bay area, artificial recharge of local runoff through spreading grounds has been carried out since the 1930s. Importation of surface water from the California State Water Project and from the Hetch Hetchy Reservoir began in the 1960s. Today, the system in the Santa Clara Valley consists of 10 reservoirs and about 400 acres of spreading ponds. This conjunctive use of groundwater and surface-water supplies has served an important role in water management in the basin over the past 70 years, as the area has shifted from a predominantly agricultural area to a major urban center. Significant benefits of the program include reduced subsidence and reduced pumping lifts. Current pumpage in the basin is about 100,000 acre-ft/yr, direct delivery of surface water is about 90,000 acre-ft/ yr, spreading of imported water averages about 40,000 acre-ft/yr, and spreading of local runoff averages about 60,000 acre-ft/yr.

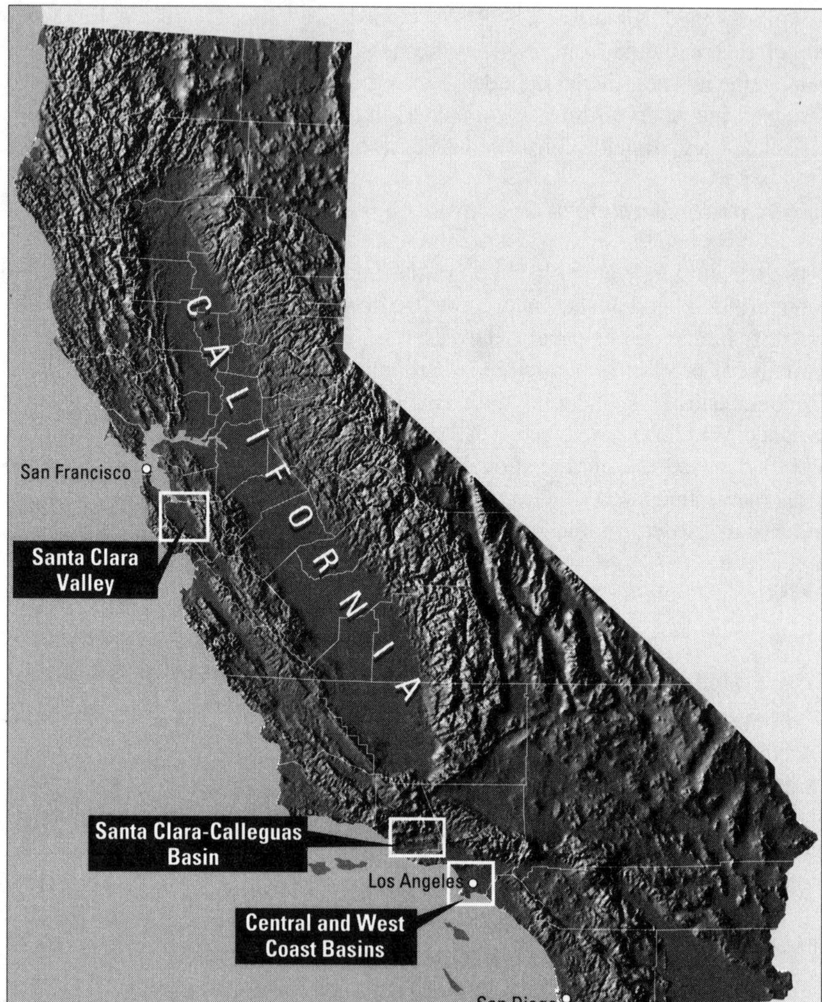

Figure 2. Locations of conjunctive use case studies.

A cost-benefit analysis of the artificial-recharge program in the Santa Clara Valley considered two alternative baselines (Reichard and Bredehoeft, 1984). The first baseline was the continued extraction of groundwater absent the increased recharge enabled by percolation ponds. In this baseline scenario, the benefits of the program included the avoided increase in the costs of pumping (due to increased pumping lifts) and subsidence (including repair of damaged infrastructure, such as sewers and bridges, as well as building/raising of levees). In other words, under the baseline of relying solely on local groundwater in lieu of conjunctive use, the water users in the region would have had to pay increased costs reflecting the consequences of mining local groundwater resources. In addition to the readily identified and estimated costs of increased pump operating costs and damages due to subsidence, other costs of over-extraction of groundwater (i.e., other benefits of conjunctive use) may be significant. For example, some of the potentially significant avoided costs that were not explicitly estimated in this case study may include deepening wells, adding water treatment to render lower-quality water suitable for potable and other uses and, ultimately, facing the longer-term consequences of water shortages. Therefore, the benefits of conjunctive use (relative to the baseline of relying on solely local groundwater) may also include the avoidance of severe water-use restrictions and (or) strict limits on the population and economic growth of the region.

The second baseline used in the study was a "without conjunctive use" scenario in which an equivalent program of surface-water facilities (reservoirs, pipelines, and treatment plants) was implemented. The benefits of conjunctive use in this scenario were the avoided cost of building all the necessary surface-water facilities. There also are opportunity costs associated with dedicating land to both conjunctive use programs (e.g., spreading ponds) and surface-water facilities.

Under either baseline scenario, the benefits of conjunctive use—the avoided costs of increased pumping lifts and subsidence damage (first baseline) or of construction and operation of new surface-water facilities (second baseline)—were greater than the costs.

The cost-benefit analysis described above was completed in the early 1980s. Currently (2002), water levels have recovered to where they were in 1900, before significant groundwater development began. While the Santa Clara Valley Water District (SCVWD) historically operated its recharge facilities with the objective of maximizing recharge in order to refill the basin, it currently considers multiple factors in determining the optimal amount artificial recharge (Jeff Micko, SCVWD, oral commun., 2001). Officials still see important benefits from artificial recharge, especially protection against future droughts—i.e., the insurance or buffer value. However, they also need to prevent water levels from getting too high and causing return flow to streams, artesian flow through unsealed wells, and increased potential of liquefaction. Once they decide on the optimal amount of artificial recharge, SCVWD then determines the combination of facilities that minimizes the marginal costs. Options include simple instream recharge, instream recharge with temporary dams, spreading in ponds with minimum maintenance, and spreading in ponds with annual or semi-annual maintenance (capacity restoration by draining and scraping of ponds). SCVWD estimates that marginal costs of artificial recharge range from near zero for the simple instream recharge to as much as $75/acre-ft for spreading in ponds with regular maintenance.

Santa Clara-Calleguas Basin, Ventura County

The Santa Clara–Calleguas Basin in Ventura County remains a predominantly agricultural basin. The United Water Conservation District (UWCD) has been diverting water from the Santa Clara River into spreading ponds since 1929. The conjunctive use system has evolved over time, with the building of an upstream reservoir and a series of pipelines that are linked to wells. One set of wells is located adjacent to the spreading facilities and in essence pumps out the recharged water, which is delivered to downgradient water users via pipeline.

The Santa Clara-Calleguas Basin provides an example of the complexities of conjunctive use in a multiaquifer system. For example, the spreading operations predominantly recharge the upper aquifers in the upgradient part of the basin, whereas the lower aquifers in the downgradient part of the basin face the most serious problems with drawdowns and increased salinity (Reichard, 1995; Izbicki, 1996a,b; Hanson et al., 2002).

Currently, the estimated average operations and maintenance costs of spreading are $10/acre-ft (Steve Bachman, UWCD, oral commun., 2001). There are benefits from reduced pumping lifts, prevented subsidence, and control of seawater intrusion and other contamination sources. As part of an analysis of a proposed 6,000 acre-ft/yr increase in the capacity of their current spreading facilities, coupled with a shift in pumpage from downgradient wells to the wells adjacent to the spreading grounds, UWCD recently estimated annual benefits from reduced pumping lift to be about $200,000 (i.e., about $33/acre-ft/yr). This assumed a unit pumping cost of $.25/acre-ft/ft. In this case, the pumping lift benefits are due to the fact that the reduction in downgradient pumpage causes a large recovery of water levels in this confined system.

Central and West Coast Basins, Los Angeles County

In the Central and West Coast Basins, water has been spread in the Montebello Forebay since the 1930s (amounts of spreading greatly increased around 1960) and injected for seawater-intrusion control since the mid-1950s. Spreading utilizes local, imported, and recycled water; injection includes imported and recycled water. Imported water also is delivered directly to users. Subsidy programs offer economic incentives for pumpers to purchase imported water instead of pumping groundwater during certain periods. These are referred to as "in lieu" purchases by local water managers. Tradeoffs between additional injection, spreading, and in lieu purchases relate to: (1) costs of different water sources; (2) impacts on seawater intrusion, pumping lifts, interbasin fluxes, and migration of inland chloride plumes; (3) operational considerations of the many water purveyors in the area; (4) regulatory restrictions on use of reclaimed water, and (5) legal issues such as water rights.

Currently, about 130,000 acre-ft/yr of water is spread in percolation ponds, about 30,000 acre-ft/yr is injected into the seawater barrier wells, and about 20,000 acre-ft/yr is delivered directly to users in the in lieu program (Theodore Johnson, Water Replenishment District of Southern California, written commun., 2001). Estimated average operation and maintenance costs of spreading are $2.6 million per year ($20/acre-ft). Estimated average operation and maintenance costs of injection are $3 million per year ($100/acre-ft). Costs of water currently range from $15/acre-ft for recycled water used for spreading to $528/acre-ft for non-interruptible imported water purchased through the West Basin Municipal Water District for injection at the barriers. In water year 2000, the total cost of purchasing about 131,000 acre-ft of imported and recycled water for spreading, injection, and in lieu use was about $27 million (an average cost of purchased water of $229/acre-ft). The capital costs of injection are considerable. Currently, the barrier projects, operated by Los Angeles Department of Public Works, consist of 230 injection wells, 4 extraction wells, and 758 observation wells. Historically, costs of installing injection wells have been about $500,000 per well. This implies a current capital investment (in current dollars) of $115 million. The estimated costs for installing new injection wells—including the costs of necessary pipeline extensions, barrier automation, and built in redevelopment equipment—are about $1million per well (Wayne Jackson, Los Angeles Dept. Public Works, oral commun., 2001).

There are several significant benefits of the conjunctive use program in the Central and West Coast Basins of Los Angeles. Since artificial recharge became significant in the early 1960's, water levels have risen by as much as 80 ft in parts of the region. This rise in water levels, which results from a combination of the increased recharge and decreased pumpage, has reduced pumping lifts and prevented subsidence. An estimate of 2.5 million dollars per year of benefits from the reduced pumping lift can be derived by multiplying current total pumpage (about 250,000 acre-ft/yr) times an average reduction in pumping lift (taken as 40 ft) times a unit pumping cost (using the same value used in the Santa Clara- Calleguas case study—$0.25/acre-ft/ft). This is an underestimate of benefits, since water levels would actually have continued to decline without the conjunctive use program. The injection program has controlled (but not eliminated) seawater intrusion. One way to quantify the benefits of this seawater-intrusion control would be to estimate the avoided costs avoided of treating intruded water prior to usage. In addition there are *in situ* benefits to controlling seawater intrusion, including the buffer or option value associated with maintaining coastal groundwater as a potable resource for future use.

CONCLUSIONS

This chapter has summarized issues and challenges for assessing economics of conjunctive use. The potential benefits of conjunctive use are large. However, there are challenges in identifying and quantifying all relevant benefits while avoiding double counting. When computing costs, both direct and indirect effects such as water-quality degradation must be considered. A rigorous economic assessment requires identification of the appropriate baseline; in some cases it may be appropriate to consider multiple alternative baselines. In order to properly incorporate the relevant three-dimensional geohydrologic complexities, it is useful to link the benefit-cost framework with a detailed groundwater model.

Although this chapter has focused on linking economics and hydrology, institutional factors also may play a key role in determining the economic viability of conjunctive use. Water markets and water-rights structures control the type of conjunctive use programs that are instituted and determine how benefits are distributed. As was noted very early by economists, the "common pool" nature of groundwater can complicate how the benefits of conjunctive use are captured. There also may be institutional barriers to implementing inter-basin programs that export stored groundwater out of a basin.

Acknowledgments. We are grateful for the helpful technical and editorial comments of Richard Bernknopf, Steven Phillips, Clark Londquist, Gerald Woodcox, Anthony Buono, and the anonymous reviewers. We thank Theodore Johnson, Wayne Jackson, Jeff Micko, and Steve Bachman for providing information for the case studies. Phil Contreras prepared the illustrations.

REFERENCES

Ahfeld, D.P. and A.D. Mulligan, *Optimal Management of Flow in Groundwater Systems*, Academic Press, San Diego, CA, 2000.

Bredehoeft, J.D. and R.A. Young, The temporal allocation of groundwater: A simulation approach, *Water Resour. Res.,* 6(1), 3-21, 1970.

Bredehoeft, J.D. and R.A. Young, *Conjunctive use of groundwater and surface water for irrigated agriculture: Risk aversion*, Water Resour. Res., 19(5), 111-1121, 1983.

Burt, O.R., The economics of conjunctive use of ground and surface water, *Hilgardia,* 35(2), 31-111, 1964.

Burt, O.R., Economic control of groundwater reserves, *J. Farm Econ.,* 48(1), 632-647, 1966.

Burt, O.R., Temporal allocation of groundwater, *Water Resour. Res.,* 3(1), 45- 56, 1967.

Chaudry, M.T., J.W. Labadie, W.A. Hall, and M.L. Albertson, Optimal conjunctive use model for the Indus Basin, *Journal of Hydraulics Division*, American Society of Civil Engineers, 100(HY5), 667-687, 1974.

Cummings, R.G. and J.W. Mc Farland, Groundwater management and salinity control, *Water Resour, Res.,* 7(6), 1415-1424, 1974.

Cummings, R.G. and D.L. Winkleman, Water resources management in arid environs, *Water Resour. Res.,* 6(6), 1559-1568, 1970.

Danskin, W.R. and S.M. Gorelick, A policy evaluation tool: Management of a Multiaquifer system using controlled stream recharge, *Water Resour. Res.,* 21(11), 1731-1747, 1985.

Donovan, D.J., T. Katzer, K. Brothers, E. Cole, and M. Johnson, Cost-benefit analysis of artificial recharge in Las Vegas Valley, Nevada, *Journal of Water Resources Planning and Management,* 128(5), 356-365, 2002.

Fisher, A. and R.S Raucher, Intrinsic Benefits of Improved Water Quality: Conceptual and Empirical Perspective, *Advances in Applied Micro-Economics,* Vol. 3, K. Smith, editor. JAI Press, 37-66, 1984.

Gisser, M. and A. Mercado, Integration of the agricultural demand function for water and the hydrologic model of the Pecos Basin, *Water Resour. Res.* 1373-1384, 1972.

Gisser, M. and D.A. Sanchez, Competition versus optimal control in groundwater pumping, *Water Resour. Res.*, 16(4), 638-642, 1980.

Gorelick, S.M., A review of distributed parameter groundwater management modeling methods, *Water Resour. Res.*, 19(2), 305-319, 1983.

Hanson, R.T., P. Martin, and K.M. Koczot, Simulation of Ground-Water/Surface-Water Flow in the Santa Clara-Calleguas Ground-Water Basin, Ventura County, California, U.S. Geological Survey Water Resources Investigations Report, 02-4136, 2002.

Izbicki, J.A., Seawater intrusion in a coastal California aquifer, U.S. Geological Survey Fact Sheet FS-125-96, 1996a.

Izbicki, J.A., Source, movement, and age of ground water in a coastal California , aquifer, U.S. Geological Survey Fact Sheet FS-126-96, 1996b.

Logan, W. and R. Allen-King. 2001. Opportunities for Sustainable Underground Storage of Recoverable Waters, *Water Science and Technology*, 18(3), 5, 2001.

Matsukawa J., B. Finney, and R. Willis, Conjunctive-use planning in Mad River Basin, California, *Journal of Water Resources Planning and Management,* 118(2), 113-131, 1992.

McClelland, G.H.; W.D.Schulze, J.K. Lazo, D.M. Waldman, J.K. Doyle, S.R. Elliott, and J.R. Irwin., Methods for Measuring Non-Use Values: A Contingent Valuation Study of Groundwater Clean-Up, U.S. Environmental Protection Agency Cooperative Agreement #CR-815183. Boulder, CO. University of Colorado. 1992.

National Research Council, *Valuing Groundwater, Economic Concepts and Approaches*, National Academy Press, Washington, D.C., 1997.

Nishikawa, T., Water resources optimization model for Santa Barbara, California, *Journal of Water Resources Planning and Management,* 124 (5), 252-263, 1998.

Philbrick, C.R., and P.K. Kitanidas, Optimal conjunctive-use operations and plans, *Water Resour. Res.*, 34(5), 1998.

Provencher, B. and O.R. Burt, The externalities associated with common property exploitation of groundwater, *Journal of environmental economics and management,* 24(2), 139-158, 1993.

Reichard, E.G., Groundwater-surface water management with stochastic surface water supplies: A simulation optimization approach, *Water Resour. Res.,* 31(11), .2845-2865, 1995.

Reichard, E.G. and J.D. Bredehoeft, An engineering econmic analysis of a program for artificial groundwater recharge, *Water Resources Bulletin,* 20(6), 929-939, 1984.

Todd, D.K., Economics of groundwater recharge, *Journal of the Hydraulics Division, Proceedings of ASCE 91(HY4),* 679-686, 1965.

Tsur, Y., The stabilization role of groundwater when surface water supplies are uncertain: the implications for groundwater development, *Water Resour. Res.*,26(5), 811-818, 1990.

Tsur, Y. and T. Graham-Tomasi, The buffer value of groundwater with stochastic surface water supplies, *Journal of environmental economics and management* 21, 201-224, 1991.

Wagner, B.J., Recent advances in simulation-optimimization groundwater management modeling, *Reviews of Geophysics, Supplement, U.S. National Report to International Union of Geodesy and Geophysics*, 991-1994, 1021-1028, 1995.

Weisbrod, B.A.1964. Collective-consumption services of individual consumption goods. *Quarterly Journal of Economics* (August 1964): 471-477, 1964.

Winter, T.C., J.W. Harvey, O.L. Franke, and W.M. Alley, Groundwater and Surface Water, A Single, Resource, *U.S. Geological Survey Circular* 1139, 1998.

Yeh, W.W-G., Reservoir management and operations models: A state-of-the-art review, *Water Resour. Res.,* 21(12), 1797-1818, 1985.

Yeh, W.W-G., Systems analysis in groundwater planning and management, *Journal of Water Resources Planning and Management,* 118(3), 1992.

Young. R.A., and J.D. Bredehoeft, Digital computer simulation for solving management problems of conjunctive groundwater and surface water systems, *Water Resour. Res.,* 8(3), 533-556, 1972.

Eric G. Reichard, U.S. Geological Survey, 5735 Kearny Villa Road, Suite O, San Diego, CA 92123, egreich@usgs.gov

Robert S. Raucher, Stratus Consulting Inc PO Box 4059, Boulder, CO 80306-4059, braucher@stratusconsulting.com

9

Water Resource Management: Science, Planning and Decision-Making

Katharine Jacobs and Roger Pulwarty

INTRODUCTION

Scientists have much to contribute to water resource management, but they may lack the socio-environmental context necessary to see their contributions fully utilized. They are often unable to provide information at geographic scales commensurate with water management[1] boundaries, and may not have a full understanding of the institutional and political ramifications of their suggestions. At the same time, water managers are frequently unaware of scientific information that could be useful to them, and may not know the constraints faced by the science community. Overcoming some of these coordination issues can improve both the usefulness of the scientific information that is developed and the quality of water management decisions.

Water management in the United States (US) occurs at multiple levels and involves substantial complexity. The degree of complexity and the number of constraints on institutional flexibility are generally not well appreciated by the public or the scientific community. In addition to issues of complexity, regulatory institutions are generally resistant to change, and there are huge public and private investments reinforcing the status quo. This means that changes to water rights systems, water supply management and decision-making approaches that may make sense from a scientific perspective are difficult to implement or may be inappropriate for a given management situation. To a large degree, private sector investments have been optimized based on the current regulatory circumstances, so even changes that are designed to gradually increase resilience and

---

[1] In this paper, we are using the term "water management" to refer to decisions regarding the use, storage, allocation, delivery or treatment of water. Just as the term "scientist" refers to individuals in broad spectrum of disciplines with multiple perspectives, water managers are involved in a wide range of activities from policy to operations.

Water: Science, Policy, and Management
Water Resources Monograph 16

10.1029/16WM11

efficiency may generate opposition. Therefore, improving the relevance of scientific products and working with water managers to identify potential applications will address only a portion of the issues in the science/decision-making context. New types of partnerships and institutional changes are needed to improve the efficiency of the water management system in the context of increasing pressures on limited water supplies.

BACKGROUND

The common expectation among scientists is that new information can help managers anticipate a range of possible outcomes and assist in designing and implementing flexible responses. Although science is influenced by socio-political factors (affecting funding priorities, research agendas, etc.), the training that scientists receive encourages development of knowledge as an objective in its own right. In addition, scientists have been trained to focus on well-defined disciplinary areas of research. This does not necessarily encourage understanding of the needs of users of the information that is generated, nor does it provide the tools needed for translation of knowledge to application by decision-makers. More importantly, it limits the ability of scientists to participate effectively in a complex world shaped by institutional, financial and political considerations.

Water managers also are trained to function in specific types of environments. Many water managers are educated as engineers and have formal training in systems analysis and complex system theory, but are still not comfortable when they are constrained by public perception, politics and other social concerns. In addition, most water managers learn a substantial proportion of their skills on the job, within the particular context of the organizations they represent. As noted by Rayner *et al.* [2000], "water management systems are heavily dependent on craft skills and local knowledge. Experience within the local system is generally valued over techno-scientific information. There is strong reliance on informal communication systems and information networks."

Water managers use scientific information such as historic flow data for managing surface water reservoirs, engineering specifications for facility design, groundwater flow models for assessing groundwater availability, and short-term weather predictions for operational decisions. However, it is the perception of many scientists that managers have a fixed view of the environmental record, such as relying on historical average values of precipitation, streamflow etc., drawn from periods that may not represent present day conditions. They question how receptive water managers are to new sources of information and changes in the "information envelope" or boundaries of their knowledge.

Some water managers may be concerned about operating outside of historic boundaries of their organizational information base, because behaving as others do is a proven or acceptable strategy. This perspective is not unique to water managers; it has been observed in other professions [see *Pulwarty and Redmond*,

1997]. New information sources and technology represent risk, and the repercussions of management error can be substantial, particularly from a liability perspective. Even small increases in real or perceived uncertainty of a negative outcome—flooding, water shortage, etc.—will often outweigh certain more positive outcomes—enhanced profitability, power generation, etc. Rayner *et al.* [2000] observe that "water resource managers are inherently conservative due to a shared and at least implicitly articulated hierarchy of values; reliability, quality and cost."

Using cutting-edge technology, such as probabilistic climate forecasts, may be intimidating to water managers who are unfamiliar with the subject area, and in some cases may be unnecessary. Failure to use new technology or sources of information does not necessarily mean that water managers are resistant to change; it may mean they have analyzed the information available and concluded it has no practical applications within their system at that time.

In many cases use of new scientific knowledge and innovative approaches requires the input of a trained professional to yield results that extract small signals out of much noise. This has been demonstrated by K. Georgakakos *et al.* [1998] and A. Georgakakos [2000] in recent work in using climate information in reservoir management. The role of such integrators is likely to expand as more applications of climate science to water management are developed.

There is currently a major focus on "usable science," providing new opportunities to develop research products that have direct applications in the "real world." Understanding the context of decision-making needs to be established as a legitimate part of scientific research. Funding agencies and current review processes tend to perpetuate the view that science should not be "contaminated" with social concerns. However, failure to appreciate the social context of decision-making has resulted in generations of scientific products that are rarely used.

A new form of scientific training may be needed to improve the "usability" of scientific products. This new approach requires different modes of communication between scientists and users of information, and possibly new types of professionals. This new type of communication moves beyond providing information (the "loading-dock mentality," where information is provided but no-one assesses its usefulness) towards developing a framework and infrastructure for ongoing relationships between scientists and practitioners. This requires more in-depth relationships and expanded communication, including communication about how new information is created and evaluated. To the extent that there is a need for improved predictive capacity to make better water management decisions, there are significant communication hurdles that need to be overcome [see *Bogardi*, 2002].

Management of water supplies is *always* political, because control of supply is closely related to political power and generation of wealth. Water is an ingredient of virtually all forms of industry and commerce, thus it is directly related to land values and growth patterns. This is particularly true in the West, where decisions to irrigate lands to "make the desert bloom" resulted in huge taxpayer subsidies that transformed major portions of the US agricultural and consumer economy, but also resulted in personal gain to some individuals and losses to others [*Reisner*, 1987]. The huge system of dams, reservoirs and canals (and other interbasin transfer mechanisms) made it possible for the West to support large-scale agriculture and burgeoning populations, but at enormous cost for aquatic habitat quantity and quality.

In this paper, there are numerous references to the water management conditions in Arizona. In particular, water issues in Tucson are used to illustrate several points made (see The Role of Science and Perception, page 12). Use of climate forecasts also receives specific attention in examples throughout the paper.

WHO MAKES WATER MANAGEMENT DECISIONS?

For consumers who receive their water from a domestic water supplier, water management may seem relatively simple: if good quality water comes out of the tap when they turn it on, the system is working. Consumers may never question the numerous management and policy decisions that ensure that a safe and reliable water supply is delivered twenty-four hours a day. Often many layers of management and regulation lie between the source of water and the ultimate users, and each of these layers functions in an environment where technical information is only one component of decision-making. This leads Helen Ingram, a political scientist at the University of California, Irvine to have the opposite experience from the typical consumer. When she turns on her tap, she "sees not water, but rather 'a cascade of builders, fishermen, farmers, environmentalists, politicians, activists, technocrats, bureaucrats and autocrats'". She also notes that "water is often thought about as an engineering and technical subject, laying pipes in the right way. People don't realize the extent to which politics and social organization determine whether water flows from their taps." [*Ingram,* UC Irvine web site].

Generally, states have the authority to regulate access to surface and groundwater, while regulation of water quality is initiated at the federal level. The legal basis of water rights administration differs substantially from state to state, and surface water is frequently regulated under a different water rights regime than groundwater. Some states are actively involved in both surface and groundwater management, particularly in regulating access to water through water rights and permits. Because of the relatively large role of the states in water management, decision-makers include state regulators, state water project managers, state legislators, and in some cases, state courts. For example, in Colorado, the majority of water rights decisions are made in the court system rather than in an administrative agency [*Nichols, et al.*, 2001].

Within states there are many local and regional water management districts, including agricultural districts, some of which have separate state authorizing legislation and different powers and duties. Local water managers may also be associated with public or private water utilities. Some public systems have been privatized, usually resulting in different management constraints than exist for systems that are run by administrators who answer to elected officials. In addition, utility regulatory agencies in most states have some authority over rate setting for private municipal water providers. This may significantly limit the management options available to such entities.

The traditional role of the federal government in water management has been building and operating massive storage and delivery systems, especially in the West. The federal regulatory responsibilities for water management have historically been limited, but federal agencies have had an expanding role in the last few decades. There are at least five major federal agencies significantly involved in water management. These agencies' activities generally are not well coordinated. In some cases, the missions and/or procedures of these agencies are in direct conflict. These agencies include three that are within the Department of the Interior (the Bureau of Indian Affairs (BIA), the Fish and Wildlife Service (FWS), and the Bureau of Reclamation (BOR)). The US Army Corps of Engineers (USACE) also has substantial management responsibilities along major river systems. Other federal agencies have a large, and potentially unappreciated impact on the water management arena. These include the Department of Agriculture, the National Marine Fisheries Service in the Department of Commerce, the Geological Survey (USGS, part of the Department of the Interior, which provides a significant component of water–related data at the federal level), and the Federal Emergency Management Agency (FEMA).

The Environmental Protection Agency (EPA) has primary authority for implementing and enforcing the Clean Water Act (CWA) and the Safe Drinking Water Act (SDWA). Some states have been delegated responsibility for enforcing components of these Acts. Most states have at least some responsibilities related to regulating domestic water supplies from a health perspective, using standards established by the Environmental Protection Agency through the SDWA.

Though these federal agencies fall under the Executive Branch, the US Congress also plays a role in water management, particularly through funding decisions. Congress can directly control water policy through passage of laws, but decisions regarding funding or defunding particular agency programs may have even more impact. Appropriations for agricultural subsidies are a good example. Changes in agricultural price support systems and set-aside programs affect cropping patterns and water use in areas of irrigated agriculture.

The watersheds of surface water and groundwater basins frequently cross state and international boundaries as well as the boundaries of Indian reservations. For more than a century, there have been conflicts between the states over access to and ownership of these water supplies, leading to multiple interstate and international compacts and commissions and court decrees. In the case of the Colorado River—which flows through seven states prior to entering Mexico—

the Secretary of the Interior is the ultimate decision maker, serving as the water master for the river. However, his or her decisions are made in the context of multiple interstate compacts and decrees, as well as international agreements with Mexico. Generally, interstate compacts negotiated under federal law take precedent over state regulation. The International Boundary Water Commission has authority over certain activities within 100 kilometers of the U.S.-Mexican boundary, pursuant to the 1983 La Paz Agreement for the Protection of the Environment in the Border Area. Due to their sovereign status, Indian tribes have authority to manage water supplies within their own boundaries and in some cases have established water quality standards that affect off-reservation users.

Water management decisions are also made directly by the public, particularly through votes regarding bonding for infrastructure. In rare circumstances, specific water management decisions have been made through initiatives and referenda. A particularly vivid example of this has occurred in Tucson, Arizona, where voters overturned plans to convert from mined groundwater to Colorado River water through the Central Arizona Project canal, (see The Role of Perception in Water Management, p 12).

Given the multiple players in the field of water management, and the significant regulatory complexity, it is not surprising that many scientists have chosen to focus on what they do best: improving the understanding of nature, within particular disciplines. Clearly there are currently engineers and other professionals who currently focus on translating scientific results into management applications. However, it is clear that new strategies must evolve to enhance the capacity of scientists and water managers to work together. It is not reasonable to expect scientists to engage in or understand the full breadth of social concerns faced by decision-makers. However, failure to appreciate the regulatory and institutional context may explain a significant component of coordination problems between scientists and water managers to date.

THE ROLE OF UNCERTAINTY IN WATER MANAGEMENT

> *"Uncertainty is not the hallmark of bad science, it is the hallmark of honest science.... This perennial question "Do we know enough to act?" – is inherently a policy question not a scientific one" (Hon. George Brown, 1997).*

There are many types of uncertainty within the realm of decision-making, ranging from technical uncertainty to uncertainties within the policy and operating environments. An initial source of uncertainty is whether the right question is being asked. Water managers may perceive a given problem in the context of past experience and fail to note that the conditions they are observing represent changes in trends, either physical, political or economic. Failure to ask the "right" question adds a layer of uncertainty that may not be recognized as a risk factor. Additional layers of uncertainty relate to the appropriateness of the methodology selected for analysis and uncertainties in the data [see *Functowicz and Ravetz*, 1990]. Uncertainties about data relate to accuracy and quality of the data, and whether the data are representative of the actual physical conditions in the area of concern.

The impact of regulatory uncertainty is frequently underestimated. With so many layers of regulators operating under different legal frameworks, alternatives that are selected at the local level may encounter "fatal flaws" when subjected to federal agency review. For instance, the lack of predictive skill is particularly problematic as it relates to the Endangered Species Act, in part because there is inadequate information about the needs of particular species relative to habitat protection but also because regulatory decisions are made by individuals who may have differing interpretations of legal requirements. It is important for scientists to recognize that improved predictive capacity is only a part of the equation in reducing uncertainty.

The final and possibly most important source of uncertainty is the role of public perception and politics in decision-making. This component of uncertainty is discussed in the Tucson example. Multiple studies [e.g., *Changnon*, 2000] indicate that water managers do not believe there is enough certainty associated with climate-related predictions to justify a change in management approach. Some of that belief may be based on an incomplete understanding of the basis and meaning of those predictions. In these instances, new tools to evaluate alternatives in the context of uncertainty and risk could help water managers "know enough to act."

Barriers to the use of climate information relate to relevance, accessibility and acceptability, and are summarized in Table 1.

HOW DO WATER PLANNERS AND MANAGERS USE CLIMATE INFORMATION IN DECISION-MAKING?

This section focuses specifically on the use of climate information in water management to illustrate points regarding the use of scientific information in water management and planning in general. Scientists have recognized the potential value of the use of seasonal climate information for water management for some time [*Changnon and Vonnhame*, 1986]. However, Pagano [2001] and others have noted that such forecasts often play only a marginal role in real-world decision-making [*Callahan et al*., 2000; *Changnon*, 1990: *Pulwarty and Melis*, 2000; *Pulwarty and Redmond*, 1997; *Sonka et al*. 1992]. There are several reasons why water managers have been reluctant to use this information.

First, water managers are not a homogenous group. They use scientific information in decision-making at different scales, based on their training, institutional culture, regulatory structure, perception of risk and stability, system constraints, and financial situation, including bonding capacity. Second, different sources of information are needed for different types of water management sys-

Table 1. Barriers to the Use of Climate Information by Water Managers
[Adapted from *Pulwarty and Redmond,* 1997; *Pulwarty and Melis*, 2001]

Certainty
- Forecasts are not seen as accurate enough to justify action
- Experts disagree, or are perceived to disagree (desire for unanimity among experts)
- Inability to verify or track information based on own experience
- Validation statistics of previous forecasts or "skill scores" are not available, or are not considered to be accurate enough to justify action
- Perception of "waffling" in successive forecasts due to new information gives impression of lack of certainty
- Role of climate information in reducing risk is unclear
- Response to climate information is viewed as more risky than using established procedures
- Manager's overconfidence in ability to manage and control risky situations
- Lack of an explicit characterization by scientists of degree of uncertainty
- Manager's overconfidence in their own knowledge, or based on heuristics covering a relatively short time span

Communication
- Overuse of disciplinary language without context (use of jargon)
- Users face new or changing definitions of terms
- Media coverage inadequate, inappropriate or inaccurate
- "Clients" are not challenged to be more precise about their needs
- Negative perceptions about the utility of climate information
- Issues in visualization of complex information

Focus
- Spatial information is too broad and non-specific or difficult to interpret
- Desired information not provided or available (the science is not yet available to meet the user's needs)
- Groundwater managers may not have the same need for climate information as surface water managers
- Water managers may be overloaded with information, not able to sift for relevant material
- Water managers' job expectations do not encourage incorporation of risk assessment and probabilistic information

Trust
- Perception of communication as "marketing" rather than based on common interest

Resources
- Failure to recognize resource limitations: time, money, staff and data may not be available to incorporate the new information
- Funding institutions may not be willing to fund applied research

Timing
- Forecast information is not available on a timely basis, relative to decision calendar of manager

Training
- Scientists are trained within specific disciplines, and rewarded for staying within their disciplines; they are not trained as integrators
- Water managers may be trained primarily as engineers, only have experience within particular job applications
- Lack of familiarity with the methodologies for analyzing climate data (and their limits)
- Inability of forecasters to recognize competing or shifting goals, need for a flexible response
- Lack of procedures for incorporating climate impacts information/models in decision-making
- Water managers may have analyzed the information available and concluded it is not applicable to their system

Boundary
- Inability to access data or apply management solutions across jurisdictional and institutional boundaries
- Entities located near places where boundary conditions change, e.g., climate divisions, watershed boundaries, etc. may have difficulty accessing information that is useful to them.

tems—surface water vs. groundwater, wholesale suppliers vs. retail, etc. Third, individual water managers in identical positions may respond differently simply due to their own experience and willingness to innovate and take risks.

At the operational level, information about temperature influences on snowpack and runoff, precipitation extremes, onset and duration of drought are frequently requested by water managers of climate scientists. More recently, interest has moved to the impacts of ENSO (El Nino-Southern Oscillation) events on the year-to-year variability of runoff conditions and on longer-term (decades or centuries) changes in supply potential. Efforts to produce appropriate information at useful scales must be framed within an understanding of the management goals at the society-environment interface.

However, decision-makers repeatedly state that climate forecasts are unreliable, and that there are no quantitative ways to evaluate their credibility [*Hartmann et al*, 2001]. Interestingly, it has been found that communication of forecast uncertainty can increase forecast credibility [*Pielke and Glantz*, 1995; *Pielke*, 1999; *Hartmann*, 2001]. It is clear that practical use of probabilistic forecasts requires a specialized type of training, and that communication issues are important. However, significant progress is being made in this arena. Forecast evaluations show that predictions are much more accurate for some regions of the country than for others. Hartmann *et al.* [2001] note:

> Forecast evaluations should focus on specific regions, seasons, and lead times of interest to different decision-makers. CPC seasonal climate outlooks clearly perform better for some users than others. From the perspective of water managers in the Southwest, winter precipitation outlooks made during fall and winter are better than climatology[2] forecasts according to all criteria. Winter and spring forecasts of summer precipitation lack skill...Compared to the Upper Colorado River Basin, not only does the Lower Basin benefit from greater storage capacity..., but from greater climate predictability as well [p.14].

One suggestion for improving the relevance of forecast and other climatic information is to identify appropriate entry points into the decision-making process, through so-called hydro-climatic "decision calendars" [*Pulwarty and Melis*, 2001]. Such calendars have been used to identify decision needs within planning and operational activities on the Upper Colorado River and at Glen Canyon Dam [see *Ray et al.*, 2001]. These calendars are simply time-frame maps of the appropriate climate-related information needed for decision-making throughout the year for developing operating plans and endangered species recovery programs. They also provide a context for discussion, a way to encourage relationships between scientists and water managers that facilitates development of common knowledge.

[2] "Climatology" refers to historic climate information, without any predictions based on models.

Multi-objective management is a tool used to optimize complex systems where there are multiple constraints, and provides a theoretical framework for decision-making [see *Schwartz*, 2000]. Unfortunately, many water management decisions are made under time and resource constraints that limit the utility of comprehensive modeling exercises. In addition, the modeling approach focuses primarily on efficiency from an economic perspective and may not be able to accommodate other management objectives such as equity.

Different types of scientific information are used if the primary source of water is groundwater rather than surface water. Although there is a growing trend toward conjunctive use, wherein surface water and groundwater are both used to maximize the water supply available, the majority of water managers focus on one supply or the other. Key inputs for groundwater managers describe aquifer conditions, such as transmissivity of the sediments (flow rate), quantity of water in storage, pumping water levels and water quality. Groundwater managers are also concerned with sources of contamination that tend to be human in origin. Changes in climate do affect the availability of groundwater, but it is long-term trends, rather than short-term fluctuations, that are of most concern. Changes in temperature and precipitation may have a stronger effect on the demand side of the equation than on the supply side.

Surface water managers are much more concerned about climate variability and short-term conditions than groundwater managers. They must also focus on longer-term (seasonal to inter-annual) climate information because of the need to manage reservoirs to maximize water supply conditions while minimizing flooding potential. Additional information is needed in cases where power production, recreation, water quality, transportation, and habitat protection objectives are part of their management program. Such activities affect timing of releases as well as the total volume of water released from reservoirs.

As noted by Yao and Georgakakos [2001] improved forecasting will not improve reservoir management unless the forecast information is used effectively. They found in a study of Folsom Lake that perfect forecasts used with traditional rule curves did not improve the performance of the reservoir, while adaptive decision systems utilizing reliable forecast ensembles to determine dynamic operational policies were found to be highly effective.

A manager's willingness to accept risk (economic, physical, or social) and to respond appropriately to changing conditions is directly related to the types of scientific information that he or she needs or requests or is willing or likely to use. As noted previously, water managers tend to be risk averse. Yet, failure to effectively use predictive information related to climate may result in poor decisions and an increase in risk within the system, due to failure to recognize drought or flood conditions in time to respond, for example. As noted by the National Research Council, 1998, 1999 and others, it is clear that although their potential has not been fully realized, hydro-climatic forecasts at inter-annual and shorter scales can be useful in improving water management decisions. However, clearly perception of risks to the individual and the system is key to understanding decision-making, not the theoretical risk as defined by an external observer.

Because of the inherently conservative nature of the water management system (both institutionally and in the perspective of individual managers), most significant changes in management and institutions result from perceived or actual crises. For example, the significant drought in Southern California in the late 1980's and early 1990's resulted in renewed interest in changing the historic water rights priority system (agriculture holds the oldest and highest priority surface water rights) and led to the establishment of the California Drought Water Bank. As is illustrated in the Tucson case (see page 12) a crisis caused by the failures in delivery of Central Arizona Project water caused changes in the management approach as well as the delivery systems used by Tucson Water.

Although many notable water management policy changes are in response to crises, significant changes also occur incrementally over a period of time. Kirby [2000] points out that:

> Political scientists suggest that substantial policy changes usually take many years (a decade or more) to occur because the policy-making game is stabilized by several system parameters. These stabilizing system parameters include current culture, laws, wealth distribution, etc. and are outside the direct control of the players in each policy subsystem ...This observation that significant policy change primarily takes place over many years provides a sobering outlook to studies expected to directly influence policy outcomes based on analytical findings. If this is true—and experience indicates that it is—analysts should pay much more attention to studies designed to improve long-term understanding among participants about the problems and their related systems [p. 20].

Institutional problems may exacerbate problems caused by inadequate science. In the case of surface water/groundwater interactions, lack of a full understanding of the role of groundwater in supporting surface water flows has led to multiple cases of unanticipated consequences and substantial habitat damage [*Glennon and Maddock*, 1997]. This outcome is more likely in cases similar to that in Arizona, where the legal framework for groundwater is separate from that for surface water. For example, there is currently no legal mechanism for considering impacts on surface water caused by new permits to pump groundwater [*Glennon and Maddock*, 1994].

Water managers have differing needs for scientific information relative to the scale of management, the type of decision being made, and the training and structure of local management organizations (e.g., elected board vs. professional managers). Decisions that have long-term implications, such as development of new infrastructure, require greater accuracy than decisions that require no capital investment. Likewise, decisions that affect millions of users, such as managing the water levels in the reservoirs along the Colorado, are made with great care because of the significant implications for both water supply and flood damage.

The difficulty of handling highly technical issues in the political arena is particularly evident in the Tucson case described below, where the primary decision makers were City Council members and the general public.

THE ROLE OF SCIENCE AND PERCEPTION: WATER MANAGEMENT IN TUCSON, ARIZONA

The Central Arizona Project (CAP) was constructed by the Bureau of Reclamation to bring Colorado River water across the deserts of Arizona to Phoenix and Tucson. The total public and private investments in the aqueduct and associated distribution systems exceed 4.4 billion dollars. The concept of bringing Colorado River water to central Arizona was first discussed in the 1920's, but the project wasn't officially approved by Congress until 1968. By 1985 deliveries were being made from the CAP to the Phoenix area; late in 1992, the Tucson area began using Colorado River water for domestic deliveries. The CAP is a marvel of engineering, involving 13 major pumping stations and 330 miles of aqueduct. However, the CAP imported controversy along with a new source of water for Tucson.

The CAP is central Arizona's most significant source of renewable water supplies. Until its arrival, Tucson Water, the City of Tucson's municipal water utility, was entirely dependent on groundwater supplies. The water table in the Tucson region had been significantly lowered since the 1940's, with drops in the water level of up to 200 feet in portions of Tucson's central wellfield by 2000. It was clear before the authorization of the CAP that Tucson's groundwater use was not sustainable over the long term.

Tucson Water went through an extensive planning process and developed a long-range plan to demonstrate how it would maximize its use of Colorado River water. Because of the public's concerns about water quality and health, the city developed a state of the art water treatment plant, designed to stay under the EPA's maximum contaminant levels for chlorine by-products by a factor of 5. The treatment system used ozonation for initial disinfection, followed by the use of chloramines for residual disinfection in the system. The plant was expected to be a showcase of the latest in treatment technology and was designed with great care to limit impacts on surrounding neighborhoods.

Prior to the delivery of CAP, Tucson had one of the largest groundwater-based distribution systems in the country, serving some 600,000 people. Tucson Water is operated by a staff of city employees, headed by an appointed director who answers to the City Manager. Ultimate authority for the management of the utility rests with the City's elected Mayor and Council.

Private consultants, working closely with Tucson Water's engineers, designed and implemented the system modifications that were required. Tucson's investment in the new treatment plant and distribution system exceeded $250 million dollars. Though CAP deliveries were initially made to half the City's customers, the transition from Tucson's dispersed well system of 140 source wells, to the single-source system fed by the new treatment plant, took place in a single day. To the amazement of many, the new source of supply was delivered through the entire system without an apparent hitch.

However, the honeymoon was over relatively quickly. Over the next few months, complaints about rusty, brown, smelly water began to escalate. Customers began to report bursting water pipes in their homes, along with substantial damage. Water mains began to rupture. At first, these complaints were dismissed by Tucson Water, which claimed that the system needed to stabilize, and that these events were normal and anticipated (though the utility had not prepared the public for such events). When the problems continued, water quality experts from all over the country were consulted. Tucson Water was given much advice by University of Arizona scientists, consultants and other water utility managers, particularly with regard to adjusting the pH and use of corrosion inhibitors. Rust in the water did not diminish, and the complaints continued to increase.

Soon, the voices of those with bursting pipes and colored water were joined by political forces that hitherto had been relatively quiet. Those who had opposed the CAP from the beginning, including those who were concerned that CAP would cause more growth in the Tucson area, developed a substantial following. The CAP delivery, originally touted as a great success, became viewed as a disaster. Credibility of the utility and its management plummeted. Anti-Tucson Water and anti-city rhetoric intensified.

The appearance of the water fed people's fears about adverse health effects. Sales of bottled water and water treatment devices soared, though there was never a violation of EPA water quality standards. The aesthetic considerations, particularly taste, were already a major concern for many people, but the media and opponents of the CAP deliveries picked up on the "brown water" issue and fed people's concerns about trihalomethanes (THM's), the by-products of chlorine and organic materials found in surface water. Experts on both sides of the THM issue confused the public about the risks, both citing what literature was available and interpreting it to their own advantage. This was particularly ironic given that the City had chosen to use chloramines as a residual disinfectant specifically to reduce THM's while most other surface water systems in the country continued to use chlorine, which has a higher THM formation potential.

Both print and TV media had a field day, with coverage of the daily catastrophes and Tucson Water's inadequate responses escalating and very few reports from objective observers. Finally, in the context of a planned outage of the CAP canal, the Tucson Mayor and Council unanimously decided to "turn off" the CAP spigot and go back to groundwater indefinitely, until the issues could be resolved.

The decision to "turn off" the CAP was made in October of 1994, a little less than two years after the CAP supply was first introduced into the Tucson system.

One of the groups that opposed Tucson Water's treatment and handling of the CAP deliveries started an initiative drive. Despite opposition from the business community, in 1995 they were successful in passing an initiative, the "Water Consumer Protection Act." This Act amazed water managers across the country because it proscribed virtually all options Tucson Water had for using its CAP water, effectively prohibiting deliveries of CAP water unless treated by membrane filtration or recharged underground and recovered first. Neither membrane filtration nor recharge and recovery prior to delivery were possible without massive additional expenditures. The costs of the CAP options began to mount, as consultants were hired and numerous special studies of alternatives were developed. The initiative also prevented Tucson Water from treating any contaminated groundwater to remove contaminants prior to delivering it to customers. Since about a tenth of Tucson's groundwater supply came from a Superfund Remediation project, this was a double blow to Tucson's water supply picture.

The damage that had been done by the initial engineering mistakes was now threatening Tucson's long-term water supply, and its designation of a 100-Year "Assured Water Supply" from the State's Department of Water Resources. Loss of this designation would mean long-term economic damage, primarily due to investors' concerns about the future of the community. Without an Assured Water Supply (AWS), no new subdivisions could be served by Tucson Water unless the individual developers were able to meet the AWS criteria themselves.

Collateral damage also occurred in the Water Utility and the City Manager's office, with two water directors, several high-level staff and two city managers losing their positions as water-related politics took its toll.

Two subsequent initiatives were considered by the Tucson voters, including one that was even more restrictive than the original. The public was now involved in making highly technical decisions about the future of the water utility. In 1998 there was a failed attempt at substituting a less restrictive initiative. In 1999 the business community, in association with the scientific community, was successful in defeating an initiative that expanded the impact of the original Act, particularly with regard to delivery of treated groundwater.

A key to the success of the 1999 initiative was developing a consensus in the scientific community about the advisability and costs of the few options that were available under the 1995 Act. Because some of the options ranged in the hundreds of millions of dollars, this was a significant consideration for Tucson voters.

In retrospect, it appeared that the majority of the brown water problems were caused by improper and too-frequent adjustments in pH, and use of the wrong corrosion inhibitor. Additional contributing causes were changes in pressure and direction of flow in the system, changes in the disinfection regime, temperature, dissolved oxygen and so forth. The condition of the pipelines prior to CAP introduction was also poor; many were thoroughly corroded and held together primarily by calcium deposits. The more aggressive CAP water dissolved the calcium, allowing older pipes to fail. The simplicity of this explanation belies the significant technical difficulties experienced in trying to stabilize the system when it was under stress.

Another reason why it was important for the scientists to sort through the facts, identify the areas of consensus and agree to speak with one voice was that the facts were misrepresented, in some cases by both sides. The media focused heavily on the apparent lack of consensus between the experts, resulting in total confusion among the voters. For many of the engineers, academic researchers and hydrologists who volunteered their time to try to help the community resolve this issue, this was their first foray into a world controlled almost entirely by perception. They all learned from this experience and few were comfortable operating in such a political environment.

The scientific community was able to make a valuable contribution in this case, but doing so required individuals to recognize institutional constraints, raise their profile (and subject themselves to considerable personal risk) to speak out. The highly polarized situation also required that new mechanisms for communication be developed, and that highly technical concepts be translated into understandable information for the public.

Since 1995, Tucson Water has substantially revised both its attitude and its CAP delivery plans [*Pearthree and Davis*, 2000]. A very large recharge and recovery project (the Clearwater Renewable Resource Facility) has been initiated near the canal, just upstream from the water treatment plant. CAP water is now being recharged and blended with groundwater prior to delivery. The pilot phase of this project began in 2001. The Clearwater Facility is anticipated to have a capacity of 60,000 acre-feet by 2003. It is not yet clear how the remaining 80,000 acre-feet of Tucson's allocation of CAP water will be utilized, though it is expected that a substantial portion will be stored in other recharge locations throughout the Tucson basin.

Tucson Water has completely overhauled its approach to handling CAP utilization issues. Public comments and complaints now are tracked and responded to immediately. Customers can log into a Web site (www.ci.tucson.az.us/water/water_quality/water_quality.htm) that gives them access to water quality data at a sampling point near their home on a real time basis. In order to ensure that the CAP blended water program would meet with maximum public approval, multiple panels of public and professional taste-testers sampled alternative blends and disinfection options. Hundreds of thousands of bottles of the selected water were distributed to the public so that they could sample the new product. Every step in the design of the new storage and recovery system has been reviewed and re-reviewed with regulators and hydrologists. A hundred miles of delivery mains have been replaced or lined. Public tours of Tucson Water facilities occur frequently, and Tucson Water regularly communicates with its customers, through brochures, newsletters and paid TV advertisements.

In short, no expense has been spared to ensure that trust re-building with the public continues.

This extended discussion of the CAP experience in Tucson illustrates the following key points. First, failure in the water management context is a real possibility, despite the perception that the manager is using the latest information that science and technology have to offer. Although most of the decisions made by Tucson Water were not evaluated at the time they were made by a panel of objective scientific observers, Tucson was working in concert with several teams of scientists and consultants throughout the design and implementation of the CAP transition. Despite this input, or perhaps because they didn't ask the right questions, Tucson Water made some serious technical mistakes. Scientists and engineers need to realize how significant the ramifications are of making the wrong decision (or, in this case, doing the right thing the wrong way). There are reasons why water managers are hesitant to make major changes in established operations. Second, it is important for scientists to understand the context in which their information will be used. As is discussed in the body of this paper, often the legal and institutional context, rather than the technical considerations, define the options available.

As has been noted in other cases where disruptive events affect the policy agenda, the media played a central role in this case. The media are driven by the need for high ratings to provide gripping stories to their customers. It is much easier for them to portray controversy through a series of one-liner quotes than it is for them to provide the substantial background information needed to understand technical material. The focus on controversy further polarized the politics in this case, making it harder for the public to get the full picture of what was going on. However, by the time the third initiative was being debated, some of the major TV and radio stations understood that part of the problem was that the facts were much harder to understand than the fiction. Several key media players supported public forums and special programs before the vote to help the public get better access to in-depth information.

The recent successes would not have been possible without substantial investments like the Ambassador Neighborhood Program, which proved through trucking recharged and recovered CAP water to neighborhoods that Tucson Water was technically capable of providing high quality, blended CAP water to customers' homes. This type of customer interaction was more intense (and risky for the water company) than standard pilot programs, and is an example of the relationship-building that is needed for applications of science in decision-making.

The Tucson example also illustrates the role of cost in water management decisions. The argument began over ways to utilize CAP water, but ended as an argument about growth and the cost of water. The Tucson community has consistently objected to water rate increases, despite the fact that they live in a desert and care deeply about the quality and quantity of the groundwater supplies in the area. Once the costs of alternative treatment methods for CAP were articulated, there was a change in the tenor of the debate.

The issue of who pays vs. who benefits has a great deal to do with the politics of water. Failure to identify the "winners" and the "losers" in evaluating particular options is a common error when focusing primarily on "facts." Information alone does not necessarily lead to better decisions.

A final major point illustrated by the Tucson case is the degree to which the public cares about the quality of their water supplies. Although the sensitivity of the Tucson community to water quality issues may be unusual, there is a national trend towards a desire for zero risk associated with drinking water quality. This is evidenced by the increasingly stringent drinking water standards and the public outcry when proposed standards are relaxed, as has been the case with arsenic regulations. Possibly more than other issues, the water quality issue is driven by perception. Because "clean" water is viewed as a fundamental right in this country, and science does not provide clear answers, people respond emotionally to information about health effects of water quality changes.

CONSTRAINTS FOR POLICY AND MANAGEMENT DUE TO THE LIMITATIONS OF THE SCIENTIFIC APPROACH

The scientific community has some problems with ensuring that its products are "usable." First, as previously described, the real world of water management is far more "messy" than any scientist would consider ideal. It is hard to accurately pin down the regulatory and social context within which scientific information is to be used. Second, scientists may have communication problems translating their findings into useful products. In some cases, scientists have not evaluated the need for their products, or what forms of product would be most useful to their intended audience. The language used by scientists may not be understood by water managers unless the managers have been trained in a similar field. Language barriers are found even between scientists in similar disciplines, (acronyms alone may be a significant stumbling block) so it is not surprising that many scientists have difficulty framing their arguments appropriately when working directly with users. Conversely, the language of bureaucrats and water managers may be nearly impenetrable by scientists. Third, to a large degree, the research programs of scientists are influenced by funding entities that may not be concerned about the ability of water managers to actually use particular products. Fourth, scientists are usually trained in a particular discipline, and may have had no experience in multi- or inter-disciplinary work. The nature of water management is extremely diverse, and requires interaction between a broad spectrum of disciplines, including atmospheric science, hydrology, economics, engineering, political science, land use planning, natural resource management, geography, etc.

In addition, water managers are not operating in a research mode, but must make decisions on a real time basis, so intuitive understanding of these inter-relationships is required. Scientists must have an appreciation of this complex web of operational considerations in order to produce useful products for decision-makers.

Further, many public values such as environmental quality, aesthetics, and the needs of future generations are not easily quantified in the context of a standard cost/benefit analysis [see for example, *NRC*, 1997]. If traditional cost/benefit techniques are used, results should be reviewed with an eye towards integration of public values into the equation.

There are substantial investments in the status quo, because people have attempted to optimize their position under the current institutional arrangement. Thus, though common sense and science would dictate that groundwater and surface water should be regulated by a common system, in Arizona, for example, current right holders have blocked any attempt to revise the current approach.

Politics at the federal, state and local levels impact fiscal policy, in turn affecting the ability to develop new infrastructure. Politics within the federal government also affect funding for science and the degree to which "pure" science vs. "applied" or applications science will be pursued. An example of politics at work can be found in the flood damage and federal disaster relief statistics. Downton and Pielke [2001] found that although there is evidence of increasing precipitation in the United States, there is no evidence that this is the primary cause of the increase in disaster declarations. They also found that presidents are more likely to issue disaster declarations during reelection campaigns.

"USABLE SCIENCE": USABLE BY WHOM?

Ensuring that scientific products are both "usable" and "useful" requires an understanding of who the "client" is—who will use the information that is produced? In some cases, the users of scientific products are academic or government researchers themselves. Generally, these users speak the same language as those who produce the information, and they are likely to communicate their needs more easily. Other users, such as the National Weather Service, might be viewed as translators, and regulatory agencies such as EPA, NMFS and FWS, as well as state agencies that interpret data for others. These translators serve as intermediaries between those who generate the science and those who make decisions on the basis of the information. A third group of users is the mission-oriented science agencies such as NASA and NOAA, regulatory agencies that use scientific information as the basis of their regulations, and state level mission agencies, such as water management agencies. Examples include the EPA, the NMFS, and the FWS. Finally, there are some scientific products that are accessed directly by the public, both private sector businesses and the general public. Examples are farmers who directly access weather data, or water managers who receive data directly that they feed into their own reservoir operating models. These users are most likely to use products only if they are tailored specifically to their needs, and some of these users already depend on the private sector for translation products.

Table 2. Differences in Perspective on the Use of Climate Information Between Scientists and Water Managers

Factor	Scientist's Perspective	Water Manager's Perspective
Identifying a critical issue	•Based on a broad understanding of the nature of water management	•Based on experience of particular system
Time frame	•Variable	•Immediate (operations) •Long-term (infrastructure)
Spatial resolution	•Defined by data availability, funding	•Defined by institutional boundaries, authorities
Goals	•Prediction •Explanation •Understanding of natural system	•Optimization of multiple conditions and minimization of risk
Basis for Decisions	•Generalizing multiple facts and observations •Use of scientific procedures, methods •Availability of research funding •Disciplinary perspective	•Tradition •Procedure •Professional judgment •Training •Economics •Politics •Job risks
Expectation	•Understanding •Prediction •Ongoing improvement (project never actually complete) •Statistical significance of results •Innovations in methods/theory	•Accuracy of information •Appropriate methodology •Precision •Save money, time •Protect the public •Protect their job, agenda or institution
Product Characteristics	•Complex •Scientifically defensible	•As simple as possible without losing accuracy •Importance of context
Frame	•Physical (atmospheric, hydrologic, etc.) conditions as drivers •Dependent on scientific discipline	•Safety, well being •Profit •Consistency with institutional culture, policy, etc.
Nature of Use	•Conceptual	•Applied

WATER MANAGER KNOWLEDGE AND INFORMATION NEEDS

This section discusses key types of scientific information that would be useful to water managers, including help in asking and answering the right questions, and dealing with information overload. Ability for water managers to test information for applicability in their own system is also important. A highly simplified summary of the differing perspectives employed by researchers and water managers on critical factors such as time frame and product characteristics is given in Table 2.

The first item in the list of water manager information needs is help in answering the right questions. In Limited by Design, Crowe and Bozeman [1998] discuss "strategy-relevant information." In responding to any decision-maker's needs, understanding and defining the problem properly are key starting points. Dissecting the components of the problem, understanding which aspects are affected by which players, and focusing on the areas that are tractable are important features of successful management in any field. These steps need to be taken before there is a discussion of the information needed to address the particular issues that have been selected.

It is common for managers to be overwhelmed with information, and to be short on time and resources. Information overload can result in difficulty separating useful from irrelevant material. As we have discussed, it is also common for water managers to use past experience to calibrate their expectations for the future. Water managers may need some assistance in framing questions in the context of potential alternative futures, within the predictive capacity of science. At the very least, scientists can help expand the surface water managers' world view by describing risk related to floods and droughts in terms of the longer historical record, and fluctuations in that record, rather than depending on the personal experience of today's water managers. Data from tree rings, ice cores, sediment and coral reefs provide new perspective on climatic conditions prior to the 20th century [*Powell Consortium*, 1995].

Another information need is the ability for water managers to test information they receive for relevance to their situation and experience. This includes the ability to provide feedback to researchers, request refinement of data or products, and allow for long-term interaction between scientists and practitioners. This long-term relationship is necessary for trust-building.

In some circumstances, the communication between scientists and decision-makers would be enhanced by the expansion of the types of professionals that specialize in translating scientific information for use by specific categories of decision-makers. There are already a number of engineers, analysts and private sector consultants who provide climate-related information for agricultural applications, power generation, etc,, but here is clearly a niche for a broader spectrum of integrators [*Kirby*, 2000; *Hartmann*, 2001] These could be consultants, scientists who are good communicators, or water managers who are familiar with scientific processes. A 1995 National Research Council Report suggested "creating programs of training for "science translators:...[which would] include exposure to the natural and social sciences, policy development and implementation, and conflict management and communication skills."

Surface water managers, in particular, need information in a timely manner. Excellent data that show up after the window for decision-making has closed are virtually useless. Scientists must understand the roles of institutions and sequence and timing of decisions (including how this calendar fits into longer term planning) in order to understand both the type and the nature of information needed at associated entry points.

Legal, institutional and data constraints as well as system and operational limitations can reduce the flexibility of water systems to respond to climatic variations and system extremes. A recent study by the consortium of western water resources institutes (the Powell Consortium) has shown that while the Lower Colorado River Basin is indeed drier than the Upper Basin, it is the Upper Basin that is vulnerable to severe, long-term drought. This is because of the legal requirement made early in the century to meet the Lower Basin requirements prior to Upper Basin needs.

Water managers also need information at the right scale, which is generally at the watershed level or smaller. Because most of the recently developed predictive capability related to climate is at the global scale, downscaling to the local levels is a key need of water managers. Significant progress has been made in downscaling from global models to watershed scale hydrologic models by researchers in the Pacific Northwest, California and the Southwest [see *US Global Change Research Program*, 2001]. However, substantial work is still needed to increase predictive capability (and appropriate applications/joint learning opportunities) at the regional scale, especially where there is substantial topographic variability.

Major trade-offs lie in the degree of accuracy versus the degree of precision (local scale information) that can be provided by climate models. Regional models can produce very precise but inaccurate numbers for small areas. It is tempting to produce such information since a consistent result across most studies of information use is that people want information pertinent to their locale (farm, stream, etc.). At the level of small watersheds it becomes extremely important not to oversell the precision of forecasts at the expense of being clear about their accuracy. Thus scaling up from local data is as important as scaling down from globally forced regional models.

Information related to the ability of existing infrastructure to adapt to changing conditions has been identified as a key information need. Most water supply and flood control features have excess capacity to deal with floods and droughts with low probability, such as the "100 year flood" or the "drought of record." However, as variability increases, it is likely that the 100-year flood (as currently defined) may become a more common occurrence and that droughts longer than those experienced in recent history will occur. There is a limit to the conditions under which existing infrastructure can continue to function, especially since much of the infrastructure in the US is aging and in need of repair. Water managers need assessments of the vulnerability of their systems, and of the design criteria currently used for flood control and water delivery. These vulnerability assessments need to be made in the context of an explicit evaluation of the uncertainties and risks of the scenarios that are identified. A common alternative—concluding that "what we can't measure we don't need to understand"—is a dangerous approach.

In the National Assessment of the Potential Consequences of Climate Variability and Change [*US Global Change Research Program*, 2001], the greatest vulnerability identified was in natural ecosystems, because there is little that can be done to help ecosystems adapt to rapid change. Very little is known about thresholds and indicators of stress in natural systems. It is clear that human activities have interfered substantially with aquatic and coastal ecosystems, but water managers could use more information about the break points and vulnerabilities of natural systems within their own water management regions.

Water managers also need more information about the likelihood of changes in storm frequency and intensity. At this time, there is no conclusive information about whether "storminess" and hurricane landfalls will increase with changes in the climate. From a risk management perspective, this information is crucial.

COMMUNICATION NEEDS FOR WATER MANAGERS

Information flows between scientists and water managers (and vice-versa) need to be carefully designed in order to be effective. As noted by Pulwarty and Redmond [1997], Miles *et al.* [2000], Rayner *et al.* [2000], "Timing and form of forecasts, and access to expertise to help implement the forecasts in decision-making processes may be more important to individual users than improved reliability." It is clear that new information pathways are needed so that decision-makers can access the "highlights" of scientific findings easily. Graphical products are particularly useful in providing a lot of information quickly, but may not be as successful in communicating the relevant caveats. In this case, the issue is not failure to communicate, but communicating complex ideas too simply. Scientists who are enthusiastic about their findings frequently encourage decision-makers to use information before it is sufficiently robust. Kirby [2000] recommends that scientists know more and say less to develop credibility. In addition, because there are so many types of water managers and so many different levels of sophistication, translations of scientific information are needed for particular audiences [*Rayner et al.*, 2000]. Information can be targeted to particular audiences to improve the accessibility for use in decision-making.

A more difficult problem is finding new modes of penetrating the water management culture and industry-wide standards. Policy development requires building and maintaining alliances of various kinds [*Kirby*, 2000]. In the absence of crisis conditions, institutional change takes time. In order to speed up the dissemination of innovation, there need to be internal "believers" who encourage the change, rather than solely external pressure. According to Rayner et al., "legitimating agencies" may be required "to disseminate information…to integrate forecasts into existing models or other decision-making tools." [*Rayner, et al.*, 2000]. Glantz [1996] suggests that "clusters of existing users can be identified and used as magnet groups to attract other potential users."

Water managers also need help in understanding levels of risk and uncertainty. Use of probabilistic predictions has been shown to be acceptable to water managers, particularly if there has been a demonstration of historic successes [*Hartmann et al.,* 2001]. However, multiple research papers on the use of climate information by water managers indicate that managers are unlikely to use climate information if they feel that the science is not certain enough to warrant action. In some cases, researchers have concluded that uncertainty is being used as an excuse for inaction [*AWWA*, 1997; *Rayner et al.*, 2000]. More importantly, the need to "address uncertainty" may be used as an evasion tactic when facing what might be a politically difficult problem.

Because the needs of water managers are relatively specialized, in many cases an impartial intermediary would be useful to sort through data and interpret and transmit key messages [*Pulwarty and Redmond*, 1997]. There are probably not an adequate number of people fluent in both the science and the context of water management who are currently qualified to perform this function. Clearly some consultants and private companies are evolving to fill this niche. However, only large water management agencies are likely to be able to afford the services of such integrators unless the government supports the cost. These equity issues are discussed by Hartmann [2001] who points out that a key to successful integration of science in decision-making is broad access to information. Expanded partnerships that serve this function may be needed among universities, federal, state, and private entities. Training programs for professional integrators should also be considered.

COMMUNICATION NEEDS FOR SCIENTISTS

Scientists often pursue research agendas based on intuitive conclusions about what types of information would be useful to water managers. Generally, the scientists are right about the utility of the tools they are developing—in theory. However, in application, there may be many considerations that have not been adequately accounted for. Key to this is understanding the difference between what is technically possible and what is practically possible. The intersection of these two sets of options is typically rather small.

A frustration for scientists is the apparent disinterest of water managers in using climate tools. It appears that scientists will need to "sell" their audience on the utility of the products. Many scientists do not know how to go about this, and they may feel that their credibility is compromised if they have to provide incentives for their audiences to use their products. Scientists need help in finding ways to encourage users to participate in research and product design, and incentives for ongoing communication about information needs. Before any kind of meaningful interaction can occur, the users need to be convinced that it will be worth their while to overcome the communication and cultural barriers. As noted by Wernstedt and Hersh [2002], "In short, producers of scientific information must ensure that such information accommodates the characteristics of the decision making process itself rather than expect that the process will eagerly embrace new input." This first step may in fact be the most difficult one to take for scientists.

Scientists may also need help in defining the types of water management problems that their science is best at addressing. Having real world demonstrations of collaborative problem solving would be useful, along with regional summaries of the key institutional, political and economic constraints that decision makers are facing. In addition, scientists may need help in identifying the key entry points for information, employing tools such as the "decision calendars" mentioned above. Further work in this area is clearly indicated.

TRENDS—CHANGING ROLE OF SCIENCE AND WATER MANAGEMENT, NEW CHALLENGES

As demands for water have changed and expanded, the costs of developing additional water sources through large-scale structural solutions have become both prohibitively expensive and socially unacceptable. In addition to rising treatment costs, increasing costs of water supply projects are inevitable because, (1) the best reservoir sites have already been developed, (2) as storage capacity on a stream increases, the quantity of water that can be supplied with a high degree of probability grows at a diminishing rate, and (3) there is a rising opportunity cost of storing and diverting water as society places higher values on instream flows [*Frederick and Sedjio*, 1991]. The limited opportunities for increasing freshwater supplies suggest that both demand management and better use of scientific information will play increasing roles in balancing the demand-supply relationship and determining the overall benefits derived.

As noted in this chapter, the barriers to information acceptability and use reflect combinations of technical, cognitive, financial, institutional, and cultural conditions. Changes in institutional priorities, strategies, and capacity may lead to unexpected outcomes. In addition, predicted events may be so complex that it is hard to know, unambiguously, how to respond. The gap between conceptual feasibility (research and planning) and practical implementation (management and operations) is immense. As argued in this paper, avenues for integration between these two frames may lie in (1) collaborative explorations of information communication and use between scientists and water managers; (2) expanding the range of professional integrators focused on the interface between science and water management; (3) developing and documenting cooperative demonstration pilot projects between scientists and water managers; (4) encouraging academic institutions to provide support for interdisciplinary, applied research including the social context of decision-making and the incorporation of knowledge in practice; and (5) developing the ability of practitioners themselves to manipulate data and to reconcile scientific claims with their own knowledge.

While there has been increasing focus on the processes by which knowledge has been produced, less time has been spent working with managers as partners to critically assess knowledge claims made in terms of their reliability and relevance. As is hopefully evident from this review, moving beyond the commonly held assumptions of information dissemination as a one-way (or even two-way) linear communication from researchers to practitioners requires a mix of partnerships and problem-solving approaches. Efforts at communicating scientific information, such as forecasts, should be viewed as processes involving ongoing evaluation of the physical conditions in the context of other decisions and information that potential users must consider throughout the year. It will thus be difficult for scientists, by themselves, to produce usable information even after the needs of stakeholders are identified.

The goal of collaborative research efforts should be to better provide scientists and managers with the capacity to: (1) follow the rate of development of new and relevant information; (2) expand the list of alternative actions and weigh competing options; (3) be clearer about their needs and better represent their uncertainties; (4) identify entry points for the application of scientific information in mitigation measures employed by water managers; and (5) know where and to whom to go for help. A key recommendation that follows from these suggestions is the need for agencies to provide support for collaborative demonstration studies, experiments in capacity building that are focused on sustained relationships between scientists and decision-makers, and training to expand the number and types of integrators who can help develop and facilitate those relationships.

Acknowledgments. This work was completed while both authors were at the NOAA Office of Global Programs. The authors appreciate the assistance of Kenneth Seasholes of the Arizona Department of Water Resources, Nancy Beller-Simms of the Office of Global Programs, Sharon Megdal and Barbara Morehouse of the University of Arizona and the suggestions of additional reviewers.

REFERENCES

American Water Works Association (AWWA), 1997. Climate Change and Water Resources. Committee Report of the AWWA Public Advisory Forum. *Journal of the American Water Works Association,* Vo. 89, No. 11, pp. 107-110.

Bogardi, J., and Z. Kundzewicz (Editors) 2002. Risk, reliability, uncertainty, and robustness of water resources systems. *International Hydrology Series*. Cambridge University Press, 220 pp.

Brown, G., 1997: Environmental science under siege in the U.S. Congress. *Environment*, 39, 13-30.

Callahan, B., E. Miles and D. Fluharty, 1999. Policy Implications of Climate Forecasts for Water Resources Management in the Pacific Northwest. *Policy Sciences*, Vol 32, pp 269-293.

Changnon, S. (ed) 2000. *El Nino 1997-1998: The Climate Event of the Century*. Oxford University Press. New York, 215 pp.

Crowe, M. and B. Bozeman. *Limited by Design: R&D Laboratories in the U.S. National Innovation System*. New York, Columbia University Press, 1998.

Downton, M. and R. Pielke, Jr., 2001. Discretion Without Accountability: Climate, Flood Damage and Presidential Politics. *Natural Hazards Review*, 2(4):157-166.

Frederick, K. and R. Sedio, (eds), 1991. Water resources: Increasing demand and scarce supplies. In *America's Renewable Resources: Historical Trends and Current Challenges*. Resources for the Future, Washington D.C.

Frederick, K., and P. Gleick, 1999. Water and Global Climate Change: Potential Impacts on U.S. Water Resources. Pew Center on Global Climate Change, 48 pp.

Functowicz, S.O., and J.R. Ravetz, 1990. *Uncertainty and Quality in Science for Policy*. Kluwer Press, 372 pp.

Georgakakos, A.P., 2000. Can Forecasts Accrue Benefits for Water Management?. *The Climate Report*, Vol. 1, No. 4, Fall 2000, pp 7-10.

Georgakakos, K.P., A. P. Georgakakos and N.E. Graham, 1998. Assessment of benefits of Climate Forecasts for Reservoir Management in the GCIP Region. *GEWEX News*, August, 1998, pp 5-7.

Glantz, M., H. 1996. *Currents of Change: El Niño's Impact on Climate and Society*.
Cambridge Press. 194 pp.

Gleick, P, 2000. The changing water paradigm. *Water International*, Volume 25, 127-138.

Glennon, R.J., and Maddock III, T., 1994. In Search of Subflow: Arizona's Futile Effort to Separate Groundwater from Surface Water. *Arizona Law Review*, The University of Arizona College of Law, vol. 36, no. 3, pp. 567-610.

_____, 1997, The Concept of Capture: the Hydrology and Law of Stream/Aquifer Interactions: Chapter 22, *Proceedings of the Forty-third Annual Rocky Mountain Mineral Law Institute*, Denver, Colorado.

Hartmann, H.C., T.C. Pagano, R. Bales, and S. Sorooshian. Evaluating Seasonal Climate Forecasts from User Perspectives. Unpublished manuscript, March, 2001.

Hartmann, H.C., 2001. *Stakeholder Driven Research in a Hydroclimatic Context*, Dissertation, Dept. of Hydrology and Water Resources, University of Arizona.

Ingram, H. "Reflecting on Water". University of California, Irvine website.

Kirby, K. W., 2000. *Beyond Common Knowledge: The Use of Technical Information in Policymaking*. Doctoral Dissertation, University of California, Davis.

Loucks, D.P., 1989. *Systems Analysis for Water Resource Management: Closing the Gap Between Theory and Practice*. IAHS Publication No. 180, IAHS Press, Institute of Hydrology, Wallingford, Oxfordshire, UK, 297 pp.

Miles, E.L., A.K. Snover, A.F. Hamlet, B. Callahan, and D. Fluharty, 2000. Pacific Northwest Regional Assessment: The Impacts of Climate Variability and Climate Change on the Water Resources of the Columbia River Basin. *Journal of the American*

Water Resources Association, vol. 36, no. 2, pp. 399-420.

National Research Council, 1995. *Science, Policy and the Coast – Improving Decisionmaking.* National Academy Press, Washington, D.C. 85 p.

National Research Council, 1997: *Valuing Groundwater: Economic Concepts and Approaches.* National Academy Press, Washington, DC.

National Research Council, 1998. *Hydrologic Sciences: Taking Stock and Looking Ahead.* National Academy Press, Washington, D.C.

National Research Council, 1999. *Making Climate Forecasts Matter.* National Academy Press, Washington, D.C.

Nichols, P.D., M.K. Murphy and D.S. Kenney, 2001. *Water and Growth in Colorado. A Review of Legal and Policy Issues Facing the Water Management Community.* Natural Resources Law Center, University of Colorado School of Law.

Pagano, T.C., H.C. Hartmann, and S. Sorooshian, Water Management and Climate Forecasts: A Case Study of the 1997-98 El Niño in Arizona. Presented at the 9th Annual Hydrology Research Exposition, Tucson AZ, April 7, 1999 (Invited).

Pagano, T. C., H.C. Hartmann, and S. Sorooshian, 2001. Using Climate Forecasts for Water Management: Arizona and the 1997-98 El Nino. *Journal of the American Water Resources Association*, Vol. 37, No. 5, pp. 1139-1153.

Pearthree M.S. and S.E. Davis. Tucson's Blended Water Delivery Demonstration Program. *Proceedings of the Arizona Water and Pollution Control Association Conference.* May, 2000.

Pielke, R.A. Jr. and M.H. Glanz, 1995, Serving Science and Society: Lessons from Large-scale Atmospheric Science Programs. *Bulletin of the American Meteorological Society*, 76, 2445-2458.

Powell Consortium, 1995. *Severe Sustained Drought: Managing the Colorado River System in Times of Water Shortage.*

Pulwarty, R.S., Redmond, K., 1997: Climate and Salmon Restoration in the Columbia River Basin: the Role and Usability of Seasonal Forecasts. *Bull. Amer. Meteorol. Soc.* 78(3), 381-397.

Pulwarty, R., and Melis, T., 2001: Climate Extremes and Adaptive Management on the Colorado River: Lessons from the 1997-1998 ENSO Event. *Journal of Environmental Management* 63, 307-324.

Ray, A.J. Webb, R., and J, Wiener, 2001. Analysis current uses and user needs for water management in the Interior West. American Meteorological Society Annual meeting January 2001, Albuquerque, NM.

Rayner, S., D. Lack and H. Ingram, 2000. Innovation in Water Agencies.

Rayner, S., D. Lack, H. Ingram and M. Houck, 2001. Weather Forecasts are for Wimps: Why Water Resource Managers Don't Use Climate Forecasts.

Reisner, M., 1987. *Cadillac Desert: The American West and its Disappearing Water.* Penguin, 582 pp.

Schwartz, S.S., 2000. Multi-objective Management of Potomac River Consumptive Use. *ASCE Journal for Water Resources Planning and Management*, Sep/Oct, pp 277-287.

US Global Change Research Program, National Assessment Synthesis Team. *Climate Change Impacts on the United States: The Potential Consequences of Climate Variability and Change.* Cambridge University Press, 2001.

Wernstedt, K. and R. Hersh, 2001. When ENSO Reigns, It Pours: Climate Forecasts in Flood Planning. Discussion Paper 01-56, Resources for the Future, Washington, D.C.

Yao, H. and A. Georgakakos, 2001. Assessment of Folsom Lake Response to Historical and Potential Future Climate Scenarios, 2, Reservoir Management. *Journal of Hydrology*, vol. 249, pp 176-196.

Katharine Jacobs, Arizona Department of Water Resources, 400 W. Congress, Suite 518, Tucson, Arizona 85701; Kljacobs@adwr.state.az.us.

Roger Pulwarty, NOAA/Climate Resources Diagnostics Center and University of Colorado, 325 Broadway Rd/CDC1, Boulder, Colorado 80305; rsp@cdc.noaa.gov.

Introduction To Section 3

This section describes the continuing central role of science and technology in water resource development. More specifically, science and technology have provided the basis for evolving a structural approach to water management, which has provided the United States at least with an effective infrastructure for irrigation agriculture, river transportation, water storage, hydropower, recreation facilities, and aquatic ecosystems. Past and future science issues and the organization of water science, particularly at the international level, are discussed as well. As time has passed, science also has shifted its focus from supporting water development and water use to assessing water quality and water's role in ecosystems, and most recently to the effects of external factors such as climate change on water resources. These large environmental issues have significance for local science as motivation for developing engineering practices to accommodate variability and uncertainty, and a scientific underpinning for risk management approaches for water systems. They also encourage the international water science community to cooperate in more integrated approaches to global water issues.

This section also describes the contributions of science to non-structural water management by supplying data, analyses, models, and information. The communication of scientific products using state-of-the-art technologies provides new opportunities for meeting the needs of a large number of water managers. Although water is a critical factor for issues such as health and ecology, this section has not focused on the needs of interdisciplinary science outside the context of discussions on the physical sciences.

Water: Science, Policy, and Management
Water Resources Monograph 16

10.1029/016WM12

10

Science and Water Policy for the United States

Neal F. Lane[1], Rosina M. Bierbaum[2], and Mark T. Anderson[3]

INTRODUCTION

Water is the most widely used natural resource in the world, consumed daily by humans and required by essentially all living things. The availability of fresh water has determined the geographic location of human settlement, underpinned local to national economies, and enriched the beauty of natural landscapes. Humans require only 2 to 5 liters of water per person per day for survival [*National Academy of Sciences*, 1977; *Gleick*, 1996]. In the United States, however, water use averages about 570 liters per day per person for domestic and municipal purposes, and an additional 5,000 liters per person per day for agriculture and industrial use [*Solley et al.*, 1998).

Water shortages have become more common in the United States in recent years. Even cities in the humid East, such as, Atlanta, Georgia, and Baltimore, Maryland, have faced the specter of water sources gone dry. For water managers, the task of securing sustainable supplies of water has become a daunting challenge, with layers of complexity added by drought and climate change, in addition to the constraints of State water rights, Federal environmental legislation, and regulatory control.

Despite its importance, the formulation of water policy in the United States is fragmented [*Hatfield*, 1994] and the supportive role for science ill defined. As our water resources become more fully developed, the short-

[1] Assistant to the President for Science and Techonology and Director, Office of Science and Technology Policy, (1998-2002), Executive Office of the President, Washington, D.C.
[2] Associate Director for Environment, Office of Science and Technology Policy, (1995-2001), Executive Office of the President, Washington, D.C.
[3] National Science and Technology Council and Office of Science and Technology Policy, (1999-2001), Executive Office of the President, Washington, D.C.

Water: Science, Policy, and Management
Water Resources Monograph 16

Published 2003 by the American Geophysical Union
10.1029/016WM13

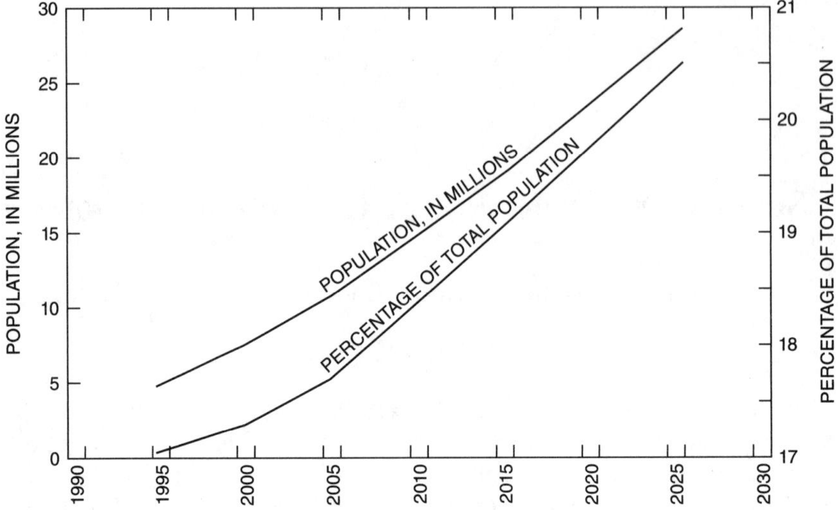

Figure 1. Projection of population growth for the Southwestern United StatesÑthe most arid region of the continent. Southwestern states include: Arizona, California, Colorado, Nevada, New Mexico, and Utah. [Source: Campbell, Paul R., 1996, Population Projection for States by Age, Sex, Race, and Hispanic Origin: 1995 to 2025, U.S. Bureau of the Census, Census, Population Division, PPL-47.]

comings of present policy become more visible. For example, we are dependent on the global water cycle to deliver water on the continents for our supply of fresh water. There is, however, no nationally coordinated program of research and monitoring to develop a better understanding of the global water cycle, much less the means to predict the impacts of climate change on the delivery of fresh water to the North American continent. Similarly, we are dependent upon ground water in many parts of the United States, especially in the West, yet there is no national program to monitor and assess these reserves or the sustainability of ground-water pumping. In fact, data on ground-water levels, potential changes, and rates of change are "not adequate for national reporting" [*The Heinz Center*, 2002]. The hydrologic sciences have defined the interconnectedness of ground water and surface water, yet these resources are still administered separately in most States. Meanwhile, the potential for water shortages, exacerbated by shortcomings in water policy, looms large on the horizon as the climate manifests evidence of change. The potential for departures from average climatic conditions to disrupt society grows.

In this paper, we stress the need for science at the Federal level to support water policy. We begin by examining a few of the overarching issues emerging over the next century that will demand our scientific attention. We point out that the science and technology of water is potentially an exportable instrument of peace, and as such it should be a valuable consideration in U.S. foreign policy. We suggest that the difficult process of adjusting the Federal science and technology portfolio to address water adequately should begin with a scientific assessment involving broad cooperation across the stakeholder community. We examine some aspects of the role science plays and will play in support of environmental decision making around water issues. Finally, we discuss a few of the anticipated, major water-monitoring and research needs for the coming century.

Water-Related Issues and Challenges of the Twenty-First Century

The supply of continental fresh water available to sustain ecosystems and human civilization on the continents is less than one percent (0.7) of the planet's total water supply—99.3 percent of the planet's water is in the oceans and polar ice. Only two one-thousandths of one percent (0.00002) is readily available in streams and lakes on land for us to drink, to bathe in, and to use to grow our crops and support industry. The fresh water in streams and lakes is the primary renewable portion of the available supply, and thus is of particular interest. The amount of fresh water on the planet is presently no greater than it was 2,000 years ago when the world's population was just 3 percent of what it is today; continued global population increase can only stress what is ultimately a finite global resource. The availability of fresh water is already a significant constraint to growth in some areas and a source of conflict between competing users [*Christen*, 2000]. As we begin this millennium, it will become increasingly difficult for the United States, and especially developing countries, to provide high-quality fresh water to their citizens at an acceptable cost. As populations increase under the current water supply and water-use patterns, the per capita availability of fresh water will inexorably decline, leading to sporadic shortages and seeds of tension that could lead to economic instability and conflict.

The American Southwest is the most rapidly growing region of the United States (Figure 1) and the most arid portion of the continent. Ground water is the source of drinking water for about half of the U.S. population, but it comprises a far greater percentage of the drinking water supply for populations in some Western States (Figure 2). Many communities in this region depend upon ground water withdrawn in unsustainable amounts in excess of aquifier recharge rates. In just the last century, submersible pumps made it possible to develop ground-water aquifers. Now, ground water is being over pumped worldwide by an estimated 160 billion cubic meters per year, mostly for agricultural use [*United Nations,* 1996; *Postel,* 1999]. This over exploitation comes with consequences beyond the obvious concern of resource depletion. Land subsidence, salt-water intrusion, and diminished water quality occur in many aquifers as the result of over pumping here in the United States and elsewhere in the world [*National Research Council*, 1991].

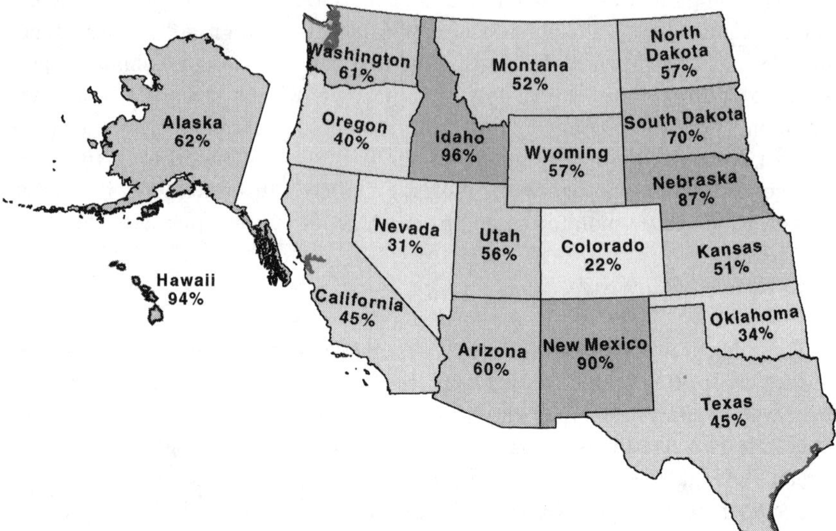

Figure 2. Percentage of population in western states dependent upon ground water for their domestic water supply [U.S. Geological Survey, 1998].

Fresh water is evaporated from the oceans and delivered to the continents by the global water cycle (GWC). Powered by the sun, the GWC delivers an estimated 46,000 cubic kilometers of water to the continents per year. Two thirds of this continental water runs off to the sea as floodwaters. In order to manage the meager fresh-water fraction in this complex setting, the fundamental processes that contribute to the temporal and spatial variability of the GWC need to be better understood. The GWC and the related forces that drive climate are responsible for delivering extreme events, such as catastrophic floods and droughts [*Hornberger et al.*, 2001]. The growing cost of natural disasters, many of them weather related, adds further economic incentives for better understanding the GWC. The scientific, technological, economic, and institutional factors underlying fresh-water management and use have grown in importance and complexity.

Ground water and surface water are a single resource; one cannot be developed without affecting the other [*Winter et al.*, 1998]. Conjunctive use, wherein ground water and surface water are managed as a conjoined resource, is a growing water-use strategy in the United States. In deference to hydrologic reality, however, many States have a bifurcated approach that treats each resource with separate laws and administrative rules [*Glennon and Maddock*, 1994; *Glennon*, 2002]. This approach arises from an era that predates the science of ground-water hydrology and will become more difficult to ignore in the next century. According to Supreme Court Justice Breyer, "there is an increasingly important need for law to reflect sound science." The judiciary has begun to look for ways to improve the quality of the science on which scientifically related judicial determinations rest [*Breyer*, 2000].

The next century will demand advances in water-treatment technology because fresh water remains a major mode of transmission of disease. Often taken for granted here in the United States, clean and safe drinking water substantially contributes to our higher standard of living and lower incidence of disease—the benefit of our past investment in the technology of water purification and wastewater treatment. Today, however, little research and development is being invested in new and improved technologies for water treatment, desalinization, or nonaqueous waste disposal [*Gleick*, 2002].

The scientific information required to administer the current water policy of the United States arises in response to both Federal and State laws and statutes. Congress defers authority to the States to allocate and administer water within their borders [*Sax et al.*, 2000]. The Federal government's role in water management has expanded over the years; initially limited to ensuring navigation along interstate rivers, it now includes instances of *de facto* appropriations of flows to fulfill mandates under the Endangered Species Act and other Federal laws. Tension has grown in recent years between the States and the Federal government regarding actions to enforce the Endangered Species Act, as well as claims of Federal reserve water rights. A less controversial role for the Federal government that has historical precedence is to conduct the monitoring and research necessary to manage water under these ever demanding circumstances.

The meaning of water availability has evolved beyond simply securing a volume of water for human use. Today, water availability encompasses additional societal goals and needs, such as the water required to sustain ecosystems, biodiversity, and sometimes even individual species. To achieve these ends, interdisciplinary science is required to monitor changes in hydrologic systems, define the physical parameters of aquatic and riparian ecosystems, and to identify the physical habitat requirements of individual threatened and endangered species. Synthesizing such interdisciplinary science and providing information in such a way that it is useful to those formulating and administering water policy is a significant and growing challenge for science.

Global Water Issues and Foreign Policy

Science and technology has become a major component of international relations for the United States. The opportunity exists for the United States to employ science and technology to reduce conflict and to promote peace through the dissemination of hydrologic expertise and of monitoring and water-treatment technologies. The availability of fresh water and related water issues frequently appears on the political agendas of countries worldwide. Several recent international conferences have highlighted water as <u>the</u> emerging, most critical environmental issue of the Twenty-first century. The European Union has adopted a fresh-water action plan that recognizes water as a strategic resource [*European Commission*, 1998].

Fresh water has become a strategic resource in which the seeds of conflict may be germinating in numerous regions of the world [*Gleick*, 1993]. At least, 250 major rivers in the world flow through more than one country, but few nations actually have treaties or other formal understandings about how water is to be shared, especially in times of shortage. Regional stability and potentially United States security and economic interests may be at stake in parts of the world where tensions run high.

The struggle for fresh water in the Middle East has been cited as a cause of the 1967 Arab-Israeli War and is a source of on-going tension in the region [*Bulloch and Darwish*, 1993; *Tessler*, 1994; *Grunfeld*, 1997]. Turkey could deprive Iraq of its water supply simply by shutting off a few crucial valves in the Anatolia Hydro Project of the Tigris and Euphrates Rivers—the source of 70 percent of Iraq's water supply [*Hoch*, 1993]. Other nations find themselves in similar strategically vulnerable positions. Egypt relies on the Nile River for 97 percent of its water. Syria depends on the Euphrates for 70 percent of its electricity production through hydro resources, and 70 to 80 percent of the total water consumed in the Middle East is used for agriculture. The Jordan River and the need for water has been a central bargaining chip in the continuing hostility between Israel and the surrounding Arab States [*Grunfeld*, 1997]. Historically, however, water has more often served as a focus for human organization and cooperation than a cause for conflict [*Postel and Wolf*, 2001; *Priscoli*, 1998].

In keeping with water's more benign and beneficial roles, U.S. foreign policy could usefully include international assistance in the science and technology of water. In this regard, opportunities for humanitarian assistance to the developing world abound. According to the most recent World Health Organization [2000] assessment, 1 billion people currently lack a clean and reliable source of drinking water, and nearly 2.5 billion people lack access to basic sanitation. An estimated 250 million cases of water-related diseases are reported each year worldwide, resulting in 5 to 10 million deaths—including one child every 8 seconds [*Nash*, 1993]. In one of the great ironic tragedies of our time, millions of people in West Bengal and Bangladesh are suffering from arsenic poisoning by drinking water from shallow tube wells installed and promoted as a safe alternative to microbe-laden surface water [*Dhar et al.*, 1997; *McArthur et al.*, 2001]. Such pain and suffering can be substantially reduced or prevented through routine water-quality analysis and simple system modifications. These examples are presented to illustrate the opportunities and ethical imperatives to employ the science and technology of water as an instrument of peace.

ENVIRONMENTAL DECISION MAKING AND SCIENTIFIC ASSESSMENTS

The belief that science will deliver a technological solution to even the most vexing problems is deeply ingrained in our society [*National Science Board*, 2000]. The determination of priority use for a natural resource, however, is more than a scientific or technological problem; such decisions must take into account various, often conflicting vested interests and values. The crucial role of scientific information in formulating water policy, and in natural resource decisionmaking generally, is widely recognized [*Holling*, 1978; *Lee*, 1993].

Environmental Decision Making

The U.S. House of Representatives, Science Subcommittee, revisited the Bush Doctrine [*Bush*, 1945] and released a report, *Unlocking Our Future: Towards a New National Science Policy* [1998]. The report reaffirmed many of the original priorities of the Bush Doctrine but proposed that a new role for science should be support for environmental decisionmaking. Today, water policy reports and management strategies often stress the need for science to aid management decisions. *Water in the West: Challenge for the Next Century* [*Western Water Policy Review Advisory Commission,* 1998], the California Bay Delta Project, the management of Glen Canyon Dam, reduction of nutrients from the Mississippi River, and the hypoxic zone in the Gulf of Mexico are just a few examples of the use of science that supports management. Despite the need, a functional model of conducting science to support resource decisionmaking is far from perfected. When legitimate uses conflict, such as hydropower generation and salmon production in the Columbia River Basin, scientific assessment should illuminate the potential consequences of different management options and reduce the uncertainty of a chosen course of action. More importantly, science needs to structure learning as experience accumulates.

Scientific Assessments

A key aspect of successfully confronting any large-scale environmental problem is the design and implementation of an integrated assessment that provides useful scientific information to decision makers along the road of policy development. An integrated assessment includes taking stock of available data and science and evaluating what the science dictates for wise policy actions now, despite remaining uncertainties. An integrated assessment also identifies what research needs to be conducted to answer key questions, both in the near and long term, with a prioritization of a research agenda [*Bierbaum*, 2002].

The National Science and Technology Council completed about 10 scientific assessments in the last 10 years, ranging from narrow science issues to large-scale integrated assessments; for example, the U.S. Global Change Research Program report entitled: *Climate Change Impacts on the United States: The Potential Consequences of Climate Variability and Change* [*United States Global Climate Change Research Program*, 2000]. Endocrine disrupters, the breadth of our environmental monitoring systems domestic energy programs, have been assessed. Specific reports include *Integrated Assessment of Hypoxia in the Gulf of Mexico* [CENR, 2000a] and *Science to Support Recovery of Pacific Northwest Salmon* [CENR, 2000b]. Each and all of these assessments figure prominently and usefully in formulating policy, setting priorities, and crafting new long-term research plans.

ESTABLISMENT OF SCIENTIFIC POLICY AND PRIORITIZATION OF RESEARCH AGENDAS

The Federal research and development (R&D) budget for environment and natural resources is stretched rather thinly across at least 10 Federal agencies. Funds tend to be allocated in response to critical and time-sensitive environmental problems, such as maintaining agricultural productivity, minimizing the cost of natural disasters, recovery of Pacific Northwest salmon, and mitigating climate change [*Bierbaum and Watson*, 1995].

Establishing Scientific Policy

The challenge of applying Federal expertise and resources to a complex set of interrelated environmental issues has increased the importance of interagency cooperation through mechanisms like the Federal Coordinating Council for Science, Engineering and Technology (FCCSET) from 1988-1993 [*Office of Science and Technology Policy*, 1993], and the National Science and Technology Council (NSTC) from 1992-2000. The Clinton administration introduced an additional priority-setting tool in the form of an annual memorandum signed by the Science Advisor and the Director of the Office of Management and Budget that provided guidance on high-priority issues that would be considered for budget augmentation. The approach stressed the importance of integrating multi-agency approaches to issues that cut across traditional boundaries.

Water science is not presently well adapted to serve societal needs in this multidimensional water-resources environment [*Naiman et al.*, 1995]. The Water Science and Technology Board (WSTB) of the National Research Council recently released a report entitled "Envisioning the Agenda for Water Resources Research in the Twenty-First Century" [2001]. The WSTB found a lack of coordination among water programs, and more significantly, that existing programs failed to anticipate the emergence of critical problems and threats. To improve coordination and responsiveness, the WSTB recommends the creation of a National Water Research Board, by Congress or the President's Office of Science and Technology Policy to establish and oversee the national research agenda [*National Research Council*, 2001].

To begin the reconsideration and redirection of water policy, a scientific assessment of the status of knowledge for the United States is needed. The assessment would examine R&D priorities as well as the role of the Federal government both domestically and internationally. This assessment could then form the foundation of a role for science to support water policy. A serious, coordinated intellectual effort needs to be invested in converting the concepts of science into effective practices for water-resources management.

Water resources are used and administered at the local level, but water cycle processes that renew and sustain fresh-water supplies operate at continental to global scales. The informed management of water resources will, therefore, require scientific information at, minimally, two scales. At the local level to regional scale, improved measurements, data collection, and earth science investigations will be needed to support water management and the resolution of conflicting uses. Above the regional scale, information is needed about problems, such as the implications of an accelerated hydrologic cycle, global climate change, and the steering mechanisms of water vapor and sea-level rise and their consequences for floods and drought.

Prioritization of Research Agendas

An exhaustive list of the priorities for research and monitoring in support of a coordinated water policy is beyond the scope of this paper. We expand, therefore, on a few of the broad and most encompassing research needs that are expected to emerge in the next century: the global water cycle, climate change, extreme hydrologic events and natural hazards, the effects of converting rivers to reservoirs, water budgets and aquifer storage change, and quantifying the physical habitat requirements of ecosystems and endangered species.

Global water cycle (GWC) and climate change. The GWC is a primary determinant of the Earth's climate, and there is some evidence that shifts in water-cycle processes are already occurring in response to climate change [*Intergovernmental Panel on Climate Change*, 2001]. The GWC encompasses many physical processes that interact to produce our climate. Even a slightly warmer global atmosphere, resulting from climate variability or global warming, is expected to accelerate the hydrologic cycle. Warmer air can hold more water vapor and, therefore, more water will be transported from the oceans onto the continents. The "atmospheric rivers" carry water vapor along somewhat predictable pathways, although we need to know more about the steering mechanisms that drive them. The current generation of general circulation models cannot determine with sufficient regional accuracy where the water vapor will be delivered. To predict changes, the

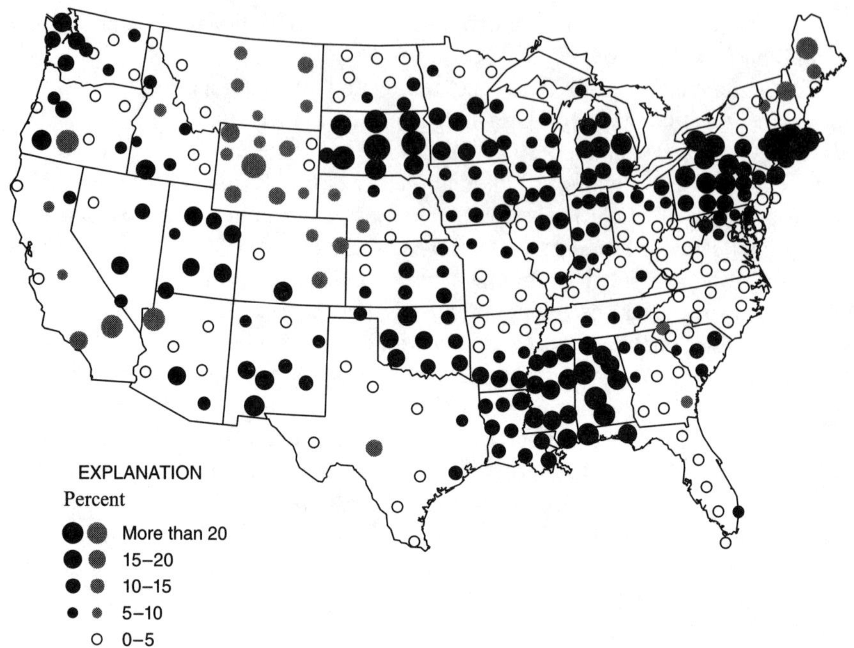

Figure 3. Precipitation trends for the conterminous United States for the period 1895-2002. The diameter of the circle centered within each climatic division reflects the magnitude of the trend. Larger circles have greater trends. Black circles reflect increasing precipitation, and gray circles reflect declining precipitation. [Source:NOAA, National Climatic Data Center].

components of evaporation, precipitation, and atmospheric circulation must be better understood. Understanding how climate change is affecting the GWC has emerged as a top research priority [*Hornberger* et al., 2001].

Extreme hydrologic events and natural hazards. There is evidence that more precipitation has fallen on the United States [*Karl and Knight*, 1998] as average temperatures have increased (Figure 3). Gleick [2000] outlined the potential consequences of climate variability and change for the United States, including changes in the timing and regional patterns of precipitation, increasing precipitation in the high latitudes, changes in flood and drought frequency, rising sea levels, and degradation of water quality.

Extreme events are not completely random in their occurrence over time but are driven by climate forces that we are just beginning to understand. This raises the problem of non-stationarity, in which the underlying assumptions of traditional engineering hydrology for flood frequency analysis are violated. In other words, our commonly accepted engineering tools for designing bridges, culverts, and flood plains are becoming less useful, at best, to obsolete at worst.

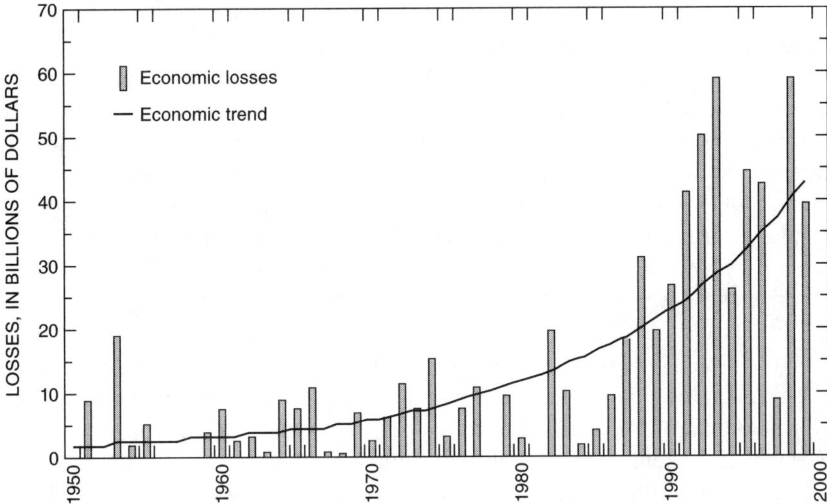

Figure 4. The costs of catastrophic weather events have exhibited a rapid upward trend in recent decades. Yearly economic losses from large events increased 10.3-fold from US$4 billion per year in the 1950s to US$40 billion per year in the 1990s (all in 1999 US$). [Modified from Figure 8.1 of Climate Change 2001:Working Group II: Impacts, Adaptation and Vulnerability, Intergovernmental Panel on Climate Change, 2001].

In terms of economic impact, the recent years have demonstrated an increasing trend in the cost of natural disasters, mostly owing to weather-related events (Figure 4). The 10 most costly weather-related disasters in our nation's history occurred in the last decade and are were mostly due to floods. The United States has spent an ever-increasing sum to recover from natural disasters. For 1998 alone, the cost of disasters to the United States was over $6 billion, with losses worldwide exceeding $90 billion, accompanied by the loss of 50,000 lives. By 2025, three–quarters of the U.S. population will live in coastal areas, making our nation even more vulnerable to disasters from hurricanes and floods.

A fundamental research question is whether climate change has produced a trend towards more extreme events and whether that trend can be observed in the historical streamflow record. Lins and Slack [1999] found significant trends of increasing minimum and median streamflow at many sites in the United States, but not high flows. McCabe and Wolock [2002] report a similar increase in minimum and median flows, but in the form of a step function that occurred in the 1970s rather than a trend. Meanwhile, Groisman and others [2001] report a significant increase in high streamflow for the contiguous United States. It is important to understand how stream flow is changing through monitoring and research.

Converting rivers into reservoirs. The nation's rivers are a crucial component of our national environment and economic infrastructure. They are critical to water availability for municipal, industrial, and agricultural supply, in addition to recreation, power generation, and transportation of goods. The history of water policy in the United States has been dominated by structures—dams, canals, dikes, and reservoirs [*Gleick*, 1998]—which were constructed with little consideration of the environmental consequences. Little opportunity exists today to increase storage along main stem rivers because few sites remain for dams and because of concern for the environmental effects. The suggestion that dams should be removed for environmental reasons evolved rapidly from an extremist idea to being given serious consideration in places such as the four lower Snake River dams (Lower Granite, Little Goose, Lower Monumental, and Ice Harbor).

In the future, decision makers will demand better scientific information when considering dam removal or remedial measures to minimize the adverse environmental impacts from the removal. Other effects of dam operation and reservoir management will need to be addressed. For instance, as fresh water is fully utilized, the evaporative loss from reservoirs, particularly in arid climates, will become less acceptable.

Water budgets and aquifer storage change. About 25 percent of the fresh water in the world is stored below the land surface. Ground water is our nation's largest reservoir of fresh water and is economically accessible in most areas of the country [*Alley et al.*, 1999]. However, broad-scale depletion of regional aquifers has occurred just in the last half of the Twentieth century due to the widespread use of turbine and submersible pumps.

A key future challenge is to prepare ground-water budgets with improved estimates of inflows, outflows and storage change. In many intensively managed areas, water managers make use of storage capacity conjunctively with surface water. In the Southwest, water is artificially recharged and banked for future use. We need to know more about our ground-water reserves and the capacity of these aquifers to store water. The effect of drought and other climatic phenomena on ground water should be better understood. At present, there is no coordinated national program to monitor changes in ground-water levels or changes in storage.

The 2000 U.S. census reveals a continued demographic shift in the nation's population to the southern and southwestern States. The most rapidly growing states in the United States are in the most arid portions of the continent—Arizona, Nevada, New Mexico, and Texas (Figure 1). An over-reliance on ground water is an unsustainable and a common situation in the Southwest. Furthermore, it will be increasingly expensive for local governments to cope with the adverse consequences of ground water overdraft, such as land subsidence and saltwater intrusion.

Sustaining ecosystems and endangered species. The U.S. Endangered Species Act (ESA) reflects society's desire to maintain or restore ecosystems and individual species. Since water is a critical habitat requirement for many endangered species, the water management alternatives in the United States are now framed by the ESA. Flow management in major Western rivers is adjusted to foster the survival of endangered species, such as the Humpback Chub in the Colorado River, Spring Chinook Salmon in the Columbia River, the Silvery Minnow in the Rio Grande River, and Winter-run Chinook Salmon in the Sacramento River. One of the great scientific challenges of sustainability will be to determine accurately the specific habitat requirements for endangered species.

CONCLUSIONS

A significant challenge of the next century will be to secure sustainable supplies of fresh water. The current approach to water policy and the science to support it, is fragmented and in need of revision. An assessment is needed to unify and bring clarity to U.S. water policy, especially to improve the scientific underpinning for environmental decisionmaking.

The problem of fresh-water availability will grow in international importance and the United States will benefit from using water science and technology as an integral part of our foreign policy. Advances in water and wastewater technology and monitoring will be exportable as demonstrations of good will. Domestically, all states face fresh-water challenges either in supply or because of degraded water quality. Basic data collection for climate parameters, streamflow, and aquifer conditions will be needed by agencies, local citizens, and watershed groups to detect changes in the environment and formulate coping alternatives. The maintenance of databases will also grow in value and importance.

Understanding the nuances of the GWC will emerge as an important research priority because these processes of the GWC deliver fresh water to the continents. Other research priorities include the scientific questions surrounding the conversion of the Nation's rivers into reservoirs and the possible consequences of the reversal of this action. Ground water is an important resource to the United States yet no nationally coordinated program of research and monitoring exists. The most rapidly expanding areas in the country are also the most arid and dependent upon ground water reserves, which is an unsustainable condition. One of the great scientific challenges in the earth sciences for the next century will be to learn how to sustain ecosystems. Furthermore, the specific habitat requirements for endangered species will need to be quantified. New ways of conducting science, more interdisciplinary and more germane to management questions, will be required in the next century.

Acknowledgments. This paper was initially prepared while we were assigned to the White House, Office of Science Technology Policy. We are grateful to the U.S.

Geological Survey and the National Oceanic and Atmospheric Administration (NOAA) for assistance with the illustrations. We thank Peter Backlund, Rick Gold, Steve Longsworth, Ken Hollett, and Carla Anderson for their helpful suggestions on the manuscript.

REFERENCES

Alley, W.M., T.E. Reilly, and O.L. Franke, *Sustainability of Ground-Water Resources*, U.S. Geological Survey Circular 1186, 79, 1999.

Bierbaum, R.M., The Role of Science in Federal Policy Development on a Regional to Global Scale: Personal Commentary, *Estuaries*, 25(4b), 878-885, 2002.

Bierbaum, R.M., and R. Watson, Recasting U. S. Federal Environmental R&D Priorities, An Editorial Comment, *Climate Change* 29, 123-130, 1995.

Breyer, S., Science in the Courtroom, in *Issues in Science and Technology*, National Academy of Sciences, 52-56, 2000.

Bulloch, J., and A. Darwish, *Water Wars: Coming Conflicts in the Middle East*. Gollancz, London, England, 224, 1993.

Bush, V., *Science, the Endless Frontier*, National Science Foundation, Washington, D.C., (reprinted 1990), 220, 1945.

Campbell, P.R., *Population Projections for States by Age, Sex, Race, and Hispanic Origin, 1995-2025*, U.S. Bureau of the Census, Population Division, PPL-47, 1996.

Christen, Kris, Global Freshwater Scarcity, *Environmental Science and Technology*, 34(15), 341a-345a, 2000.

Committee on Environment and Natural Resources (CENR), *Integrated Assessment of Hypoxia in the Northern Gulf of Mexico*, National Science and Technology Council, Washington, D.C., 2000a. Available online at: http://www.nos.noaa.gov/produc/spubs_hypox.html.

Committee on Environment and Natural Resources (CENR), *From the Edge: Science to Support Restoration of Pacific Salmon*, National Science and Technology Council, Washington, D.C., 2000b.

Dhar, R.K., B.K. Biswas, G. Samanta, B.K. Mandal, D. Chakraborti, S. Roy, A. Jafar, A. Islam, G. Ara, S. Kabir, A.W. Khan, S.K. Ahmed, and S.A. Hadi, Groundwater arsenic calamity in Bangladesh, in *Current Science*, 73, 48-59, 1997.

European Commission, *Freshwater: A Challenge for Research and Innovation: A Concerted European Response*, European Joint Research Center, EUR 18098 EN, 1998.

Gleick, P.H., Water and Conflict, in *International Security*, 18(1), 79-112, Summer 1993.

Gleick, P.H., Basic Water Requirements for Human Activities: Meeting Basic Needs, in *Water International*, 21(2), 83-92, 1996.

Gleick, P.H., *The World's Water 1998-1999: The Biennial Report on Freshwater Resources*, Island Press, Washington D.C., 200, 1998.

Gleick, P.H., *Water: The Potential Consequences of Climate Variability and Change for the Water Resources of the United States*, U.S. Global Change Research Program, Washington D.C., Sept. 2000.

Gleick, P.H., *The World's Water 2002-2003: The Biennial Report on Freshwater* Resources. Island Press, Washington, D.C., 334, 2002.

Glennon, R.J., *Water Follies: Groundwater Pumping and the Fate of America's Fresh Waters*, Island Press, Washington, D.C., 304, 2002.

Glennon, R.J., and T. Maddock, *In Search of Subflow: Arizona's Futile Effort to Separate Groundwater from Surface Water*, 36 Arizona Law Review 567, 1994.

Groisman, P.Y., R.W. Knight, and T.R. Karl, Heavy precipitation and high streamflow in the contiguous United States: Trends in the 20th century, in *Bull. Amer. Meteor. Soc.*, 82, 219-246, 2001.

Grunfeld, L., *Jordan River Use, Inventory of Conflict and Environment (ICE)*, American University, Washington, D.C., 1997. Available online at: http://www.American.edu/projects/mandala/TED/ice/JORDAN.HTM.

H. John Heinz III Center for Science, Economics and the Environment. *The State of the Nation's Ecosystems: Measuring the Lands, Waters, and Living Resources of the United States*, Cambridge University Press, Cambridge, United Kingdom, 270, 2002.

Hatfield, Mark O. Point/Counterpoint: the Long's Peak Work Group and River Basin Trusts, Northwestern School of Law, Lewis and Clark College, in *Environmental Law*, 24, 145, 1994.

Hoch, G., Will the Next War Be Over Water? in *Moment*, August, 34, 1993.

Holling, C.S., ed., *Adaptive Environmental Assessment and Management*, New York, John Wiley & Sons, 377, 1978.

Hornberger, G.M., J.D. Alber, J. Bahr, R.C. Bales, K. Beven, E. Foufoula-Georgiou, G. Katul, J.L. Kinter III, R.D. Koster, D.P. Letenmaier, D. McKnight, K. Miller, K. Mitchell, J.O. Roads, B.R. Scanlon, and E. Smith, *A Plan for a New Science Initiative on the Global Water Cycle*, U.S. Global Change Research Program, Washington, D.C., 2001

Intergovernmental Panel on Climate Change, *Climate Change 2001—Third Assessment Report,* Cambridge University Press, London, England, 1–3, 2644, 2001. Available on-line at: http://www.grida.no/climate/ipcc_tar/wg2/001.htm.

Karl, T.R., and R.W. Knight, Secular trend of precipitation amount, frequency, and intensity in the United States, in *Bulletin of the American Meteorological Society*, 79, 231-242, 1998.

Lee, Kai N., *Compass and Gyroscope: Integrating Science and Politics for the Environment*, Island Press, Washington, D.C., 243, 1993.

Lins, H.F., and J.R. Slack, Streamflow trends in the United States, in *Geophys. Res. Lett.*, 26, 227-230, 1999.

McArthur, J.M., P. Ravenscroft, S. Safiullah, and M.F. Thirlwall, Arsenic in Groundwater: Testing Pollution Mechanisms for Sedimentary Aquifers in Bangladesh, in *Water Resources Research,* 37(1), 109-117, 2001.

McCabe, G.J., and D.M Wolock, A Step Increase in Streamflow in the Conterminous United States, in *Geophys. Res. Lett.*, 29(24), 2185, 2002.

Naiman, R.J., J.J. Magnuson, D.M. McKnight, and J.A. Stanford, (eds.), *The Freshwater Imperative: A Research Agenda*, Island Press, Washington, D.C., 165, 1995.

Nash, L., Water Quality and Health, in *Water in Crisis: A Guide to the World's Fresh Water Resources*, edited by P.H. Gleick, Oxford University Press, New York, 25-39, 1993.

National Academy of Sciences, *Drinking Water and Health*, National Academy Press, Washington, D.C., 948, 1977.

National Research Council, *Mitigating Losses from Land Subsidence in the United States*, National Academy Press, Washington, D.C., 58, 1991.

National Research Council, *Envisioning the Agenda for Water Resources Research in the Twenty-First Century*, National Academy Press, June 2001. Available online at: http://books.nap.edu/books/0309075661/html/

National Science Board, Science and Engineering Indicators—2000. Arlington Virginia: National Science Foundation, (NSB-00-01), 2000. Available online at: http://www.nsf.gov/sbe/srs/seind00/start.htm and: http://www.nsf.gov/od/lpa/news/press/00/pr_indicators.htm.

Office of Science and Technology Policy, Federal Coordinating Council for Science, Engineering, and Technology *(FCCSET) Initiatives in the FY 1994 Budget*, White House, Washington D.C., April 8, 1993.

Postel, S.L., *Pillar of Sand: Can the Irrigation Miracle Last?*, Worldwatch Institute, Washington, D.C., 313, 1999.

Postel, S.L., and Wolf, A.T., Dehydrating Conflict, in *Foreign Policy*, September-October, 60-67, 2001.

Priscoli, J.D., Water and Civilization: Using History to Reframe Water Policy Debates and to Build a New Ecological Realism, in *Water Policy*, 1(6), 623-636, 1998.

Sax, J.L., B.H. Thompson, J.D. Leshy, and R.H. Abrams, *Legal Control of Water Resources*, Third Edition, American Casebook Series, West Group, St. Paul, Minnesota, 956, 2000.

Solley, W.B., R.P. Pierce, and H.A. Perlman, *Estimated Water Use in the United States in 1995*, U.S. Geological Survey, Circular 1200, 1998.

Tessler, M.A., *A History of the Israeli-Palestinian Conflict*, University Press, Bloomington, Indiana, 928, 1994.

U.S. Geological Survey, *Strategic Directions for the U.S. Geological Survey Ground-Water Resources Program—A Report to Congress*, U.S. Geological Survey, 18, 1998. Available online at: http://water.usgs.gov/ogw/gwrp/stratdir/stratdir.pdf

U.S. Global Change Research Program, National Assessment Synthesis Team, *Climate Change Impacts on the United States: The Potential Consequences of Climate Variability and Change*, Dec. 2000. Available online at: http://www.usgcrp.gov/usgcrp/nacc/default.htm.

U.S. House of Representatives, Science Subcommittee, *Unlocking Our Future: Toward a New National Science Policy*, A Report to Congress by the House Committee on Science, Sept. 24, 1998. Available online at: http://www.house.gov/science/science_policy_report.htm

United Nations, Groundwater: A Threatened Resource, UNEP/GEMS Environmental Library No. 15, UNEP, Stock No. 001810, 1996.

Western Water Policy Review Advisory Commission, *Water in the West: Challenge for the Next Century*, National Technical Information Service, Springfield, Virginia, 1998.

Winter, T.C., J.W. Harvey, O.L. Franke, and W.M. Alley, *Ground Water and Surface Water a Single Resource*, U.S. Geological Survey, Circular 1139, 79, 1998.

World Health Organization (WHO), Global Water Supply and Sanitation Assessment 2000 Report, World Health Organization and United Nations Children's Fund Report, 2000. Available online at:
http://www.who.int/water_sanitation_health/Globassessment/GlobalTOC.htm.

Neal F. Lane, Rice University, Department of Physics and Astronomy MS108, 6100 Main St., Houston, TX 77005; neal@rice.edu

Rosina M. Bierbaum, University of Michigan, School of Natural Resources and Environment, Dana Building, 430 E. University, Ann Arbor, MI 48109-1115; rbierbau@umich.edu

Mark T. Anderson, U.S. Geological Survey, 520 N. Park Ave., Ste. 221, Tucson, AZ 85719; manders@usgs.gov

11

International Hydrologic Science Programs and Global Water Issues

R.G. Lawford, J. M. Landwehr,
S. Sorooshian, and M. P. L. Whitaker

1. INTRODUCTION

This chapter provides an overview of the purpose and role of international science programs that contribute to the resolution of water resource issues. Some of these issues are inherently global or at least multinational in nature, whereas others are of regional or local scale, but nevertheless are of global concern or have long-term future relevance and implications. The orientation for international hydrology and water resource programs may be toward either social or scientific goals, but all require participation by hydrological scientists to ensure the central issues are successfully addressed. This chapter considers the factors that are causing the hydrologic sciences to shift their focus from studies at the basin scale to studies at continental and global scales to more effectively address these larger scale issues. It also examines international mechanisms for conducting and applying science to water resource issues. Appendix A provides a listing of the numerous international programs referred to in this chapter that address some aspect of water resources. The coordination of science within an international policy framework can be challenging because it relies on trust and the development and pursuit of a shared vision. Even though the theoretical approaches are primarily apolitical, and the coordination frameworks are provided by international bodies, the scientific efforts are generally funded by individual nations, each with different priorities. The United Nations (UN) structure tends to be the primary forum for much of this coordination. However, an additional international, nonpolitical and non-governmental scientific perspective is afforded by the International Council of Science Unions.

Water: Science, Policy, and Management
Water Resources Monograph 16

10.1029/016WM14

2. BACKGROUND

2.1 Trends in Water Issues Affecting Water Research Priorities

An increasing global population, coupled with a widespread desire for improved living standards, leads to increased demands for water of high quality. Consequently, the world is faced with increasing limitations to the availability of water for development and even for the maintenance of ongoing social and economic activities. The limitations on economic development caused by the lack of access to adequate useable water supplies in many developing countries are well known [See for example, *Boehmer et al.,* 2001]. In many developed countries, economic growth has proceeded rapidly over the last century, facilitated by a regional abundance of fresh water. However, serious constraints to growth are occurring, due to inadequate water supplies, even in some basins in these developed countries, particularly because of the rapid growth of urban centers and industrial activities and failure to account for year-to-year or even decade-to-decade variability in water availability arising from global climate variability. Nations without sufficient water resources are becoming aware that they may be able to overcome this limitation to their development only if they obtain increased access to water from other areas. There is a growing awareness that access to water can be secured through negotiated transboundary agreements or through a free market system if water is treated as a commodity. Science also has a critical role to play as needs for water increase and paradigms for water management shift. As *Rodda* [2001] eloquently states, "at a time when the demand for water is rising faster than at any previous moment in human history and more death and destruction are being caused by floods and droughts, it is vital that the scientific basis for action to alleviate these problems is strong and well founded."

2.2 Trends in Scientific Research

Over the past few decades, a view of hydrologic systems as a global concern, rather than just a mosaic of local or regional problems, has emerged. Following World War II, the development of international networks for the observation of global weather conditions led to an integrated perspective of climate systems and, in turn, encouraged the monitoring of water and energy cycle variables over the world's land areas. Since the 1960s, these capabilities have advanced in parallel with the development of increasingly sophisticated satellite systems for measuring these variables over land and ocean. In addition, the advent of advanced computing systems has accelerated the development of data assimilation and prediction capabilities.

Satellites, in particular, have changed our view of the water cycle. The ability to look at the Earth from space has given us a more holistic appreciation of the interactions between the elements of the Earth system. In a sense, Earth System science has developed as a new discipline that comprises an integrated understanding of the functioning of the various elements of the Earth system. For example, the role of water in the Earth system can be considered as an integrated unit. Although, historically, local and even national needs have dominated the requirements for hydrologic information, the science community is developing the theoretical framework needed for a dynamic global understanding of the water-cycle through the implementation of national and international water cycle research initiatives.

3. INTERNATIONAL POLICY ISSUES AND FRAMEWORKS

A number of issues are addressed in international discussions and activities pertaining to water resources and the contributions of the international hydrologic sciences. First, to what extent are current water resources of appropriate quality to satisfy the near-term social and economic concerns of the countries of the world? Second, are the relative quantities of available fresh water reserves changing, either because of short-term social conditions (e.g. uncontrolled increasing consumptive demand or removal from use by contamination) or long-term physical limitations (e.g. climate change with water balance impact)? Third, what technical or scientific capabilities, information and data are needed to address questions concerning adequate supplies? Fourth, how does the interaction between policy and the mechanisms of governance make use of scientific information to reach decisions regarding the management of resources on all timescales. Often this dialogue is hampered by a lack of understanding of the kind of information that scientists can provide to policy makers and of the types of information that are needed by policy makers.

Historically, societies have tended to treat water as a free public good to which everyone had a right to access. However, even early civilizations such as the Romans or Incas had water works projects, thereby introducing social controls (i.e. engineering works) into water-supply management and distribution. Facilitated by national policies and institutions, engineering capabilities during the last century rapidly increased the amount of water that was subject to control. [See, for example, *L'vovich*, 1979; *Beaumont*, 1979; *Linsley et al.*, 1982). More recently, environmental and community concerns have re-examined the reliance on storage development, and have led to a new emphasis on better information and more timely decisions as more effective ways to manage water. Accordingly, numerous international agencies now recognize the need to understand water use in a more socially and ecologically integrated way and to use that knowledge in the development of coordinated management programs.

Examples abound of the enlightened use of information to manage environmental problems within international frameworks. For example, successful efforts to curb ozone levels and, more recently, to develop a comprehensive understanding of the effect of the emissions of greenhouse gases to the atmosphere, provide some guidance on how society could address water issues. To deal with cross-boundary and international water management needs and their implications for regional economies, the global environment and world peace, it is essential to coordinate relevant issues and necessary supporting research in an international framework. The United Nations system provides a framework to develop initiatives for many of these coordinated efforts. The policy aspects of water issues are addressed in the United Nations Educational Scientific Cultural Organization (UNESCO) framework as well as through other UN initiatives. Many aspects of water hazards are addressed through the United Nations Environmental Programme (UNEP). In some cases, two international bodies coordinate the responses to a specific problem. For example, the Intergovernmental Panel on Climate Change (IPCC) is a coordinated effort between the World Meteorological Organization (WMO) and UNEP to undertake periodic assessments of the scientific, technical and socioeconomic information relevant for understanding the risks, including risks to water supply systems, arising from human-induced climate change.

The effective use of information about water resources is commonly constrained by factors such as legal agreements and requirements, conflicting issues affecting the management of the resource, and values attached by different communities to long-term environmental conditions and water security. In the case of tactical year-to-year decisions, information is often very critical during some seasons but of less value during other seasons because of the nature of the water management decisions that must be made. In order to understand the use of information in policy and management decisions, studies are needed of the role of institutional and policy factors in decision-making related to water resources. The International Human Dimensions Program (IHDP) seeks to understand the institutional and human factors that affect society's response to changes in water availability.

Water use is highly dependent on water quality, which, in turn, is closely tied to land use and land-use changes. Consequently, an effective water policy will need to include a land-use policy. For example, the drainage of wetlands and the gradual conversion of grasslands to managed agriculture have significant implications for the seasonal nature and magnitude of evaporation into the atmosphere. In addition, it can affect the quality of water percolation through the soils in areas where fertilizer is used extensively. These issues can be addressed only by a program whose primary focus is water.

The World Water Council (WWC), the UNESCO World Water Assessment Program (WWAP), and the Global Water Partnership (GWP) are all attempting to draw international and national groups together for dialogues that will facilitate the structuring of policy perspectives and the identification of issues that may transcend national boundaries. Some of these issues are short-term, such as flood warning systems requiring inputs and tactical water management decisions. Other considerations are long-term and depend on several evolving factors, including the increasing depletion of groundwater reserves, the increasing dependence of water systems on surface water of diminishing quality, multinational resource allocation issues, and the options for communities to deal with natural limitations (such as droughts) or social limits (such as those imposed by the development of certain types of industry). Long-term consequences can result from the cumulative effects of poor short-term water management decisions, and from the degree to which societies allow themselves to become more vulnerable to risks such as floods or water-borne diseases, as well as uncertain, but potentially important effects of climate change.

In order to address the complex issue of water planning at large scales, a number of socio-economic strategies need to be considered:

- more effective and integrated supply-side management;
- demand-side management, in accordance with a social evaluation of competing uses, including environmental protection;
- the development of regional and international markets among competing users;
- recycling of used water;
- the engineering of new technologies to increase local or regional supplies such as harvesting of cloud and fog water, desalinization, etc.
- conjunctive planning and use of surface and groundwaters.

The optimum mix of these strategies will depend on the regional and local patterns of supply and demand. Choices will be influenced by the values held by the national and regional governments overseeing the decision process. Both the assessment of the options and the enhancement of the efficiency of water use require inputs from the water science community to understand the vulnerabilities of sources and fluxes. Techniques are needed to allow for comparisons that show the advantage of one strategy over another. The next section reviews the progress of the hydrologic sciences as they have moved from addressing basin-scale issues to global-scale issues.

4. INTERNATIONAL HYDROLOGIC SCIENCES AND PROGRESSION FROM BASIN TO GLOBAL SCALES

Research pertaining to the hydrologic cycle typically examines either: (1) the physical processes governing the hydrologic cycle; or (2) the assessment and management of water resources, including hydrologic applications. This distinction is helpful in understanding the internationalization of research and development related to the hydrologic cycle as well as water resources management. Internationalization of research related to water resources management and hydrologic applications has achieved only limited success. The primary reason for this slower development is that many of the water resources problems of acute interest to society are experienced primarily at local, regional, and/or national scales. Depending upon the nature of problems in water resources (hazard mitigation, water supply, etc.), the issues often are viewed as sensitive and central to sociopolitical and legal disputes. For example, rivers that flow through two or more countries are sometimes at the center of disputes and controversy.

For reasons of sensitivity described above, the hydrologic information required for a thorough analysis at a basin scale may not be easily attainable. In this case, a few programs, such as the Flow Regimes from International Experimental and Network Data (FRIEND) program of UNESCO have shown some degree of success. FRIEND has provided a forum in which to share the necessary tools (such as models and new instrumentation technologies) and knowledge (how to use them) required to obtain a more complete understanding of regional conditions. Accordingly, the specific hydrologic application becomes the responsibility of a given country, and this approach reduces the likelihood of reluctance on the part of some nations to participate in international activities.

The area of physical hydrologic research has progressed more rapidly where international collaboration has increased. There are three critical functions in physical hydrology research: *observations*, *process studies,* and *system modeling*. Relatively speaking, a strong degree of collaboration and sharing of information at the international level occurs in the areas of process studies and modeling. The same cannot be said to date about the sharing of in situ measurements or other observations regarding specific water systems.

Historically, an understanding of the water cycle has been derived from observations at watershed or catchment scales; and the research was overseen by engineers. This research has been particularly important to individual communities and specific locales, because it pertains to preventing or mitigating damage from hydrologic hazards such as flooding and soil erosion, and ensuring adequate water supplies in times of drought. More recently, with the advent of satellite technology and the ability to observe large-scale hydrologic phenomena, hydrologic research has become internationalized, with the development of multinational hydrologic research programs interfacing with large-scale climate modeling projects. While satellite observations have been increasingly prolific, it has been a challenge to accumulate adequate corresponding ground-based measurements with which to calibrate these space-based data. In particular, many countries have very sparse, if any, hydrometeorological measurements, and other countries cannot or will not release their data because of national policy restrictions. This is a significant barrier to the international science community in many areas, including climate modelers [e.g. *International Association of Hydrological Sciences'* Ad hoc *Group on Global Water Data Sets*, 2001].

Various international activities, such as the Global Observing Systems, are addressing the information needs of both scientists and policy makers. A critical goal of international science is to construct a global data set of hydrometeorological information combining continental-scale and regional-scale process studies with improved and more expansive observational data sets available from satellites and in situ networks. The development and coordination of such a data set is one of the goals of the World Climate Research Programme's Global Energy and Water Cycle Experiment (WCRP-GEWEX) as well as several other programs discussed below.

In order to gain a better understanding of the global hydrologic cycle, comprehensive continental-scale and even regional-scale process studies are needed. If properly coordinated and planned, process studies can potentially provide readily transferable observational data in a standardized format, and can also improve hydrologic understanding of those aspects of the hydrologic cycle that contribute to a global perspective. In particular, hydrologists can discern how different aspects of the hydrologic cycle influence the accuracy and precision of land-surface/atmosphere models. Examples of successful process studies include: the First ISLSCP Field Experiment (FIFE; ISLSCP is the International Satellite Land Surface Climatology Project); Monsoon 90 Multidisciplinary Experiment; the Hydrology-Atmosphere Pilot Experiment in the Sahel (HAPEX-SAHEL); and the Semi-Arid Land-Surface-Atmosphere program (SALSA).

WCRP focuses on the development of climate prediction techniques that will allow prediction on seasonal to century-long timescales, and GEWEX deals with global and regional data set development to support climate model development and climate process research, as well as the use of these data in improving our understanding of the global water cycle. In particular, GEWEX coordinates a number of global data set development activities, including ISLSCP, the Global Precipitation Climatology Product, and the Global Runoff Data Center, to name a few. Through its Global Hydrometeorology Panel (GHP) and the Continental Scale Experiments (CSEs), GEWEX was the first to internationally coordinate the establishment of nine continental-scale or regional-scale river basins on five continents through six Continental-scale Experiments (see http://www.gewex.com /cseslocation.html). Over the past decade, scientists involved with the GEWEX program have contributed to intensive studies of specific hydrological regions through these CSEs. These continental-scale studies provide improved observations and coupled land-atmosphere models for data assimilation and prediction purposes. Some of these experiments, such as the GEWEX Continental-scale International Project (GCIP), have reached maturity and are being followed by extensions such as the GEWEX Americas Prediction Project (GAPP), which focuses on seasonal timescales.

International research programs support applications efforts such as UNESCO's International Hydrologic Program (IHP) and its two primary study programs, FRIEND (described above) and the Hydrology for Environment, Life and Policy (HELP) initiative. The IHP's purpose is to improve the scientific and technological basis for the development of methods and expertise for rational management of water resources including environmental protection, and to integrate developing countries into worldwide ventures of research and training. UNESCO is placing a growing emphasis on the use of knowledge and forecasts by local communities. FRIEND is primarily an international research program whose goal is to develop an understanding of the hydrologic variability among regions of the world. This is accomplished in a cooperative manner through mutual exchange of data, techniques and knowledge among FRIEND scientific partners. In contrast, HELP focuses in depth on individual regions seeking to characterize both environmental and anthropogenic/policy concerns. HELP has identified more than 25 basins for study. These basins cover a range of socioeconomic and climatic conditions. In general, the scientific developments from this research find their way into operational services through the national agencies responsible for flood prediction and resource management. A number of countries have hydrologic services that provide users with forecasts for river levels and volumetric flows.

UNESCO has also recently accepted the Dutch Government's offer to transform IHE-Delft (International Institute for Infrastructure, Hydraulic and Environmental Engineering in Delft) into the UNESCO-IHE Institute for Water Education to make the most recent knowledge on water management accessible to water managers throughout the world [*UNESCO*, 2001]. Through the efforts of the World Meteorological Organization's (WMO's) Hydrology program, new technologies are made known to these national hydrometeorological operational agencies. WMO programs that facilitate the transfer of models, technologies and data among nations also make an important contribution to these efforts.

Ultimately, many of these international research programs, which receive significant support from space agencies, contribute to improvements in the ability of satellites to provide more accurate hydrologic information at relevant spatial and temporal scales. This will be helpful in bridging the gap between the definition of hydrologic fields at global scales and what can be provided by the regionally limited and sometimes hard-to-obtain hydrologic data systems. Scientists will then be able to continue the work toward understanding processes and developing improved prediction models for the benefit of water resources management.

The International Council of Scientific Unions (ICSU) also fosters international research pertaining to hydrology on a global scale, both through the activities of its 25 scientific member unions and through special committees. Directly relevant to the promotion of hydrologic sciences is the ICSU member International Union of Geodesy and Geophysics (IUGG) and its semi-independent member association, the International Association of Hydrologic Sciences (IAHS) that serves as a communications forum for a number of international process studies. Various ICSU special committees also have objectives touching on water resource issues. The Scientific Committee on Problems of the Environment (SCOPE) fosters international research pertaining to hydrology on a global scale, albeit indirectly, through its focus on global environmental and ecological issues. It includes a number of projects with hydrologic components because of the importance of water to the functioning of ecosystems. For example, the SCOPE project, "Nitrogen Transport and Transformations" [see http://www.nceas.ucsb.edu/public/scope-n/wgl.html] has multiple objectives, including the development of a regional and ultimately a global understanding of nutrient export to coastal areas. Indeed, the globalization of hydrologic research has come about not only because of the need to encourage more nations to share hydrologic data to improve models, but also because many nations share regional ecological challenges and all nations may be affected by the global-scale consequences of anthropogenic alteration of the Earth's ecosystems. One of the major environmental issues forcing hydrologists to develop a global perspective is climate change and its implications for water resources in every part of the globe.

5. GLOBAL CLIMATE ISSUES: DEVELOPING HYDROLOGY TO ADDRESS GLOBAL EARTH SYSTEM SCIENCE NEEDS

The uncertainties surrounding climate change and climate variability issues cannot be understood without a systematic understanding of the global water cycle. For example, processes whereby water moves from the land and ocean through transpiration and evaporation, from the atmosphere to the land through precipitation, and from the land to the ocean by runoff and groundwater outflow need to be understood and quantified. A number of important water cycle issues have been raised in the context of climate change. Recently an IPCC report stated that the inability to model precipitation over as large an area as the Amazon basin "reflects our poor understanding of the interaction of convection and land surface processes" (IPCC, 2001). According to the US National Assessment report on the water sector, reliance on traditional management mechanisms may be untenable in view of the anticipated effects of climate change on water resources. [*Gleick* and *Adams*, 2000]. In addition, the consequences of potential changes in precipitation and temperature patterns for nations and communities are also important issues that are not likely to be adequately addressed in traditional water resources management strategies.

Due to the global nature of atmospheric processes, hydrology must be able to address the climate-hydrology interface at all scales up to global. Some water programs have adopted climate issues as a priority theme; accordingly, they give emphasis to the consequences of changes and variations in temperature and precipitation patterns. Furthermore, changing patterns of use must be monitored to ensure that changes in climate do not threaten the ability of the supply systems to meet their existing commitments.

An important, relatively new component of the WCRP, initiated by GEWEX, is the Coordinated Enhanced Observing Period (CEOP). As discussed in a later paragraph, the Committee on Earth Observing Satellites (CEOS) is also a partner in this initiative. CEOP is a cooperative effort being carried out in the 2001-2004-time period to compile global-scale continental data sets derived from a new generation of satellites. By collecting data on different continents, all of which represent different hydrologic ecosystems, modelers aim to test the transferability of data estimates to other, unsampled but hydrologically similar locations. The ultimate goal is to improve the predictability of the climate affecting water resources planning and management in various regions of the world.

Another program that studies hydrologic and land surface processes pertaining to but extending beyond climate-related concerns is the Biospheric Aspects of the Hydrologic Cycle (BAHC) program within the International Geosphere-Biosphere Program (IGBP). Studies conducted through this initiative provide a basis for studying the effects of land use changes on runoff, and assessing the consequences of climate change for water resources by downscaling climate model scenarios to assess the consequences for water resources in a particular basin.

The involvement of the hydrologic sciences in climate issues has introduced new requirements for, and approaches to, global observations. Global and regional data sets are needed for the development of global and regional hydrological models that are needed, in turn, to support climate studies. Furthermore, macro-scale policies and management rely on continued access to data and data products to provide baseline information on resource inventories and the factors that may influence future resource availability. Some data programs are maintained for reasons in addition to climate because they provide a basis for assessing effects of land use, urbanizations, etc. In order to address data issues, many international organizations have programs that are focused on water-related issues (see Appendix A). Some of these programs promote work that addresses the social dimension of water resources (e.g. FAO, UNESCO, IHDP, HELP); others directly pertain to anticipating questions of change (e.g. UNFCC). Still others are directed toward developing scientific capabilities pertaining to these water-related questions (WCP, WCRP, and GEWEX), while some promote the sharing of data and technology transfer on a global basis for climate and other purposes (GTN-H, GCOS, GTOS, GOOS, GEMS, GPCC, GRDC, WHYCOS, FRIEND).

Three interconnected international activities, known as the *Global Observing Systems,* were established in the 1990s specifically to address the needs of both scientists and policy makers to understand the Earth system as a whole and to provide a capability to monitor changes in the Earth's climate. These three systems are the Global Climate Observing System (GCOS), Global Terrestrial Observing System (GTOS) and Global Ocean Observing System (GOOS). The purpose of GCOS is to ensure that the comprehensive, long-term observations and information (atmospheric, oceanic, terrestrial and cryospheric) needed to improve our capability to detect, predict and assess climate change, are obtained and made available to all potential users. Like many international activities, GCOS does not itself directly make observations nor generate data products; rather, it facilitates the coordination and collaboration of national and international observational programs. GTOS was established to provide necessary data to policy makers, resource managers, and researchers in order to detect, quantify, and give early warning of changes in the capacity of terrestrial ecosystems, with the ultimate goal of supporting sustainable development and improvements in human welfare. GCOS and GTOS share a joint scientific advisory panel called TOPC, Terrestrial Observation Panel for Climate, which has produced an analysis of information and actions needed to concurrently address GCOS's and GTOS's climate-related goals. The key variables discussed in this plan are listed in Appendix B.

The Integrated Global Observing Strategy (IGOS) is an important mechanism for coordinating global observational programs. In particular, the IGOS Partners, their working groups, and their Committee on Earth Observing Satellites (CEOS) were formed to provide more effective coordination between the international science and resource management organizations and the national agencies that implement observing systems. IGOS-Partners recently approved the development of a Global Water Cycle theme that will provide a framework for water-related observations. Coordination will be directed at three priorities that have been identified: precipitation, surface hydrology and water resource applications, including irrigation. The theme also provides an umbrella for the ambitious WCRP/CEOS CEOP initiative described earlier in this chapter.

6. APPLYING INTERNATIONAL HYDROLOGIC SCIENCE TO GLOBAL POLICY ISSUES

The major policy framework for addressing international water issues is provided by the UN. The UN approach is based upon the principles embodied in its charter, namely the promotion of peace and equality among the people of the world, and the protection of societies from threats that may lead to widespread loss. Access to natural resources, including water, is correlated with the economic activity of the country. The world-wide adoption of the principle of sustainable development as outlined in what is generally known as *"The Brundtland Report" [World Commission on Environment and Development*, 1987] is widely regarded as a feasible path for achieving a system of long-term global stability. In order to clarify its approach to water and other renewable resources, the UN has structured many of its new initiatives around the theme of sustainable development. Obstacles that prevent the achievement of these goals include different cultural practices, limited access to educational capabilities, differences in technologies between developed and developing countries, and the inability to predict medium and long-term variability of water supplies. The contributions of science are greatest when the research provides techniques and information that deal with these obstacles. International science programs that have an element devoted to capacity building and technology transfer, in addition to improving scientific understanding, are most effective in the international arena.

Water is a critical factor in the war on poverty. Water for human consumption is a primary social concern because clean water is critical for life and is important for health. Contaminated water accounts for more than 3 million deaths per year worldwide [*World Health Organization* (*WHO*)—see http://w3.whosea.org/wwd/waterdises.htm]. The provision of clean water is a problem particularly in poorer countries where the sanitation systems are commonly woefully inadequate. This is a primary concern of the Water, Sanitation and Health Program of WHO. Water is also important for food production, notably livestock support and agricultural production, especially in areas where rain is inadequate to support crops without the use of irrigation. Water is also needed for industrial productivity and the production of energy. The Global Environment Monitoring System Freshwater Quality Programme (GEMS/Water), sponsored by UNEP, is an international science programme that focuses on providing information for freshwater quality issues throughout the world. It seeks an understanding of these issues through sharing monitoring and assessment information for both surface and groundwaters, and fosters capacity building in sixty-nine countries around the world.

International science supports these global objectives in a number of ways and is partitioned into programs that focus on one or more of these priority areas. In general, the goals of these programs also reflect the interests of the scientists who developed them and the agencies that fund them, leading to a rather open-ended mosaic of programs rather than a monolithic coherent structure. From a management perspective, however, it is important that the overarching policy questions be addressed through a combination of research, development and public education. It is critical that scientific programs such as UNESCO/IHP and WCRP inform exercises such as the World Water Forum to ensure that they provide substantive and reliable input to the broader sustainable development agenda. A number of case studies to demonstrate the importance of information sharing and transfer are also underway through the World Water Assessment Program (WWAP) and other studies within the CSEs are being coordinated by the GEWEX Water Resources Applications Project (WRAP). Another important coordination mechanism for the hydrological sciences is the International Association of Hydrological Sciences (IAHS), which has established a number of working groups to enable it to influence both the governmental science programs as well as the policy issues involving water quantity and quality.

Specific scientific programs have been developed to examine the global policy issues and to explore the potential links between the physical sciences and the policy agenda. The International Human Dimensions Programme (IHDP) on Environmental Change also places an emphasis on studies of human interactions with the fresh water cycle. In particular, IHDP seeks to achieve water and environmental security through a well-integrated socio-hydroecological approach to water problems. This programme supports a number of projects aimed at understanding how human interactions with the global water cycle can be addressed through integrated catchment management or superstructures (e.g. legislation, financing) to allow for certain actions to be taken. IHDP interests in security issues, vulnerability and global environmental change, and sustainability science all create opportunities for researching the links between the user needs and the abilities of science to support this development. While the physical sciences address problems such as the lack of the resource, the Human Dimension program is needed to assess water scarcities that are the result of political and social challenges to manage the resource.

In order to develop a more integrated approach to problems related to environmental science and global water issues, a Global Water System Project (GWSP) involving WCRP, IGBP, IHDP and Diversitas has been established. The GWSP addresses two major questions: "how are humans changing the global water cycle, the associated biogeochemical cycles, and the biological components of the global water system, and what are the social feedbacks arising from these changes?" [*Hoff*, 2002]. The science questions underpinning this project relate to inventorying changes in the global water system, determining mechanisms whereby human activities affect the global water system and determining the degree to which the global water system is resilient to global change. The challenges posed by these questions are addressed through a number of the activities of the Global Environmental Programmes, most notably WCRP and IGBP.

The Global Water Partnership (GWP) is an important policy framework that was formed in 1996 to promote and implement integrated water resources management through the development of world-wide networks of partnerships that could coordinate financial, technical, policy, and human resources to address the critical issues related to sustainable water management, such as domestic water services for the poor and maintaining water for ecosystems [*van den Heuvel* and *Willemse*, 2001]. Although it is not feasible to consider a world-wide water management policy, water must become an important element of major global environmental programs. Partnerships like the GWP rely on influencing the development of global approaches by contributing to awareness activities within programs such as the World Summit on Sustainable Development. The GWP has established a number of technical committees covering all continents except North America. It uses cross-sector dialogue on common water problems and develops action plans based on integrated water-resources management to resolve these problems. In part, this is done through the promotion of an Integrated Water Resource Management (IWRM) package that facilitates the coordinated development and management of water, land and related resources. Technology-oriented projects must draw upon the scientific capabilities of participating nations in the relevant international programs.

7. MECHANISMS AND CHALLENGES IN THE COORDINATION OF INTERNATIONAL SCIENCE

7.1 Mechanisms

Science efforts at the international level have tended to involve large projects that entrain a number of scientists from different regions. Because of their inherently large size and the opportunities they provide for international collaboration, these projects often attract some of the best and most visionary researchers from each nation. International project leaders tend to be facilitators who draw together international scientists to develop science plans, coordinate implementation and discuss research results. In many cases, the activities funded by individual nations are coordinated through UN special agencies, such as WMO and WCRP that establish project offices for specific initiatives. These offices oversee the development of science plans that are subsequently funded by various nations. Individual nations are encouraged to come forward with their contributions to these international efforts and to outline how they plan to contribute. Commitment conferences have been found to be an effective means for securing investments from various countries for specific programs. Generally, the largest contributions for water programs come from the United States, Europe, and Japan. There appears to be a second tier of funders, including Brazil, China, Canada and Australia, with other countries contributing as they can. The three major funders all maintain strong space programs and consider global water cycle issues as part of their observations and research related to the Earth system, although Europe and Japan provide more support for international programs dealing with water demand issues (such as UNESCO, WWAP).

International research programs, such as WCRP and its components, have progressed by coordinating, influencing and guiding national research activities to achieve the goals set out in their respective international science plans. This approach has been successful where strong leadership is provided at the international level. Frequently, this scientific leadership is provided through a scientific committee and a project office or a secretariat to coordinate the actions. In addition, nations or agencies make specific commitments to carry out certain aspects of the research program and these international offices must track their progress and ensure that their work contributes to the projects carried out by other countries. One challenge for this approach arises from the lack of suitable scientific experts who are willing to fill coordination roles at both national and international levels.

National and international programs tend to be implemented in an iterative and symbiotic way. International programs are commonly developed in coordination with national programs that provide the initial intellectual framework and, within limits, funding. National programs provide critical support for international programs, while international programs place national programs in a broader context that allows for these programs to achieve more comprehensive objectives and aid in developing a broader base of support within their own countries. National programs are generally stronger when they consider global objectives and linkages in their formulation.

In many international environmental programs, targets are frequently established at the international level, and nations and programs are asked to commit to helping achieve those targets. Earth Summit Agenda 21, crafted in Rio in 1992, represented one such overarching agreement for advancing the sustainable development agenda on a broad scale. National commitments to these goals are uneven, however, as periodic reviews of progress toward this agenda indicates. In the section of the Agenda 21 report [*UN*, 1993] dealing with the protection of the quality and supply of fresh water resources, priority activities are identified for: 1) integrated water resources development and management; 2) water resource assessment; 3) protection of water resources, water quality and aquatic ecosystems; 4) drinking water supply and sanitation; 5) water and sustainable urban development; 6) water and sustainable food production and rural development; and 7) impacts of climate change on water resources. In an ideal world, international research programs would address these issues in setting their priorities.

7.2 Challenges

Progress in understanding the global water cycle requires observations, process studies and modeling studies. As is evident in this paper, there are a number of international programs that address process and modeling studies, where needs exist for improved coordination. In addition, through the efforts of the space agencies and related international activities, the availability and use of satellite data are growing. However, in view of the need to validate these satellite data, the single most pressing international science issue involves the diminishing availability of in situ information due to the erosion of traditional observing networks, the loss of historical data sets because of a lack of adequate quality control and archiving policies, and the lack of incentive for nations to share data across international boundaries. In many countries, routine hydrological observations are made that could be used for climate research and monitoring purposes and for freshwater assessments. However, there is no international hydrological network that operates on a standardized set of procedures for data collection, dissemination, analysis and use. Even many international hydrological research programs, such as FRIEND, limit mutual exchange of data to the community of their scientific partners.

Many factors are contributing to the shrinkage of hydrologic information globally, both as a result of the physical erosion of monitoring systems and the tendency of countries to be less willing to share global data [*International Association of Hydrological Sciences'* Ad hoc *Group on Global Water Data Sets*, 2001].

Nonetheless, this is also a particularly optimistic time for the data exchange issue in the light of WMO Resolution 25 (CgXIII) on the free and unrestricted exchange of hydrological data and products adopted by the Thirteenth WMO Congress in June 1999. In the summer of 2000, a panel of experts met to discuss the establishment of a global hydrological observation network for climate. [*Cihlar et al.*, 2001]. This report recommends a mechanism or network for sharing hydrologic information that would meet policy and science requirements in the area of climate research and management goals. This network, known as the Global Terrestrial Network for Hydrology (GTN-H), would satisfy many requirements for information beyond those arising from climate questions. The GTN-H has recently been adopted as an active international project under the sponsorship of GTOS, GCOS and WMO/HWRP.

8. RECOMMENDATIONS FOR A WAY FORWARD

Over the last two decades there has been a growing awareness of the need to link the distribution of water to globally teleconnected patterns. Today, a substantial international effort is directed at developing an understanding of the global dynamics of hydrologic systems and their physical and societal links to water resource issues. Considerable progress has occurred in the last two decades toward developing science programs that help solidify our understanding of continental-scale water cycle processes and their links to climate processes, as well as the mechanisms that transfer energy and materials from the land to the ocean. However, the approach to global water problems is not seamless, as weaknesses exist in the scientific efforts on some topics and in some regions. More basin-scale science programs are needed to test the knowledge from smaller- scale studies in order to address global change issues and to resolve the world's large-scale water issues. In addition, there are obstacles and impediments to conducting international water research in support of global policy issues. Possible ways to strengthen international hydrologic endeavors are suggested below.

Currently, a dominant theme in earth resources studies is the quantification of global change at all scales in the Earth system. In addition, assessments of the consequences of climate change for water resources have encouraged global views of changing patterns of water availability. Studies promoted by the IPCC and the UNFCC have contributed to a consensus on the importance of climate change as an issue for international attention. Perhaps even more importantly, these studies have drawn attention to the need to more fully address the potential impact of climate change on water resources. However, even apart from any consideration of climate change, a strong argument can be made for establishing an independent program to facilitate research and the sharing of information about water systems within a global context, both for social and scientific reasons. While the distribution of water resources could be affected by climate change, other critical factors are known to influence the quantity and quality of water, and thus affect the total resources available. A focused effort should be directed at developing the intergovernmental cooperation needed to achieve Agenda 21 objectives for water.

The erosion of national hydrometric observational networks is a trend that is disruptive to developing a comprehensive understanding of the water cycle. The resulting gaps in these networks can only partially be filled by remote sensing data. Furthermore, in situ measurements are needed for the calibration and validation of remotely sensed data as well. An overall strategy for hydrometric measurements over land must be developed, and nations should be encouraged to make commitments to support it. The free and open access to hydrometric data should be adopted as a basic principle for this network. WMO Resolution #25, which provides a basis for more liberal national data exchange policies, should be applied to these networks.

Since World War II, the United States and other developed countries have tended to be strong players in international programs in which there is a major atmospheric component. Despite recent precautionary restrictions on public domain hydrologic knowledge as a result of post-September 11, 2002 security considerations, the USA has been and continues to be a leader in making its hydrologic information available to all users [see http://www.water.usgs.gov] at a time when even some developed countries will make their data available only for a cost, if at all. However, the lack of US involvement in some of the mainstream application programs such as UNESCO has led to a reduction of US international contributions in these areas. Despite this disconnect, the USA has some excellent applications programs at the regional and national levels (e.g. NOAA's Regional Integrated Science and Assessment projects). In addition, satellite agencies, especially NASA, have contributed to a global perspective of surface water distribution. The challenge is to forge a unified approach for the study of global hydrologic systems. A broader and more open US approach to water issues discussed in certain fora, such as UNESCO, could strengthen national programs and ensure that the views of the United States are effectively introduced to the full international community.

REFERENCES

Beaumont, P., Man's impact on river systems: a world-wide view, Area, v.10, p. 38-41, 1979.

Boehmer, K., A. Memon, and B. Mitchell, Towards Sustainable Water Management in Southeast Asia: Experiences from Indonesia and Malaysia, Water Resources Journal of the UN Economic and Social Commission for Asia and the Pacific, March 2001, p.1-30, 2001.

Cihlar, J., W. Grabs, and J. Landwehr, Establishment of a Global Hydrological Observation Network for Climate: Report of the GCOS/GTOS/HWRP Expert meeting, GCOS Report 63, GTOS Report 26, WMO/TD-No.1047, WMO, Geneva, Switzerland, 93pp,

2000. Available at http://www.wmo.ch/web/homs/geisenheim.pdf and http://www.fao.org/GTOS/gtospub/pub26.htm

GCOS/GTOS, Plan for Terrestrial Climate-Related Observations, version 2, June 1997, GCOS Report 32, GTOS Report 12, WMO/TD-No796, WMO, Geneva, Switzerland, 130 pp. Available at http://www.wmo.ch/web/gcos/pub/topv2_1.html and http://www.fao.org/GTOS/gtospub\pub12.htm, 1997.

GCOS/GTOS Terrestrial Observation Panel for Climate, fifth session, Report of Meeting at Birmingham, U.K, July 27-30, GCOS Report 59 and GTOS Report 22, FAO, Rome, Italy, 1999. Available at http://www.fao.org/GTOS/gtospub\pub22.htm and through http://www.wmo.ch/web/gcos/publications.htm

Gleick, P.H. and B. Adams, Water: the Potential Consequences of Climate Variability and Change for the Water Resources of the United States, USGCRP, 151 pp, 2000.

Global Water Partnership, Making Every Drop Count, Sustainable Development International: Strategies and Technologies for Agenda 21 Implementation, ICG Publishing Ltd., 127-128pp, 2000.

Hoff, 2002. Personal communication with Richard Lawford.

Intergovernmental Panel on Climate Change (IPCC), Climate Change 2001: The Scientific Basins. Cambridge University Press, p. 443, 2001.

International Association of Hydrological Sciences' Ad Hoc Group on Global Water Data Sets of the International Association of Hydrological Sciences (C. Vorosmarty, A. Askew, W.Grabs, R.G.Barry, C.Birkett, P.Doll, B.Goodison, A.Hall, R.Jenne, L.Kitaev, J.Landwehr, M.Keeler, G.Leavesley, J.Schaake, K.Strzepek, S.Sundarvel, K.Takeuchi, F.Webster), Global Water Data: A Newly Endangered Species, EOS Transactions of the American Geophysical Union, v. 82 no. 5, p. 54-58, 2001.

IPCC TAR, Part 2: Impacts, adaptation and vulnerability is equally important for water resources.

Linsley, R.K., M.A. Kohler, and J.L.H. Paulhus, Hydrology for Engineers, McGraw-Hill Book Company, 508 pp, 1982.

L'vovich, M.I., World Water Resources and their Future, English translation by R.L. Nace, 1979, American Geophysical Union, 415 pp, 1994.

Rodda, J.C., Water Under Pressure, Hydrological Sciences Journal 46(6), December 2001, p. 841-854, 2001.

United Nations (UN), Earth Summit Agenda 21, United Nations Publications: Sales No. E.93.I.11, United Nations, Geneva, 294 pp, 1993.

van den Heuvel, M. and E. Willemse, Achieving Water Security: Making Water Everybody's Business, (In Sustainable Development International: Strategies and Technologies for Agenda 21 Implementation), 123-125, 2001.

World Commission on Environment and Development, Our Common Future, Oxford University Press, Oxford and New York, 1987.

Richard Lawford, NOAA Office of Global Programs, 1100 Wayne Avenue, Suite 1210, Silver Spring, MD 20910; richard.lawford@noaa.gov

Jurate Maciunas Landwehr, U.S. Geological Survey, National Center-MS431, 12201 Sunrise Valley Drive, Reston, VA 20192

S. Sorooshian and M.P.L. Whitaker, University of Arizona, Department of Hydrology and Water Resources, Harshbarger Building, Room 122, Tucson, AZ 85721

APPENDIX A: TABLE OF INTERNATIONAL ACTIVITIES CONCERNED WITH WATER RESOURCES

I. ICSU Based

ICSU - International Council for Scientific Unions
http://www.icsu.org/
IGBP- International Geosphere-Biosphere Programm
http://www.igbp.kva.se
(ICSU)

BAHC - Biospheric Aspects of the Hydrological Cycle
http://www.PIK-Potsdam.DE/~bahc/
(IGBP)

IUGG - International Union of Geodosy and Geophysics
http://www.iugg.org/eoverview.html
(ICSU)

IAHS - International Association of Hydrologic Sciences
http://www.cig.ensmp.fr/~iahs/index.html
(IUGG)

IHDP - International Human Dimensions Program
http://www.uni-bonn.de/ihdp
(IGBP, WCRP & DIVERSITAS)

II. UN Based

UN - United Nations
http:// www.un.org and http://www.unsystem.org

FAO - Food and Agriculture Organization
http://www.fao.org
(UN)

IAEA - International Atomic Energy Association
http://www.iaea.org/
(UN)

GNIP and ISOHYS - Global Network for Isotopes in Precipitation and Isotope Hydrology Information System
http://isohis.iaea.org
(IAEA & WMO)

UNEP - United Nations Environment Programme
http://www.unep.org
(UN)

GEMS-Water - Global Environment Monitoring System - Water
http://www.cciw.ca/gems/
(UNEP)

IPCC - Intergovernmental Panel on Climate Change
http://www.ipcc.ch
(UNEP & WMO)

UNESCO - United Nations Environment, Science, and Cultural Organization
http://www.unesco.org/
(UN)

IHP - International Hydrologic Programme
http://www.unesco.org/water/ihp/index.shtml
(UNESCO)

FRIEND - Flow Regimes from International and Experimental Network Data
http://www.nwl.ac.uk/ih/www/research/bfriend.html
(IHP)

HELP - Hydrology for Environment, Life and Policy
http://www.nwl.ac.uk/ih/help/index.html
(IHP)

IWE - Institute for Water Education
http;//www.ihe.nl/vmp/articles/News/NEW-Unesco-GC_decisions.html
(UNESCO & IHE Delft)

IOC - Intergovernmental Oceanographic Commission
http://ioc.unesco.org/iocweb
(UNESCO)

WWAP - World Water Assessment Program
http://www.unesco.org/water/wwap/index.shtml
(UNESCO)

UNFCCC - United Nations Framework Convention on Climate Change

http://www.unfccc.de
(UN)

WHO
World Health Organization
http://www.who.int/home-page
(UN)

WSH - Water, Sanitation and Health
http://www.who.int/water_sanitation_health/index.html
(WHO)

WMO
World Meteorological Organization
http://www.wmo.ch/
(UN)

HWRP - Hydrology and Water Resources Programme
http://www.wmo.ch/web/homs/hwrphome.html
(WMO)

WHYCOS - World Hydrologic Cycle Observing Systems
http://www.wmo.ch/web/homs/whycos.html
(HWRP)

III. Organizations Sponsored by ICSU and UN

GCOS - Global Climate Observing System
http://www.wmo.ch/web/gcos/gcoshome.html
(ICSU, IOC, UNEP, WMO)

GTOS - Global Terrestrial Observing System
http://www.fao.org/GTOS
(ICSU, FAO, UNEP, UNESCO & WMO)

TOPC - Terrestrial Observation Panel for Climate
http://www.wmo.ch/web/gcos/topc.htm
(GCOS, GTOS)

GOOS
Global Oceanographic Observing System
http://ioc.unesco.org/goos
(ICSU, IOC, UNEP, WMO)

GOSIC - Global Observing Systems Information Center
http://www.gos.udel.edu
(GCOS, GTOS, GOOS)

WCP - World Climate Programme
http://www.wmo.ch/web/wcp/wcp_prog.htm
(ICSU, IOC, UNEP & WMO)

WCRP - World Climate Research Programme
http://www.wmo.ch/web/wcrp/wcrp/home.html
(WCP)

GEWEX - Global Energy and Water Cycle Experiment
http://www.gewex.com
(WCRP)

GHP - GEWEX Hydrometeorolgy Panel
http://www.usask.ca/geography/MAGS/GHP/ghp.html
(GEWEX)

GPCC - Global Precipitation Climatology Center
http://www.dwd.de/research/gpcc
(WCRP)

GRDC - Global Runoff Data Center
http://www.bafg.de/grdc.htm
(WCRP)

ISLSCP - International Satellite Land Surface
Climatology Project
http://www.gewex.com/islscp.html
(GEWEX)

FIFE - First ISLSCP Field Experiment
http://www-eodis.ornl.gov/daacpages/fife.html
(ISLSCP)

HAPEX-SAHEL-Hydrology - Atmosphere Pilot
Experiment in the Sahel
http://www.orstom.fr/hapex/index.html
(GEWEX, CESBIO, ORSTOM, CNES)
CEOP - Coordinated Enhanced Observing Period
http://www.usask.ca/geography/MAGS/GHP/ceop.issues.html

(WCRP, IGOS, CEOS)

IV. OTHER
IGOS-P - Integrated Global Observing Strategy – Partners
http://ioc.unesco.org/igospartners/igoshome.htm
(UNESCO, IOC, WMO, WCRP, IHDP, IGBP, FAO, CEOS, GOOS, GCOS, GTOS)

WWC - World Water Council
http://www.worldwatercouncil.org
(UNDP, UNESCO, World Bank, CIDA, CIHEAM, ICID, IUCN, IWA, IWRA & WSSCC)

WWF - World Water Forum
http://worldwaterforum.net
(World Water Council)

GWP
Global Water Partnership
http://www.gwpforum.org/
(Membership Organizations)

11. APPENDIX B

Nine key hydrospheric, cryospheric and biospheric variables for the assessment of climate impacts on water resources were identified in the GCOS/GTOS Plan (1997), namely:

surface water flow - discharge;
surface water storage fluxes;
ground water storage fluxes;
precipitation;
evapotranspiration,
relative humidity;
snow water equivalent;
soil moisture; and
biogeochemical transport from land to oceans.

A tenth variable, water use, was added by the Terrestrial Observation Panel for Climate (1999) because this flux is needed not only for sociological and economic studies of water resources, but also to assess the dynamics of the terrestrial water cycle of which anthropogenic impacts are now understood to be a key component. A perusal of these variables will clearly indicate that they provide information that is useful for far more than just climate related issues.

12. APPENDIX C: ACRONYMS

BAHC – Biospheric Aspects of the Hydrologic Cycle
CEOP – Coordinated Enhanced Observing Period
CEOS - Committee on Earth Observing Satellites
CSE - Continental Scale Experiment
FAO – Food and Agriculture Organization
FRIEND – Flow Regimes from International and Experimental Network Data
GCOS – Global Climate Observing System
GEMS – Global Environmental Monitoring System
GEWEX- Global Energy and Water Cycle Experiment
GHP – GEWEX Hydrometeorology Panel
GOOS – Global Oceanographic Observing System
GPCC – Global Precipitation Climatology Center
GTN-H – Global Terrestrial Network Hydrology
GRDC – Global Runoff Data Center
GTOS – Global Terrestrial Observing System
GWP - Global Water Partnership
GWSP – Global Water System Project
HELP – Hydrology for Environment, Life and Policy
HWRP - Hydrology and Water Resources Program
IAHS – International Association of Hydrologic Sciences
IGOS – Integrated Global Observing Strategy
IHDP – International Human Dimension Program
IHP – International Hydrology Program
IPCC – Intergovernmental Panel on Climate Change
IWRM – Integrated Water Resource Management
TOPC – Terrestrial Observation Panel for Climate
UN – United Nations
UNEP – United Nations Environmental Program
UNESCO – United Nations Environment, Science and Cultural Organization
UNFCC – United Nations Framework Convention on Climate Change
USGCRP – United States Global Change Research Program
WCP – World Climate Program
WCRP – World Climate Research Program
WHO – World Health Organization
WHYCOS – World Hydrology Cycle Observing System
WMO – World Meteorological Organization
WRAP – Water Resources Application Project
WWAP – World Water Assessment Program
WWC – World Water Council

12

The Role of Climate in Water Resources Planning and Management

Dennis P. Lettenmaier

INTRODUCTION

Water is essential for human life. Since the earliest days of civilization, man has attempted to deal with variations in climate through physical adaptations. For instance, the ancient Egyptians diverted water from the Nile River during the spring and summer flood, and grew crops on the land as the flood receded in winter. Among early efforts to develop engineering works to provide reliable water supplies, the Roman aqueduct systems, which brought water to the major population centers from surrounding sources, are perhaps the best known. Indirectly, most if not all, early civilizations were susceptible to the vagaries of climate through the effects of drought on agriculture, or other food sources. The response of the ancient civilizations of the Middle East was irrigation, a prime example of which is the major water delivery system of the Mesopotamians, which once supported a prosperous Baghdad from what has for over 600 years been a barren desert.

Although the availability of disease-free sources of water is now an expectation in developed countries, this situation has only pertained over the last century or so. The modern era of water resources planning and management can be argued to date to the mid-19th century, when piped water distribution and sewage removal systems were introduced to the large cities of western Europe and North America. Much of the motivation for the development of such systems was the eradication of waterborne diseases like yellow fever and typhus. For instance, in the late 1700s, yellow fever outbreaks in Philadelphia, then the capitol of the fledgling U.S., occurred nearly every summer, and most elected officials, including the President, vacated the city from spring until late fall. As a result, Philadelphia constructed the first piped water system in the U.S in the early 1800s. Other major eastern U.S. cities soon followed suit, with New York's Croton Aqueduct completed in 1842 and Boston's Cochituate Aqueduct completed in 1848.

Water: Science, Policy, and Management
Water Resources Monograph 16

10.1029/016WM15

However, inherent in the development of piped water systems was the necessity to provide reliable water supply sources, and methods for estimating water supply reliability in the 1800s were essentially nonexistent. In general, the early municipal systems were greatly oversized, so reliability in the early years was not an issue, particularly given the relative abundance of water in the semi-humid eastern U.S. Later development of water sources in the western U.S., where water was less available, was more problematic. In the West, the main issues revolved around the sustainability of agriculture. At the time of the first wave of western migration in the 1800s, little was known about the reliability of streamflow, and the nature of drought. Examples abound of failure of settlements that were fortuitously established during times of abnormally wet conditions, only to fail in subsequent years of drought (or even more normal precipitation).

In the late 1800s, irrigation systems evolved as a means of reducing the vulnerability of agriculture to drought. Early systems were mostly small dams, which were built without the benefit of formal estimates of reliability. In the early 20th century, though, the era of large water management systems began with the construction of reservoirs like Roosevelt Dam on the Salt River and Lahontan Reservoir, fed by Nevada's Carson and Truckee Rivers. The modern stream gaging network of the U.S. Geological Survey began about the same time, with the first U.S. gage installed on the Rio Grande near Embudo, New Mexico in 1889. By the late 1800s, the gage network had grown to include some locations on most major U.S. rivers. Nonetheless, as will be shown below, modern water resources planning is critically dependent on the availability of records long enough to characterize the natural variability of streamflow (a common, although arbitrary, record length that is often cited is 50 years), and early in the evolution of the network all streamflow records were short.

Modern water managers are faced with two characteristic problems. The first is to plan the constructed and managed systems that will provide water for a specified need. In most cases, the planning or design problem is a variant of what is called in operations research the capacity expansion problem—i.e., an existing system will, or is projected to be, inadequate to meet demands, and some expansion will be required. The second characteristic problem is operations. Given an existing system, how can it be operated to provide the best performance as measured by some set of metrics—such as maximum reliability for a fixed cost. Although these problems have often been approached using different methods, they are in fact linked in an obvious manner, because the required size of a water supply system depends on the manner in which it is operated.

The purpose of this paper is to track the evolution of modern water management methods, and their relationship to climate—especially new and evolving information about climate variability and change—and to indicate the potential and role of new methods for predicting climate variability and change. The importance of climate change to water planning has been highlighted in the Hydrology and Water Resources chapters of the Intergovernmental Panel on

Climate Change (IPCC) 1995 *Second Assessment Report* and 2001 *Third Assessment Report* [*Kaczmarek et al.*, 1996; and *Arnell and Liu*, 2001], as well as the Water Sector Report of the U.S. National Assessment of the Potential Consequences of Climate Variability and Change for the Nation [*Gleick*, 2000]. These reports, and many papers referenced therein, suggest the potential sensitivity of water resources to climate change, and, in the *Third Assessment Report*, the potential consequences of acceleration of the hydrological cycle. What is generally not acknowledged in the IPCC reports is that water resources planning methods as currently used in practice already embed methods that reflect the role of climate variability, if not climate change [*Lins and Stakhiv*, 1998]. Nonetheless, this paper will show that essentially all methods used in current practice assume, either directly or indirectly, that the future will resemble the past. The nature and consequences of this assumption are evaluated, and methods that might be used to incorporate dynamic information about climate in both planning and operations contexts are suggested.

BACKGROUND

Mass curve analysis was developed in the 1800s as a means of determining the required size of a reservoir to meet a fixed demand. Rippl [1883] formalized the procedure, which, as explained in most texts [e.g., *Loucks et al.*, 1981] inverts the cumulative sum of a sequence (or times series) of reservoir inflows less demand to solve for the required storage, which is related to the maximum minus the minimum of the cumulative mass curve. There is a good deal of discussion and debate in the literature about implementation of the Rippl method—the commonly cited method utilizes what is known as the Sequent Peak Algorithm, which essentially uses a repetition of the observed time series of inflows to handle the problem of shortages which occur near the end of the observed record. Klemes [1997] contends that the Rippl method as commonly explained, beginning with Maass et al. [1962] is incorrect, as the original reference is based on analysis of the mass curve of differences between inflow and draft, rather than the mass curve of inflow as discussed in most texts. However, the methodological differences may well pale in comparison with issues related to the length and quality of observations used to estimate required reservoir storage. In the late 1800s, streamflow networks in the U.S. were embryonic, and as noted above, the U.S. Geological Survey stream gaging network was not started until 1889 (although some stream gauge records predate the USGS—for instance, gauging of the Columbia River at the Dalles, OR by the railroad dates to the mid-1800s). The role of climate in water resources planning is apparent from the nature of mass curve analysis—quite clearly, the required storage determined by mass curve analysis depends on the characteristics of the sequence of inflows used. In general, the range of the series of cumulative sums depends not only on the demand, but also on the variability of the inflow sequence, as well as its persistence characteristics—i.e., the likelihood of anomalously low flows persisting from year to year or season to season. Work in the 1970s [e.g., *Wallis and Matalas*, 1972; and *Lettenmaier and Burges*, 1977) explored these relationships in detail through the use of Monte Carlo analysis. Nonetheless, notwithstanding the evolution of modern methods like those described in the following section, reservoir design is still largely based on methods like mass curve analysis applied to observed streamflow time series—the implicit assumption being that the range of variability represented in historic observations is representative of the future.

The first reservoirs were mostly built to provide water supply, and were operated solely for that purpose. Generally they were relatively small in size relative to their inflows (a commonly used measure of reservoir size is the usable storage capacity divided by the mean annual inflow). For instance, within the Columbia River basin, it is estimated that by the early 1900s, over 100 reservoirs (mostly for irrigation) had already been constructed, some 40 years prior to the completion of the first large storage reservoir in the system, Grand Coulee Dam [*NRC*, 1996]. Operating considerations for these small reservoirs were simple—inflows during high flow periods or seasons were stored, and withdrawn as needed during lower flow periods. Where reliability estimates were needed, e.g., to determine reasonable rates of withdrawals, these were determined largely by trial and error.

Nonetheless, early methods of reservoir design were based on variations of mass curve analysis, as noted above originally attributed to Rippl. It was only later, as reservoirs became larger and reservoir systems more complex (e.g., multiple reservoirs were constructed within river basins so that the effects of operations interacted), and reservoirs began to be constructed to serve multiple purposes that more careful attention began to be given to reservoir operating policies. Nonetheless, as with reservoir design, modern reservoir operations remain largely tied to a sequence of observed inflows, and quite often operations are based on a particular "critical period," usually the drought of record, which is taken as a worst case.

EVOLUTION OF MODERN WATER MANAGEMENT

Modern water management methods date to the development and widespread use of digital computers in the 1950s and 1960s. The Harvard Water Program of the early 1960s was the nucleus of much of the early activity—both in the representation of reservoir inflows through the evolution of the field of stochastic hydrology, and in the development and application of operations research methods applied to reservoir operation. Both of these fields are relevant to the interaction of climate and water management, and will be addressed separately in the next two subsections.

Stochastic Hydrology

As indicated in the previous section, the early development of methods for sizing reservoirs was based on mass curve analysis applied to observed sequences of reservoir inflows. In the 1960s, two related problems with this approach were recognized. The first was that most observed sequences of reservoir inflows were at that time quite short. In the U.S., the longest streamflow records mostly were less than 50 years in length, and the many USGS gages installed in the late 1940s offered record lengths of only 20-30 years. Furthermore, it was recognized that the nature of the variability represented by the historic observations was not necessarily indicative of conditions that might be experienced in the future. For instance, the worst drought observed in a historic record might be far less severe than could be experienced in the future.

The approaches that were developed recognized the historic record of observations as being only one realization of an underlying stochastic process. Although the underlying process could never be known, under an assumption of stationarity (essentially meaning that the probability distributions that characterize the process are unchanging in time), parameters of the probability distributions could be estimated from the record of historical observations. The simplest model that could represent the observed persistence in streamflow records (the tendency of anomalous conditions to persist), the lag one Markov model, was at the heart of the Thomas-Fiering model (see Fiering and Jackson, 1971, for a description):

$$Q_t - \mu = \rho Q_{t-1} - \mu + \varepsilon_t \tag{1}$$

where Q_t and Q_{t-1} are the annual streamflows in years t and t-1, respectively, μ is the mean annual flow, ε_t is a random error term, and ρ is the lag one correlation coefficient. Generalizations to this model make it applicable to seasonal (e.g. monthly) streamflows, as well as to multiple locations. A good deal of work in the 1960s and 1970s addressed the question of whether the persistence structure of this relatively simple model (streamflow at time t depends only on the previous (e.g., year's) streamflow, and not on earlier time steps) was consistent with observed streamflow records [e.g., *Mandelbrot and Wallis*, 1968; and 1969a]. Mandelbrot and Wallis [1969b] and Mandelbrot [1971] proposed more complex models, based on the work of Hurst [1951; 1956; and 1965], who had analyzed very long records of Nile River streamflows and found that the Thomas-Fiering type models did not capture certain observed variations in streamflow that they characterized as having long-term persistence. In a simplified sense, the concept of long-term persistence is related to the tendency for a variable, such as streamflow, to have excursions that can take it well away from the long-term mean for extended periods. For example, Figure 1 shows long realizations of time series generated from a model with a Hurst coefficient (a measurement of long-term persistence) of 0.7, as compared with a similar series with no long-term persistence. In a formal sense, long-term persistence is characterized by autocorrelation coefficients whose time integral is unbounded, which is

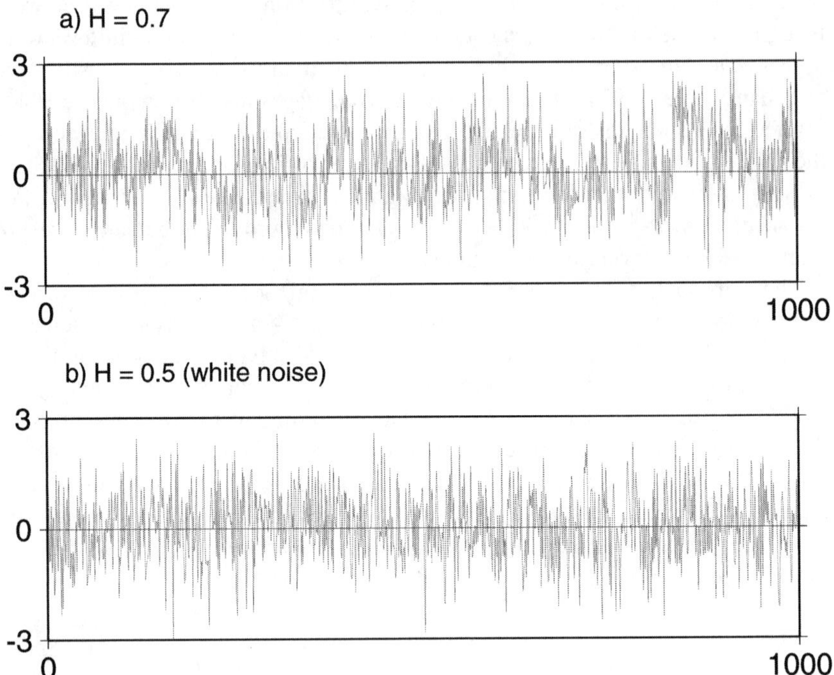

Figure 1. Simulated 1000-year time series of white noise (upper panel) and fractional Gaussian noise with Hurst coefficient H = 0.7 (lower panel). Low frequency associated with higher H values is apparent from long excursions from the mean, arguably typical of observed streamflow series. Figure adapted from Mandelbrot and Wallis [1969a].

equivalent to an unbounded spectrum at zero frequency. The practical implications are that in long-term persistent models, droughts of much greater duration and severity can occur than with short-term persistence models. A good deal of work in the 1960s and 1970s [e.g., *Wallis and Matalas*, 1972; and *Wallis and O'Connell*, 1973] examined the implications of such models when used for estimating required reservoir storage to meet a fixed demand.

A weak link in the sometimes heated debate over the appropriateness of short versus long-term persistence models for streamflow was the problem of parameter estimation. O'Connell [1974], McLeod and Hipel [1978] and Hipel and McLeod [1978] proposed that a generalization of the Markov models of Fiering and Thomas, known as autoregressive moving average (ARMA), could be used to represent long-term persistence. The advantage of ARMA models was that Box and Jenkins [1970] had developed estimation methods that could be used both for model identification and parameter estimation. O'Connell [1974] showed that ARMA models could yield persistence behavior essentially indistinguishable from that displayed by the more complex long-term persistence models proposed by Mandelbrot [1971], which were parameterized using the so-called Hurst coefficient. McLeod and Hipel [1978] and Hipel and McCleod [1978] showed, for record lengths typical of observed streamflow, that estimators of the Hurst coefficient tended to have an upward bias (toward greater inferred long-term persistence). The result was that it was practically impossible, absent very long streamflow records, to distinguish between short and long-term persistence. Although the debate was never truly resolved, after the explosion of effort in the 1960s and 1970s, work in stochastic hydrology migrated toward practical problems of simulating seasonal and annual streamflows at one or more sites with compatible statistics, using disaggregation methods [e.g., *Valencia and Schaake*, 1973; *Mejia and Rousselle*, 1976; and *Stedinger and Vogel*, 1984). Stochastic hydrology can now be considered a mature field, although it has to be said that its impact on water resource system design and management has not been as great as might have been hoped. Notwithstanding examples to the contrary where stochastic methods have been used in practice [see, e.g., *Frevert et al.*, 1989], the norm in the applied water management community remains to base design and management decisions on analyses of an historical record of streamflow observations.

In a sense, the stochastic models of long-term persistence, which challenged the notion of stationarity of hydrologic records, can be viewed as a precursor of climate change assessments. For instance, Matalas and Fiering [1977] in a National Academy of Sciences report, stated that "Recent climatic literature has pointed out that the past several decades have been a period of rather mild and stable climate, but that the future may be less so. If this is indeed true, then more severe floods and droughts may be expected in the relatively near future. Whether or not the climate is changing is subject to debate concerning the time scale over which change is defined; if it is changing, the nature of the change and how and when it would impact on hydrology are uncertain." The authors then laid out a framework for robust design of water resource systems in which they acknowledged that future climate was so uncertain that it was best for planners to evaluate the robustness of their designs by using statistical parameters characterizing reservoir inflows (e.g., mean and variance) over plausible ranges, rather than the "best estimate" from the record of historic streamflow observations.

From the standpoint of climate change, it is now clear that the critical shortcoming of stochastic streamflow synthesis methods is the assumption of statistical stationarity. This assumption is equivalent to assuming that the statistics underlying the streamflow model will be the same in the future as in the past. Quite clearly, climate change implies that this will not be the case [see e.g., *Matalas*, 1997]. However, notwithstanding numerous studies demonstrating the sensitivity of hydrology and water management to climate change, there are at present no practical methods for incorporating such information into management and design decisions. Furthermore, the applied water management community remains skeptical that climate change is a "real problem", and for the most part, climate change information is not incorporated into their decision making process.

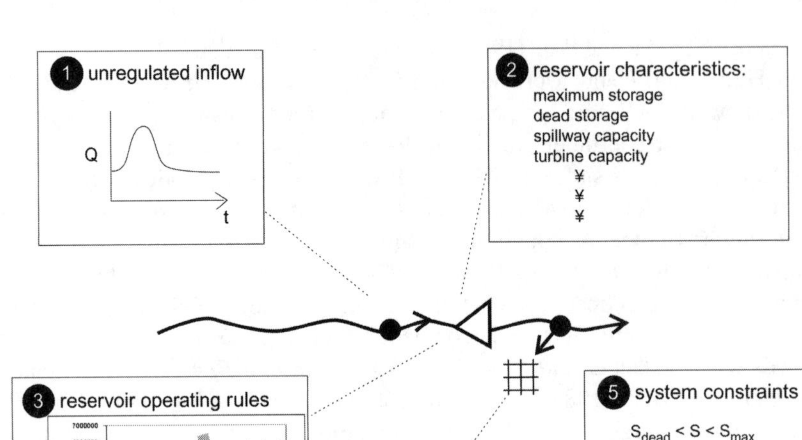

Figure 2. Components of a water resources (reservoir) system as abstracted by system simulation models.

Operations Research Methods in the Context of Water Management

In addition to its role as the genesis of subsequent interest in stochastic streamflow simulation methods, the Harvard Water Program [*Maass et al.*, 1962] was responsible for developing early algorithms for optimizing the design of water management systems, and/or allocation of water from existing systems. In broad generalities, two types of approaches evolved: simulation and optimization. Water resource simulation models attempt to capture the main elements of a water management system, which consists of natural and built components, and rules by which the combined system is operated. A typical abstraction of a water management system is shown in Figure 2. The abstracted system consists of 1) a set of streamflows (generally assumed to be at locations far enough upstream so that there is minimal effect of water management), 2) the built and/or natural storage system, which might consist, for instance, of surface reservoirs and a managed groundwater system, 3) a set of rules for operating the managed (or manageable) part of the system, 4) a set of water demands or uses, and 5) prescribed system constraints. In most cases, the system inflows are taken from a record of observations, often averaged over time increments that range from a few days to a month or more, depending on the size of the system. These inflows usually are (or should be) adjusted via an external process to remove effects of upstream water management. The storage system typically is simplified depend-

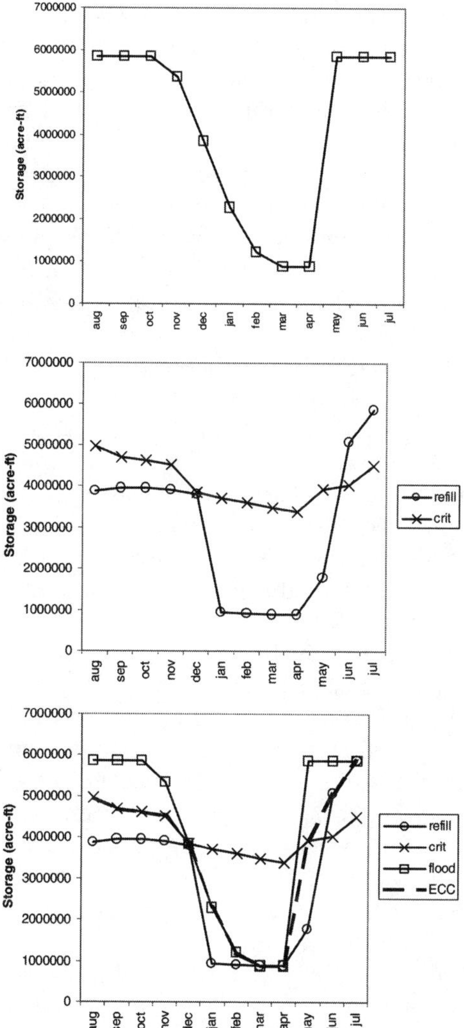

Figure 3. Reservoir operating rules for the Columbia River system. Panel a) shows critical and assured refill curves; critical rule curve represents reservoir elevations under the most severe condition in the historic record; assured refill curve indicates level from which system could refill following reoccurrence of most severe year in the historic record. Panel b) shows flood control rule curve, which defines drawdown required to regulate streamflow to prevent damage if the most severe flood in the historic record were to reoccur; Panel c) shows variable energy content curve, which is used as the target for nonfirm energy generation (usually the lowest of the four curves in winter and early spring; thereafter determined by streamflow forecasts. System operation is governed by flood rule curve until early winter; thereafter variable energy content curve or assured refill curve, whichever is lower).

ing on the required detail of the analysis. For instance, storage and run-of-river reservoirs may be aggregated to reduce the complexity of the modeled system.

A key distinction between simulation and optimization methods is that in simulation models, the operating rules are prescribed, usually in a manner similar to the way the reservoirs are operated in practice. For instance, Figure 3 shows the current reservoir operating rules for the Columbia River system, as abstracted from the ColSim model of Hamlet and Lettenmaier [1999a]. These rules dictate releases from the reservoir system (operated primarily for hydropower generation, water supply, and flood control, subject to environmental constraints) depending on time of year, current reservoir storage, and forecasted future inflows.

System constraints typically are physical capacity limits (e.g., on storage, turbine or bypass canal capacity, and the like). In some cases, environmental considerations, such as minimum low flows, may be treated as constraints as well. Simulation models are well suited to testing sensitivity of the performance of a water management system to changes in the various component factors, and/or combinations thereof. For instance, Lettenmaier et al. [1999] examined the performance of several water management systems across the U.S. to changes in inflows that might be associated with climate change using simulation models developed for each of the systems.

Simulation models are routinely used in performance studies by water management agencies. As compared with optimization models, simulation models generally have had greater acceptance in practice because they can, to varying degrees of accuracy, reproduce as a starting point the current operating conditions of a system. Simulation models, however, are not very well suited to the exploration of potential changes in reservoir operating policies to cope with changing operational priorities or external forces (such as climate change), and for these uses optimization models, which in effect allow exploration in a systematic manner of the effect of alternate operating policies on reservoir system performance, may be preferred.

Optimization models represent all but one of the same elements of a water management system as are used by simulation models. The exception is that system operation is not defined by a set of operating rules, rather it is determined by specifying an objective function or functions, the maximization (or minimization, as appropriate) of which defines a unique set of system variables (reservoir storage and/or releases) for each time step. For instance, in a simple case where a system has two objectives, hydropower production and minimum instream flows, the objective function might be to maximize hydropower production subject to physical system constraints (e.g., storage capacity) and minimum flows. If tradeoffs between several objectives are to be explored, the use of penalty functions for each management objective may be used instead of "hard" constraints [see e.g. *Sylla*, 1995]. Numerous variations of optimization methods are described in the literature [see e.g. *Hipel*,

1992; *Yeh*, 1992; and *Yang et al.*, 1995]. Among these are linear and dynamic programming techniques, and deterministic and stochastic formulations (deterministic formulations take the historic inflows as fixed, whereas stochastic formulations attempt to characterize from the observations the joint probability distributions of inflows in the present and future time periods). Likewise different modeling formulations treat single or multiple reservoirs and/or groundwater storage [*Yeh*, 1992; and *Belaineh et al.*, 1999].

Historically, the challenge faced in modeling large (multiple reservoir) systems, particularly using stochastic formulations, was that the dimensionality of the mathematical problem rapidly exceeded available computing resources as the number of system elements increased. In recent years computer resources and solution algorithms have improved dramatically, and appropriate tools are now available as off-the-shelf software packages [see e.g. *Andreu et al.*, 1996; and *Zagona et al.*, 2001]. However, in practice, the use of optimization techniques is still limited by two key problems. The first is associated with the difficulty in representing intentionally sub-optimal water resources objectives and policies that are the rule rather than the exception in most water resources systems. Optimizing such systems, while technically feasible, generally has reduced value, because the optimal solutions are dominated by the constraints imposed by the sub-optimal management objectives. For instance, these constraints are often imposed by water law or allocation conventions (e.g. the prior appropriation doctrine in western water law, or prescribed flood control evacuation requirements), and are frequently very difficult to change. The second problem is the absence of sufficient lead time and/or skill in streamflow forecasts to permit meaningful optimization at the seasonal to interannual time scales most relevant to water resources management.

As a result of these limitations, optimization methods have found their greatest acceptance in practice in systems where a) an economic objective function can fully describe the water resources objective (e.g. a system of reservoirs designed to produce maximum revenue from hydropower), and b) deterministic elements of the system to be optimized dominate over stochastic elements. This usually means relatively short lead times, so that the sequence of inflows over the decision period is reasonably well known. Depending on the size of the physical system, this can imply time horizons as short as a few hours, to as long as a month or so. For longer time horizons (e.g. seasonal to interannual), system managers usually have been more comfortable with simulation methods, which utilize prescribed operating rules for the reasons discussed above. A common combination is to use simulation methods to make long-range "planning" decisions (from a month or so out to a year or more), while shorter term "scheduling" decisions (e.g. allocation of hydropower releases between multiple projects) are made using optimization methods, which are nested within the framework of a simulation model for the longer term (e.g., by constraining end-of-run storage to be that determined by simulation). Optimization is also sometimes used to attempt to search for improved heuristic management rules. In these kinds of studies, reservoir releases are optimized using a set of observed inflows, and characteristics of the optimal solution are then generalized to produce a new heuristic management rule (e.g. a reservoir rule curve). Recent work by Yao and Georgakakos [2001], Hamlet and Lettenmaier [1999a] and others on long-lead streamflow forecasts based on climate information may lead to more efficient reservoir operation. However, Maurer and Lettenmaier [2003] show that the potential is limited for very large reservoir systems, e.g., those where the ratio of storage to mean annual reservoir inflow is greater than one.

Deterministic Streamflow Simulation Models

Deterministic streamflow simulation models were initially developed for engineering purposes, as a more generalized and flexible method of estimating characteristics of streamflow needed for design purposes than could be done with pre-computer era manual methods. For instance, estimation of design floods in the absence of lengthy streamflow records (from which direct statistical estimates might be made) utilized so-called design storms, and simple methods of transforming rainfall to runoff like the unit hydrograph, which essentially is a (linear) convolution method. Although tedious, such methods could be performed by hand. Critical limitations included the required assumption of linearity, which could only be defended if "initial abstractions" were removed from the observed precipitation forcings to account for the effects of initial soil moisture. Furthermore, such methods, sometimes now characterized as "lumped", require considerable spatial, if not temporal, aggregation, the error characteristics of which were unknown. Finally, these methods were entirely "black box", that is, the unit hydrograph coefficients must be estimated from coincident observations of precipitation and runoff, and the relationships extracted from the estimation period are assumed to be applicable to the prediction period.

With the advent of the digital computer, hydrologists began to develop rainfall-runoff models, which represented at some level the processes by which rainfall is transformed to runoff. The best-known of the early models was the Stanford Watershed Model of Crawford and Linsley [1966], which was the predecessor of the still widely used HSPF (Hydrologic Simulation Package—FORTRAN) and the National Weather Service River Forecast System (NWS-RFS) [*Burnash et al.*, 1973]. The Stanford Watershed Model and its successors are continuous simulation models — that is, they account for the hydrologic states (soil moisture and snow water storage) that affect subsequent runoff, as well as the retention of moisture at and near the land surface that controls runoff and streamflow generation. Continuous simulation avoids the need to specify initial moisture conditions. Furthermore, it makes possible the simulation of lengthy streamflow time series, to which statistical tools can be applied as if they were observed sequences. Such methods are particularly useful in cases where streamflow records are short, but longer records of model forcings (typically precipitation and temperature) are available.

Parameter estimation is a major issue in the application of streamflow simulation models. These models usually have a number of parameters (in the case of NWSRFS, sixteen), which are not readily related to physically measurable quantities. These parameters must be estimated through either manual (trial and error) or automated methods. A great deal of work has been done on development of automated parameter estimation for streamflow simulation models using various search procedures [*Sorooshian and Dracup*, 1980; *Sorooshian and Gupta*, 1983; and *Duan et al.*, 1992, among others]. Nonetheless, the state of practice remains to use manual methods, perhaps augmented by automated search procedures for exploratory purposes [see e.g. *Boyle et al.*, 2000].

Another limitation of the now-traditional streamflow simulation models, like NWSRFS and HSPF, is that they do not represent the effects of vegetation on hydrologic processes directly. Although vegetation rather obviously affects evaporative demands, which are represented by these models, this typically is done by prescribing potential evapotranspiration (in practice, potential evapotranspiration often is adjusted in the calibration process to achieve the best match of simulated and observed streamflow). Although these procedures do not impact the utility of this type of model for applications like flood forecasting, or extension of streamflow records, they clearly limit their usefulness for applications like assessing the hydrologic effects of changing land cover. Another key limitation of these early generation rainfall-runoff models is that they are spatially lumped—that is, they simulate runoff over a watershed using parameters that represent the aggregate effects of soil moisture and surface conditions. Furthermore, the precipitation forcings are usually taken as a spatial mean over the watershed area. The motivation for these abstractions lies largely in computing limitations, which were a critical concern at the time the models were developed. In some cases, like the USGS PRMS (Precipitation-Runoff Modeling System) model [*Leavesley et al.*, 1983], which originally was spatially lumped, partially distributed models have been developed which treat watershed subareas (which may or may not be physical subcatchments) as spatially lumped, but aggregate the hydrologic responses over subareas. More recently, the trend in hydrology has been toward the development of spatially distributed hydrologic models that represent the effects of local topographic, land cover and vegetation, and soil conditions on runoff generation [*Wigmosta et al.*, 1994; and *Ivanov et al.*, 2003]. Although these models are considerably more computationally intensive than are spatially lumped models, improved computing performance has made feasible simulations using models that discretize river basins into hundreds of thousands, or even more, pixels that represent explicitly spatial variations in topography, soils, and vegetation within a river basin. Computing requirements are now much less a concern with such models than are parameter estimation issues, which remain a topic of research [e.g. *Beven and Feyen*, 2002].

A final limitation of "traditional" rainfall-runoff models is that they do not perform a surface energy balance. In contrast, land surface schemes used in numerical weather prediction and climate models must, by construct, close both the surface energy and water budgets. In fact, in contrast to hydrologic models, the primary role of land surface schemes in coupled land-atmosphere models is to partition net radiation at the land surface into sensible, latent, and ground heat fluxes. Representation of the surface water balance (aside from evapotranspiration, and its control via soil moisture) has been somewhat less a concern. Over the last decade, however, there has been a convergence of land surface models usable in coupled applications. Those models that come from a hydrologic heritage (like the Variable Infiltration Capacity, or VIC, macroscale hydrologic model of Liang et al. [1994]) have evolved to include more sophisticated representations of surface energy balance processes, whereas some Soil-Vegetation-Atmosphere Transfer Schemes (SVATS), which tend to focus more on vertical exchange processes between the land and atmosphere, have incorporated improved runoff generation schemes. Two recent experiments of the Project for Intercomparison of Land-surface Parameterization Schemes (PILPS) for the Arkansas-Red River basins of the central U.S., and the Torne-Kalix River basin of Sweden and Finland, have shown that the better-performing land surface models are able to reproduce reasonably well the observed hydrographs for these large rivers [*Wood et al.*, 1998; and *Bowling et al.*, 2003].

TOOLS FOR WATER RESOURCE PLANNING WITH CLIMATE CHANGE

Hamlet and Lettenmaier [1999b] outline a general approach for evaluating the sensitivity of water resource system management to climate change, which is summarized in Figure 4. The main elements are: a) future climate scenarios, which typically come from a General Circulation Model of the global land-atmosphere-ocean system, downscaled using either dynamical or statistical methods (see Arnell [2002] for a discussion of downscaling methods as applied to hydrology and water resources climate change assessments); b) a hydrologic (precipitation-runoff) model, which is driven by downscaled climate scenarios, typically as a time series over a period of several decades to a century or so; c) a water management model, which is driven by the output of the hydrologic model; and d) a set of system performance metrics, used to summarize the overall performance of the system, and how it might change for different climate scenarios.

A key aspect of this assessment structure is its sensitivity analysis context—that is, the focus is on evaluation of the consequences for water management implied by a given climate scenario. Of course, the only thing that can be said with certainty about any given climate scenario is that the future will be different from it. Nonetheless, essentially all work reported in the literature that evaluates hydrologic and water resource consequences of climate change is cast in such a sensitivity analysis context—including, for instance, the over 40 papers

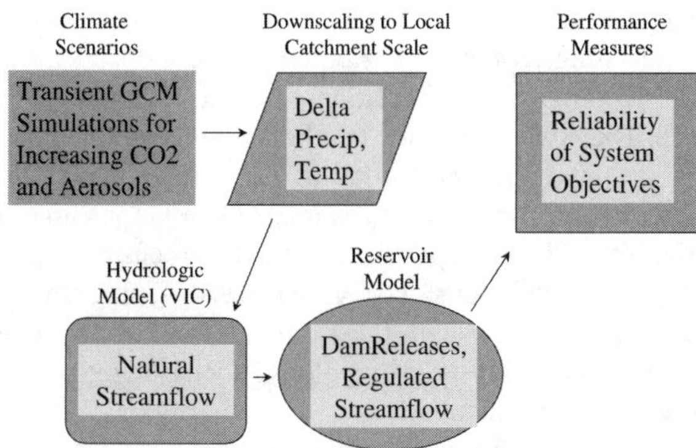

Figure 4. General approach for evaluating the sensitivity of water resource system performance to climate change

included within two special issues of *Water Resources Bulletin* (December, 1999 and April, 2000) produced as part of the Water Sector activity of the U.S. National Assessment of the Consequences of Climate Change [*Gleick*, 2000], and the numerous papers cited in recent IPCC reports [*Kaczmarek et al.*, 1996; *Arnell et al.*, 1996; and *Arnell and Liu*, 2001].

On the other hand, despite the evolution of an entire sub-field of hydrology that was motivated by the need to simulate streamflows for purposes of assessing reservoir system performance, the acceptance of these methods in practice has been slow at best. Furthermore, there is a nearly complete disconnect between the work reviewed above, which attempted to characterize the long-term properties of streamflow persistence and its effect on the performance of reservoir systems, and more recent climate change (sensitivity) assessments like those reviewed in the IPCC and U.S. National Assessment reports. Furthermore, there has been relatively little examination of the extent to which existing management practices are inherently robust to climate and other changes in future operating conditions.

THE PATH FORWARD—RESEARCH NEEDS

When one considers the historical development of methods for hydrologic design, it is quite clear that climate is, and always has been, the primary driver. The development of stochastic hydrology, and reservoir simulation and optimization methods, have all had the intent of providing system designs and operating methods that buffer water users from the vagaries of climate. On the other hand, documents like the Hydrology and Water Resources chapter of the 2001 IPCC report [*Arnell and Liu*, 2001] and the U.S. National Water Assessment Group [*Gleick*, 2000] leave one with the impression that potential vulnerabilities of water resources systems to climate is a discovery of the climate change community. Why is this? I suggest that there are two reasons, resolution of which must be at the heart of any viable strategy to deal with climate change issues in the context of water.

First, work undertaken in the 1960s on development of time series methods that could reflect long-term persistence in hydrologic observations (which can be viewed as a hydrological manifestation of climate variability) lacked a sound physical basis. There was never any real link with the climate community, and in fact most of the modeling work took as its starting point observed series of streamflow, so there was no effort to examine the relationship of streamflow and climate variability. Absent a physical basis, the focus was primarily methodological, and by the late 1980s, work in the area had stagnated. Those who had been involved in this area sought other pursuits, and by the time climate issues had begun to be taken seriously in hydrology, stochastic hydrology was no longer an active or viable research area.

Second, academic interest in water management has always had a strong engineering focus. The engineering community has been (and for the most part continues to be) skeptical of climate change as a serious design issue. Notwithstanding that methods for design of structures and systems that account for environmental uncertainty are a legitimate engineering problem, climate issues have not been taken very seriously by the civil engineering community.

Furthermore, with the release of the NRC report *Opportunities in the Hydrologic Sciences* [*Eagleson et al.*, 1990] the water resources sector of the hydrologic community was more or less cast adrift from the science sector. For instance, the NSF Hydrological Sciences Program in general does not consider proposals for development of water management tools, which are not considered "scientific." Likewise, other U.S. agencies in which there has been funding growth for research related to hydrology and climate, like the NOAA Office of Global Programs, and the NASA Land Surface Hydrology Program, in general have a predominantly scientific focus. The result has been that funding opportunities for those working on water management problems have become scarce within the federal research context, and researchers in these areas have tended to migrate toward funding sources associated with applications problems faced by operating agencies, or state, local, and private funding sources. Insofar as these operating agencies have been slow to recognize climate change issues, little in the way of new or innovative research has resulted.

Resolution of these issues arguably will require fostering and/or development of interest by an interaction of the water resources engineering and climate science communities. Opportunities for development of procedures to account for climate change in water resources planning and management exist in several areas, but two that deserve attention are:

- Development and use of stochastic methods that account for the possibility of future climate change through use of estimation methods that are weighted more toward recent observations than those from the distant past. Work that relates hydrologic forcings (especially precipitation) to large area circulation features that may be undergoing change shows particular promise (see for instance recent work applying hidden Markov processes by Charles et al. [1999]).

- Rejuvenation of methods for planning water management systems (especially system expansion) under uncertainty. Like stochastic hydrology models, classical capacity expansion methods that incorporate information about hydrologic uncertainty do so in the context of statistical stationarity. Methods that are adapted to ensemble simulation, for instance, could be a productive area for future research, but would need to be fostered by research managers whose focus and interests presently lie elsewhere.

Acknowledgments. The assistance of Mr. Alan Hamlet, Ph.D. student in the Department of Civil and Environmental Engineering (CEE), University of Washington, who reviewed the manuscript and assisted in preparation of figures, is greatly appreciated. Dr. Chunmei Zhu and Mr. Hyo Seok Park, also University of Washington CEE graduate students, assisted in preparation of figures. This publication was supported in part by the Joint Institute for the Study of the Atmosphere and Ocean (JISAO) at the University of Washington, funded under NOAA Cooperative Agreement number NA17RJ11232, Contribution 942, as part of the GEWEX Continental-Scale International Project (GCIP) and the GEWEX Americas Prediction Project (GAPP).

REFERENCES

Andreu, J, J. Capilla, and E. Sanchis, AQUATOOL, A generalized decision-support system for water-resources planning and operational management, *Journal of Hydrology*, 177, 269-291, 1996.

Arnell, N., *Hydrology and global environmental change*, Prentice-Hall, Harlow, England, 346 pp., 2002.

Arnell, N., B. Bates, H. Lang, J. Magnuson, and P. Mulholland (convening authors), Hydrology and freshwater ecology, in *Climate Change 2001: Impacts, Adaptation, and Mitigation of Climate Change, Contribution of Working Group II to the Second Assessment Report of the Intergovernmental Panel on Climate Change*, pp. 325-363, Cambridge University Press, Cambridge, UK, 879 pp., 1996.

Arnell, N., and C. Liu (convening authors), Hydrology and Water Resources, in *Climate Change 2001: Impacts, Adaptation, and Vulnerability, Contribution of Working Group II to the Third Assessment Report of the Intergovernmental Panel on Climate Change*, pp. 191-233, Cambridge University Press, Cambridge, UK, 1032 pp., 2001.

Belaineh, G, R.C. Peralta, and T.C. Hughes, Simulation/optimization modeling for water resources management, *J. of Water Res. Planning and Mgmt.-ASCE*, 125, 154-161, 1999.

Beven K, and J. Feyen, The future of distributed modelling - Special issue, *Hydrological Processes* 16, 169-172, 2002.

Box, G.E.P., and G.M. Jenkins, *Time series analysis forecasting and control*, Holden-Day,San Francisco, 553 pp., 1970.

Bowling, L.C., D.P. Lettenmaier, B. Nijssen, L.P. Graham, and others, Simulation of high latitude hydrological processes in the Torne-Kalix basin: PILPS Phase 2(e) 1:

Experiment description and summary intercomparisons, *Global and Planetary Change*, in press, 2003

Boyle, D.P., H.V. Gupta, and S. Sorooshian, Towards improved calibration of hydrologic models: Combining the strengths of manual and automatic methods, *Water Resources Research*, 36, 3663-3674, 2000.

Burnash, R.C., R.L. Ferral, and R.A. McGuire, A generalized streamflow simulation sytem, Conceptual modeling for digital computers, Federal-State River Forecast Center, Sacramento, 1973.

Charles S.P., B.C. Bates BC, and J.P. Hughes, A spatio-temporal model for downscaling precipitation occurrence and amounts, *Journal of Geophysical Research* 104, 31,657-31,669, 1999.

Crawford, N.H. and R.K. Linsley, Digital simulation in hydrology: Stanford Watershed Model IV, Stanford University Department of Civil Engineering Technical Report No. 39, 1966.

Duan, Q., V.K. Gupta, and S. Sorooshian, Effective and efficient global optimization for conceptual rainfall-runoff models, *Water Resources Research* 28, 1015-1031, 1992.

Eagleson, P.E. (chair), and others, *Opportunities in the Hydrologic Sciences*, National Academy Press, 348 pp., 1990.

Fiering, M.B., and B. Jackson, Synthetic streamflows, *Water Resources Monograph* 1, American Geophysical Union, 1971.

Frevert, D.K., M.S. Cowan, and W.L. Lane, Use of stochastic hydrology in reservoir operation, *Journal of Irrigation and Drainage Engineering-ASCE*, 115, 334-343, 1989.

Gleick, P.H., Water: The potential consequences of climate variability and change for the water resources of the United States, edited by P.H. Gleick, Pacific Institute, Oakland, CA, 2000.

Hamlet, A.F. and D.P. Lettenmaier, Columbia River streamflow forecasting based on ENSO and PDO climate signals, *J. of Water Res. Planning and Mgmt.*, 125, 333-341, 1999a.

Hamlet, A.F. and D.P. Lettenmaier, Effects of climate change on hydrology and water resources objectives in the Columbia River basin, *Water Resources Bulletin*, 35, 1597-1623, 1999b.

Hipel, K.W., and A.I. McLeod, Preservation of the rescaled adjusted range 2. Simulation studies using Box-Jenkings models", *Water Resources Research* 14, 509-518, 1978.

Hipel, KW, Multiple objective decision-making in water-resources, *Water Resources Bulletin*, 28, 3-12, 1992.

Hurst, H.E., Long-term storage capacity of reservoirs, *Transactions American Society of Civil Engineers*, 116, 770-806, 1951.

Hurst, H.E., Methods of using long-term storage in reservoirs, *Proceedings Institute of Civil Engineers*, 1, 519-543, 1956.

Hurst, H. E., R. P. Black and Y. M. Simaika, *Long-term Storage, an Experimental Study*, Constable, London, 145 pp., 1965.

Ivanov, V.Y., E.R. Vivoni, R.L. Bras, and D. Entekhabi, Development of a TIN-based distributed hydrologic model, *Journal of Hydrology*, in press, 2003.

Kaczmarek, Z., N.W. Arnell, and E.Z. Stakhiv, Water resources management, in *Climate Change 2001: Impacts, Adaptation, and Mitigation of Climate Change, Contribution of Working Group II to the Second Assessment Report of the Intergovernmental Panel on Climate Change*, pp. 469-486, Cambridge University Press, Cambridge, UK, 879 pp.,

1996.
Klemes, V., Water storage: Source of inspiration and desperation, in *Reflections on Hydrology, Science and Practice,* edited by N. Buras, pp. 287-314, American Geophysical Union, 1997.
Leavesley, G.H., R.W. Lichty, B.M. Troutman, and L.G. Saindon, Precipitation-Runoff Modeling System user's manual, U.S. Geological Survey Water Resources Investigations Report 83-4238, 207 pp., 1983.
Lettenmaier, D.P., and S.J. Burges, Operational assessment of hydrologic models of long-term persistence, *Water Resources Research* 13, 113-124, 1977.
Lettenmaier, D.P., A.W. Wood, R.N. Palmer, E.F. Wood, and E.Z. Stakhiv, Water resources implications of global warming: A U.S. regional perspective, *Climatic Change*, 43: 537-579, 1999.
Liang, X., D.P. Lettenmaier, E.F. Wood, and S.J. Burges, A simple hydrologically based model of land and energy fluxes for general circulation models, *Journal of Geophysical Research*, 99, 14, 415-14,428., 1994.
Lins, H.F., and E.Z. Stakhiv, Managing the nation's water in a changing climate, *Journal of the American Water Resources Association* 34, 1255-1264, 1998.
Loucks, D.P., J.R. Stedinger, and D.A. Haith, *Water Resource System Planning and Analysis*, Prentice-Hall, Englewood Cliffs, NJ, 559 pp., 1981.
Maass, A., M.M. Hufschmidt, R. Dorfman, H.A. Thomas, Jr., S.A. Margolin, and G.M. Fair, *Design of Water Resource Systems, New Techniques for Relating Economic Objectives, Engineering Analysis, and Governmental Planning*, Harvard University Press, Cambridge, MA, 620 pp., 1962.
Mandelbrot, B.B., and J.R. Wallis, Noah, Joseph, and operational hydrology, *Water Resources Research* 4, 909-918, 1968.
Mandelbrot, B.B., and J.R. Wallis, Some long-run properties of geophysical records, *Water Resources Research* 5, 321-340, 1969a.
Mandelbrot, B.B., and J.R. Wallis, Computer experiments with Fraction Gaussian Noises Part 1, Averages and variances, *Water Resources Research* 5,.228-241, 1969b.
Mandelbrot, B.B., A fast fractional Gaussian noise generator, *Water Resources Research* 7, 543-553, 1971.
Matalas, N.C., and M.B. Fiering, Water resources system planning, in *Climate, Climatic Change, and Water Supply*, pp. 99-110, National Academy of Sciences, 1977.
Matalas, N.C., Stochastic hydrology in the context of climate change, *Climatic Change* 37, 89-101, 1997.
Maurer, E.P. and D.P. Lettenmaier, Potential effects of long-lead hydrologic predictability on Missouri River main-stem reservoirs, *J. Climate*, in press, 2003.
McLeod, A.I., and K.W. Hipel, Preservation of the rescaled adjusted range 1. A reassessment of the Hurst phenomenon, *Water Resources Research* 14, 491-508, 1978.
Mejia, J.M., and J. Rousselle, Disaggregation models in hydrology revisited, *Water Resources Research* 12, 185-186, 1976.
National Research Council, Committee on Protection and Management of Pacific Northwest Anadromous Salmonids, *Upstream: Salmon and Society in the Pacific Northwest*, National Academy Press, Washington, D.C., 472 pp., 1996.
O'Connell, P.E., Stochastic modeling of long-term persistence in streamflow sequences", Ph.D. thesis, Imperial College, London, 1974.
Rippl, W., The capacity of storage reservoirs for water supply, *Minutes, Proceedings of*

the Institute of Civil Engineers, 71, 270-78, 1883.

Sorooshian, S., and J.A. Dracup, Stochastic parameter estimation procedures for rainfall-runoff models: Correlated and heteroscedastic error cases, *Water Resources Research* 16, 430-442, 1980.

Sorooshian, S., and V.K. Gupta, Automatic calibration of conceputal rainfall-runoff models: The question of parameter observability and uniqueness, *Water Resources Research* 19, 251-159, 1983.

Stedinger, J.R., and R.M. Vogel, Disaggregation procedures for generating serially correlated flow vectors, *Water Resources Research* 20, 47-56, 1984.

Sylla, C., A penalty-based optimization for reservoirs system management, *Computers and Industrial Engineering*, 28, 409-422, 1995.

Valencia, R.D., and J.C. Schaake, Disaggregation processes in stochastic hydrology, *Water Resources Research,* 9, 580-585, 1973.

Wallis, J.R., and N.C. Matalas, Sensitivity of reservoir design to the generating mechanism of inflows, *Water Resources Research* 8, 634-641, 1972.

Wallis, J.R., and P.E. O'Connell, Firm reservoir yield—how reliable are hydrological records, *Hydrological Sciences Bulletin*, 39, 347-365, 1973.

Wigmosta, M.S., L.W. Vail, and D.P. Lettenmaier, A distributed hydrology-vegetation model for complex terrain, *Water Resources Research*, 30, 1665-1669, 1994.

Wood, E.F., D.P. Lettenmaier, X. Liang, D. Lohmann, and others, The Project for Intercomparison of Land-Surface Parameterization Schemes (PILPS) Phase-2(c) Red-Arkansas River experiment: 1. Experiment description and summary intercomparisons, *Global and Planetary Change*, 19, 115-135, 1998.

Yang, X.L., E. Parent, C. Michel, and P.A. Roche, Comparison of real-time reservoir-operation techniques, *Journal of Water Resources Planning and Management*, ASCE, 121, 345-351, 1995.

Yao, H, and A. Georgakakos, Assessment of Folsom Lake response to historical and potential future climate scenarios 2. Reservoir management, *Journal of Hydrology*, 249, 176-196, 2001.

Yeh, W.W.-G., Systems-analysis in groundwater planning and management, *Journal of Water Resources Planning and Management*, ASCE , 118, 224-237, 1992.

Zagona, E.A., T.J. Fulp, R. Shane, Y. Magee, and H.M. Goranflo, Riverware: A generalized tool for complex reservoir system modeling, *Journal of the American Water Resources Association*, 37, 913-929, 2001.

Dennis P. Lettenmaier, Box 352700, Department of Civil and Environmental Engineering, University of Washington, Seattle, WA 98195

13

Identifying Hydrologic Variability and Change for Strategic Water System Planning and Design

F. Russell Walker, Jr. and Ellen M. Douglas

INTRODUCTION

Most engineering projects must accommodate some level of uncertainty in design and some level of failure risk (however small) during operation. This is especially true in hydrologic engineering, where natural variability can result in fluctuations as large as or larger than the magnitude of the design parameters. Hydrologic risk has typically been estimated by conventional frequency analysis, which entails fitting a probability function to historical data (i.e., streamflow, precipitation) and then using the fitted function to estimate the magnitude of an event with a specified risk of occurrence (i.e., the 100-year flood or the 50-year storm). Figure 1 illustrates the results of this method applied to historical flood flows. Classical frequency analysis used in infrastructure design and operation requires an assumption of stationarity. A time series is stationary if it is free of trends, shifts, or periodicity, implying that the statistical parameters of the series (e.g., mean and variance) remain constant through time [Salas, 1993].

Classical frequency analysis, as illustrated by Figure 1, has evolved without much consideration of climatic variability or changes in watershed response [Stedinger et al., 1993]. It is reasonable to speculate that on a planet with a dynamic atmosphere such as that possessed by our Earth, the hydrologic cycle will respond to changes in climatic factors. In addition, massive changes in land use and land management practices can greatly affect a basin's hydrologic response, perhaps even offsetting some of the impacts of climate change [Potter, 1991; DeWalle et al., 2000]. Non-stationarity in hydroclimatic processes is of interest to scientists and policy-makers alike [Baldwin and Lall, 1999], and has

Water: Science, Policy, and Management
Water Resources Monograph 16

10.1029/016WM16

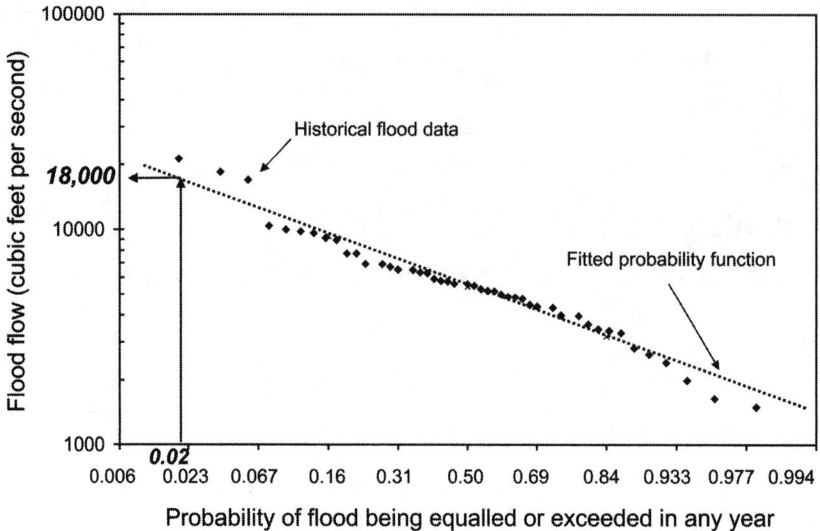

Figure 1. Example of classical frequency analysis applied to annual flood flows. A theoretical probability function is fitted to historical data, and is used to estimate the magnitude of a flood event (18,000 cfs) with a specified exceedence probability (0.02).

profound implications for strategic water system planning, operation and design. Hydrologic risk is not constant (as assumed by conventional methods) but may vary significantly over extended periods of time. A great deal of literature has been devoted to debating the merits and pitfalls of the assumption of stationarity in hydroclimatic time series. Most recently, Jain and Lall [2001] assert that the traditional assumption of stationarity for flood frequency analysis may translate into sub-optimal decisions on flood control design. Improving frequency analysis to include non-stationarity requires a new paradigm. Likewise, it is important to address the impact of hydroclimatic connections (or forcings) on the trends and inter-decadal variability observed in hydrologic data. Therefore, it is important to develop statistical techniques that can adequately address such issues, while providing the engineer and hydrologist with improved methods for detecting and accommodating changes in the hydrologic record and the resulting impacts on hydrologic risk.

CHANGES IN THE HYDROLOGIC RECORD

First, let us address the nature of changes in hydrologic records. A trend is generally defined as an observed increase or decrease in a hydrologic variable over time. Trends may be identified over differing time steps (monthly, seasonally, annually) and in different statistical parameters (mean, maxima or minima, ranks). Trends observed over long time periods imply an ongoing process of change that may result in hydrologic characteristics unlike those in prior historical records. Hydrologic records may also change significantly due to periods of increased and decreased variability, known as regimes. Such regimes are believed to be the manifestation of climatic connections or forcings, which are also variable but over larger spatial and temporal scales within the dynamic Earth system.

Examples of Hydrologic Regimes

In a study of fluvial responses to small-scale climatic changes, Knox [1984] examined the magnitude of floods on the Mississippi River at St. Paul, Minnesota. His work was directed at sedimentary processes, which are known to vary with large flood events. Knox hypothesized that large-scale atmospheric patterns contribute to changes in hydrology and thus fluvial responses. He performed a traditional flood-frequency analysis on four distinct subseries of the St. Paul flood record and provided estimates of flood magnitudes for several return periods (see Figure 2). The subseries were characterized by identifiable, yet slow, changes in atmospheric conditions that demonstrated long-term climatic variability over multiple decades. Figure 2 illustrates that changing climatic condi-

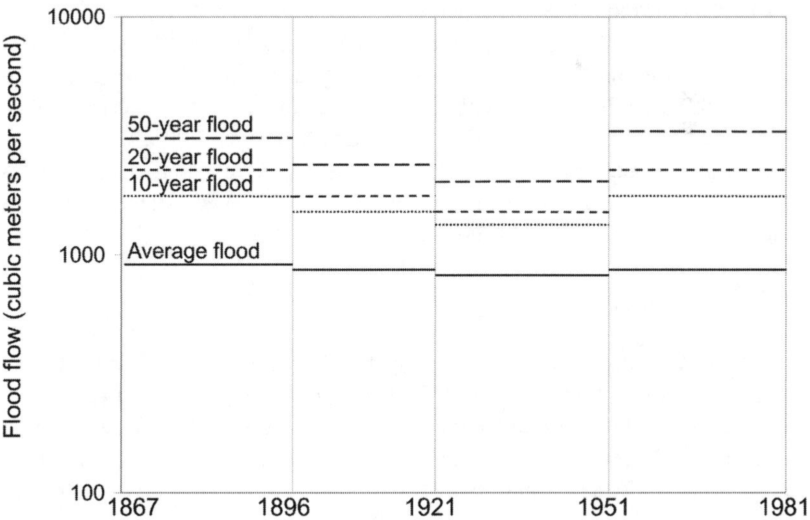

Figure 2. Varying flood risk on the Mississippi River. Flood flows estimated using the log Pearson type III probability density functiondistribution and plotted in logarithmic format to account for differences in magnitude. Vertical lines separate climate regimes [from Knox, 1984].

tions had little effect on the mean flood with only an 8% decrease between 1867 and 1950 and 6% increase between 1921 and 1981. However, impacts were amplified in the more extreme events: decreases of greater than 30% in the 20- and 50-year flood estimates between 1867 and 1950 and increases of 50 to 60% in these floods between 1921 and 1981. Even if one cannot reject the hypothesis that all subsets of the record are characterized by a common distribution, the implications are profound for flood frequency analysis: flood control structures and operational plans based on the pre-1950 flood record would have been subject to much larger failure risk than anticipated or desired.

In a recent paper, Barros and Evans [1997] examined the climatic coincidence of superfloods (defined as floods with return periods on the order of 100 years in basins exceeding 4,000 km^2) within the Upper Mississippi and Missouri River basins. Figure 3 [from Barros and Evans, 1997] shows that the occurrence of superfloods is linked to the combination of both large precipitation anomalies and the negative phase Southern Oscillation Index (SOI) associated with El Niño events. The implication is that, for a given short time interval of up to several years, extraordinary floods could be better predicted by including information on large scale climate phenomena.

A reduction in the time between extreme events, also known as clustering, is another indication of a change in hydrologic or climatic regime. The flood record of the American River indicates that 10 of the top 13 flood flows in its 1905-1998 record have occurred since 1950, with the two largest occurring in the 1990s (NRC, 1995; USGS, 1999]. While not necessarily verifiable by hypothesis testing, this curious increase in large floods in recent decades suggests some possible long-term variability in the arrival rate of large floods on the American River. Walker and Stedinger [Walker, 1999; Walker and Stedinger, 2000; Walker et al., in review] investigated the issue of flood clustering and concluded that statistically significant flood clustering is evident in the Mid-Atlantic and the Southeast-Gulf Coast regions of the U.S.

Hydrologic Trends

A number of recent studies have investigated the presence of trends in streamflow data. Westmacott and Burn [1997] reported a large proportion of streams with decreasing trends in annual flood flow data from West-Central Canada. Genta et al. [1998] examined 30-year running averages for four major rivers in southeastern South America and found that streamflow has been increasing nearly monotonically since the mid- to late-1950s. Lettenmaier et al. [1994] detected strong increases in monthly streamflow across the U.S. with the largest trend magnitudes occurring in the north-central region (Michigan, Illinois, Wisconsin, and Minnesota). Knapp [1994], Baldwin and Lall [1999], and Olsen et al. [1999] demonstrated trends in the record resulting in larger floods on the Mississippi River. Douglas et al. [2000] found that, after accounting for spatial correlation

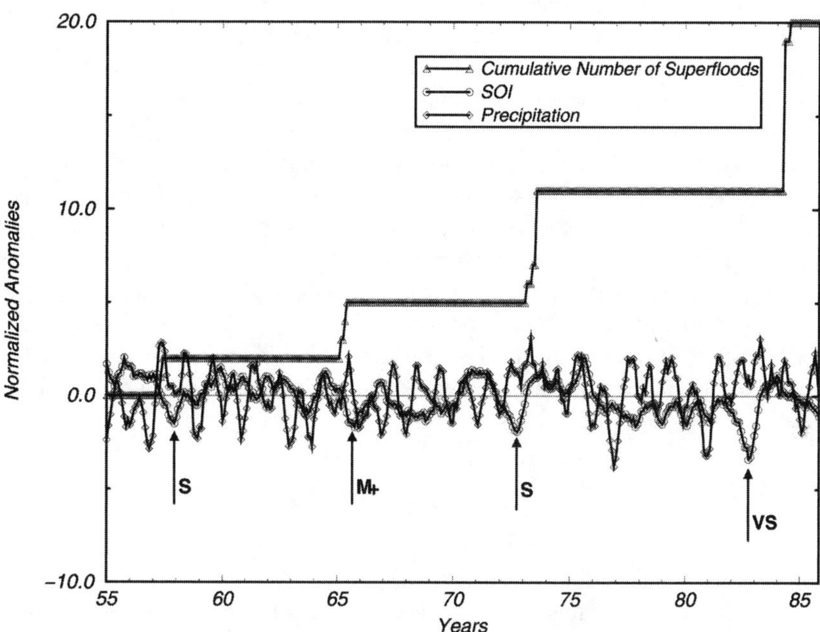

Figure 3. The influence of El Niño-Southern Oscillation (ENSO) on the occurrence of extreme floods in the Upper Mississippi River Basin. Superfloods are defined as floods with return periods on the order of 100 years for basin areas exceeding 4,000 km^2 (10,000 mi^2). Precipitation and SOI are 5-month running averages of anomalies. Arrows indicate strength of ENSO events:

VS — very strong, S — strong, M+ — moderate/strong. [from Barros and Evans, 1997, used with permission].

in the streamflow records, there was no evidence of regional trends in annual flood flows, but there was evidence of upward regional trends in low flows in the upper Midwestern U.S. Lins and Slack [1999] also reported increasing trends across the U.S. in lower magnitude streamflow percentiles (annual minimum through the 70th percentile) but not at higher percentiles (90th percentile and annual maximum).

IDENTIFYING HYDROCLIMATIC NON-STATIONARITY

Non-stationarity can potentially take several forms: a shift in the mean with constant variability, constant mean but changing variability, or changing mean and changing variability. Olsen et al. [1998] argue that the major impact of the non-stationary behavior of a random variable, such as streamflow, is manifested in the extremes. However, because extreme events are by definition rare events, detecting non-stationarities in extremes requires special tests and often significantly more data to achieve desired levels of confidence in the results of the analysis. It should be noted that some researchers [e.g., Matalas, 1997] have asserted that the assumption of stationarity should not totally be disregarded and that a "wait and see" strategy has and may continue to serve water managers well. Even so, Matalas recommends (1) making "what if" scenarios a part of the design process, and (2) incorporating trend and persistence into hydrologic analyses, thereby introducing a higher degree of variability to at least partially account for uncertainties. The former can be accomplished through the use of stochastic modeling. The latter approach will provide statistically based engineering factors of safety in the design of water systems and can serve as an interim measure while new methods and modeling linking climatic variations and hydrologic risks are developed.

Short-term Persistence

Short-term persistence (or autocorrelation) in hydrologic records refers to the similarity of observations in the near-term. For example, rainfall has a high short-term persistence; that is to say, rainfall accumulation in the previous hour tends to resemble rainfall accumulation in the hour following. Annual minimum streamflows also tend to be persistent, in other words, the annual minimum streamflow in a given year is likely to be similar in magnitude to that of the previous year. This is usually attributed to the influence of ground water inflows which contribute to base streamflow. By contrast, annual maximum (flood) flows generally possess very low persistence and are usually considered to be independent in time. Identification of the presence or absence of persistence in hydrologic variables is essential in selecting the proper model for design, prediction and operational planning.

When evaluating persistence in a continuous time series, it is convenient to consider autoregressive (AR) or autoregressive moving average (ARMA) processes. In general, AR processes are considered to be short-memory processes and ARMA processes are considered to be long-memory processes [Salas, 1993]. Hydroclimatic processes are often modeled with low-order AR(k) models, such as the AR(1) model which takes the form

$$y_t = \mu + \phi_1(y_{t-1} - \mu) + N_t$$

where y_t is the time series, μ and ϕ_1 are model parameters and N_t is a normally distributed "noise" term with mean zero and variance σ_ε^2. Given a time series of observations, μ can be estimated by the sample mean and ϕ_1 by the sample estimate of the lag-one autocorrelation coefficient, r_1. Statistical tests for autocorrelation much greater than lag-one require strong persistence, which is not often found in hydrologic data.

Long-term Persistence

Whereas short-term persistence can be expected to impact hydrologic records for a handful of years (interannual variability), it is probable that climatic forcings can result in hydrologic regimes on the order of decades (interdecadal variability). Interdecadal variability can be more difficult to detect and more problematic to interpret, especially if the available hydrologic record is short. And yet changes in hydrologic risk due to interdecadal variability can impact the design and operation of major hydrologic projects, such as flood control dams and water supply reservoirs.

Hurst [1951] studied the behavior of the observed range of the cumulative departures, R, from the annual mean streamflow into a reservoir. Hurst showed that the mean of the scaled random variable (R/s) is equal to Cn^H where C is a constant, s is the sample standard deviation, n is the record length, and H is known as the Hurst coefficient. Many mathematical approaches support this theory and show that H should approach 0.5 as n approaches infinity [Klemes, 1974]. Hurst further showed that many natural time series exhibit H significantly greater than 0.5, usually between 0.6 and 0.8. This behavior is known as the Hurst phenomenon and is often considered evidence of non-stationarity in the mean of a time series [Klemes, 1974; Potter, 1976; Mandelbrot and Wallis, 1969].

Clustering

Some hydrologists and climatologists have asserted that different regimes exist during which floods tend to occur more frequently or droughts tend to last longer. In this case, we are not interested so much in the magnitude of extreme events as we are in the interarrival times (time between events) or durations (length of time over which events, such as droughts, take place). Identification of changes in flood clustering or drought durations is particularly important in the management of multiple use reservoir systems; if a period of extended droughts is anticipated, water managers would be wise to reduce the reservoir volume set aside for flood storage in favor of increased water supply storage.

Walker and Stedinger [Walker, 1999; Walker and Stedinger, 2000; Walker et al., in review] examined various statistics based on run-length theory to measure flood clustering, including the variance of interarrival times, the mean time to the closest flood, and the variance of the time to the closest flood. They examined the empirical estimator of the 5-year flood (i.e., the flood with a 20% probability of occurrence each year). The 5-year flood was studied as it is a reasonable flood to study with record lengths of 40-50 years, typical for annual flood records in the U.S. Their findings indicate that the variance of flood interval times (expressed as the number of years between floods) was the most valuable cluster statistic. The variance of flood interval times was found to be the best measure of the persistence of floods, particularly when floods of interest are not consecutively ordered in the time series. All the clustering statistics were able to measure persistence on the order of decades, while changes in the arrival rate of floods were not detected with other methods such as autocorrelation, the Hurst phenomenon, or trend analysis (to be discussed later).

Single-site Versus Regional Estimates

Although based on work underlain by the stationarity assumption, Hosking and Wallis [1997] found that more accurate frequency analysis results from analyzing all the data samples within a suitably defined region rather than by analyzing the samples at one site. Vogel et al. [1998] point out that regional streamflow statistics may contain more information about the persistence and variability of streamflow than do single-site statistics. Walker [1999] took a regional approach in evaluating flood cluster statistics, using (1) a simple average of the standardized form of the cluster statistics for all sites in each region, and (2) an average of these standardized cluster statistics weighted by the record length of each site. The significance levels of the observed regionalized cluster statistics were resolved using a bootstrapping method that preserved the concurrent cross-correlations.

RISK ESTIMATION IN THE PRESENCE OF PERSISTENCE

Hydrologic risk is often expressed in terms of the expected time for an extreme event to occur, and is commonly referred to as the "average return period" for that event. For example, a flood with a one percent annual exceedence probability (the flow or depth that has a one in one hundred chance of being equaled or exceeded in any given year) has an average return period of 100 years and is known as the "100-year flood." Gumbel [1941] and Thomas [1948] defined the average return period as the expectation of the time between a sequence of events (the "recurrence" interval) and Lloyd [1970] showed that the average return period defined in this way is insensitive to streamflow persistence. Vogel [1987], Fernandez and Salas [1999a, b], Sen [1999], and Douglas et al. [2002] defined the average return period as the expected number of years before the first occurrence of a system failure, (the "occurrence" interval). They demonstrated that the average return period defined in this manner is sensitive to short-term persistence and they developed general formulations for estimating the average return period of design events when such persistence is present in the streamflow record. Douglas et al. [2002] showed that accounting for short-term persistence in the flow record increases the occurrence interval for a design event (a flood or low flow of a specified magnitude) and reduces the risk of failure over the design life of a system.

IDENTIFYING HYDROLOGIC TRENDS

Trends in the hydrologic record can be observed over many different time scales. Trends in ground water levels or streamflow can be observed at a daily or weekly time scale in response to fluctuations in usage. Trends at the seasonal and interannual scales can be observed in response to unusual weather patterns or short-term climate variability and are important in forecasting near-term water availability. Trends observed at the interdecadal to century scales may be indicative of long-term climatic shifts that must be anticipated and accommodated during water system planning and design.

Most trend analysis of hydrologic data are performed using statistical hypothesis tests for detecting a general increase or decrease in the data over time. The Mann-Kendall trend test [Kendall, 1962] is a non-parametric, rank-based method that makes no assumptions about the probability distribution of the random variable but does assume that, if a trend exists, it is monotonic. This method is insensitive to issues of non-normality and missing data. Another popular method for testing trend is one in which a simple linear regression is performed on the random variable against time. For details on these and other trend tests, see Helsel and Hirsch [1992].

Compensating for Hydrologic Persistence in Tests of Trend

When a trend in a time series is perceived, one needs to question whether this is actually an indication of an underlying trend or whether the apparent trend is due to long-term persistence in the data or in a process upon which the data are dependent [Hirsch and Slack, 1984]. One must consider either selecting a trend test that can account for persistence or "pre-whitening" the data prior to testing for trend. Pre-whitening is the term used for removing persistence from the time series, thereby reducing it to independent residuals ("white noise"). Von Storch [1995], Krzyscin [1997], and Douglas et al. [2000] offer details on the use of pre-whitening.

Bence [1995] evaluated methods for correcting for autocorrelation in short time series and found that most methods undercorrected due to bias in the estimation of the autocorrelation coefficient. Hamed and Rao [1998] showed that positive autocorrelation can result in detection of trends by the traditional Mann-Kendall trend in randomly generated stationary data, and developed a modified Mann-Kendall test for autocorrelated data. Hirsch and Slack [1984] developed a similar test for seasonal data. Hirsch et al. [1982] and Harcum et al. [1992] did the same for monthly and seasonal water quality data, respectively. Berryman et al. [1988] proposed a procedure for selecting amongst the Mann-Whitney, Spearman, and Kendall trend tests.

Incorporating Spatial Correlation in trend tests

Spatial dependence (cross-correlation) creates an overlap in the information contained in each datapoint. For example, if flood flows are cross-correlated and a trend is found at one site, one is more likely to find trends at nearby sites as well. From a statistical perspective, correlation reduces the effective sample size of the dataset. This results in a more "liberal" hypothesis test, meaning that, if correlation is ignored, the null hypothesis (of independence) will tend to be rejected more frequently than it should be. This was dramatically demonstrated by Douglas et al. [2000], who found no evidence of regional trends in flood flows but did find evidence of upward regional trends in low flows in three Midwestern regions. When the same analyses were performed ignoring regional cross-correlation, statistically significant trends were indicated in low flow data in all regions across the U.S. and in flood flow data in two-thirds of the regions. They also quantified a dramatic reduction in the effective sample size (from hundreds to less than ten) due to cross-correlation.

In order to evaluate trends at a regional scale, one must be able to interpret the statistical significance of a collection of individual tests. The statistical significance of each individual test (known as the p-value) is defined as the local significance. Field significance (denoted as α) is the collective significance of a group of hypothesis tests and may be defined in a number of different ways. Vogel and Kroll [1989] present one method for determining field significance by utilizing a uniform probability plot correlation hypothesis test. Livezy and Chen [1983] demonstrated another approach using a binomial distribution for a number of significance tests. When cross-correlation between flow series exists, the hypothesis tests are no longer independent and the binomial distribution no longer describes the probability distribution of the floods. Therefore, a simulation technique must be performed to assess overall field significance (α). Studies of trends in climatological data using field significance include Lettenmaier et al. [1994], Wilks [1996], Shabbar et al. [1997], Chu and Wang [1997], Suppiah and Hennessy [1998], and Douglas et al. [2000]. However, of recent studies that have evaluated trends in monthly and annual floods and low flows, only two [Lettenmaier et al., 1994; Douglas et al., 2000] accounted for the spatial dependence of the flow records.

Lettenmaier et al. [1994] performed regional trend tests on monthly and annual hydroclimatic data using the seasonal Mann-Kendall test [Hirsch et al., 1982]. They preserved the regional cross-correlation structure by developing parametric correlation functions from product-moment correlation estimates for station pairs within 15° of the north-south and east-west axes and correlation distances estimated for each variable and season. They evaluated the field significance of the regional hypothesis tests by Monte Carlo simulations that incorporated the appropriate regional correlation function. Douglas et al. [2000] developed a new test statistic, which was termed the regional average Kendall's S, computed as the average of the Kendall statistics for each site in the region [see also Kendall, 1962; Helsel and Hirsch, 1992]. For independent, identically distributed flow records, a normalized test statistic can be computed and corrected for ties in the same way as the original Mann-Kendall test [Kendall, 1962; Helsel and Hirsch, 1992]. A resampling method such as the bootstrap method [Efron, 1979] or the analytical expression presented by Douglas et al. [2000] can be used to preserve the spatial correlation structure of regional time series.

THE IMPORTANCE OF UNDERSTANDING THE CAUSES OF HYDROCLIMATIC CHANGE

There is often disagreement as to whether or not understanding the causes of change in the hydrologic record is important. This was recently illustrated by members of a focus group of consulting engineers (Annual Meeting of the Water Resources Planning and Management Division, American Society of Civil Engineers, 8 June 1999, Tempe, Arizona), who clearly stated their disinterest in the non-stationarity implied by paleohydroclimatological data without a demonstrable link between climate and watershed processes of the past and present [H. Hartmann, University of Arizona, personal communication, 2001]. Tools for understanding hydroclimatic variability are still evolving and yet the responsibility of hydrologists and engineers to provide the best quantification of hydrologic risk remains clear. Recent research has begun to unravel the complex relationships between basin-, regional- and global-scale hydroclimatology. However, a great deal of work has yet to be done before these relationships can be incorporated into standard engineering practice. We present a few studies to elucidate some of the questions and the methods to help answer them.

Lins [1993] defined coupled models of monthly variability between regional U.S. streamflow and mid-tropospheric circulation during the period 1951-1988 using canonical correlation analysis. He found that monthly streamflow variations were most influenced by five generic circulation modes: the Pacific/North American (PNA), Mixed Pacific (MP), Asian/Eurasian (AE), East Pacific (EP), and North Atlantic (NA). The streamflow responses to these forcing regimes was observed to vary from month to month, although some regimes preferentially affected hydrologic conditions in particular areas of the U.S. Anomalous streamflow conditions during winter, for example, were most affected by PNA forcing, in spring and summer by MP forcing, and in autumn by AE forcing. This study demonstrated the enormous spatial and temporal complexity and subtlety of the relations between streamflow and only one component of the climate system, i.e., atmospheric circulation. It indicates that a more complete accounting of climate system effects on streamflow for use in water system planning and design, including ocean temperature and circulation, is an important, albeit difficult, task.

El Niño is an example of a global factor that would be expected to affect hydrologic behavior on an interannual basis as it diminishes and reappears within 5-8 years [Piechota and Dracup, 1996]. In recent years, strong El Niño patterns have reminded the world of the global teleconnections that exist between the phenomena measured by the SOI and various hydrologic phenomena, including floods in Peru, increased rainfall in Florida and California, and ice storms in the U.S. Northeast [Piechota and Dracup, 1996; Ropelewski and Halpert, 1996; Sun and Furbish, 1997]. Rajagopalan and Lall [1998] used spectral analysis to evaluate interannual variability in precipitation records from the western U.S. They examined the connection between monthly precipitation and widely used atmospheric indices and found statistically significant signals related to El Niño-Southern Oscillation (ENSO) indices. However, in his analysis of the great Mississippi River flood of 1993, Rodenhuis [1996] found that the origin of flood-producing rain conditions in 1993 could not be assigned to any one atmospheric mechanism, and that the influence of ENSO, while important, was indirect.

Other climatic factors, such as the Pacific Decadal Oscillation (PDO) and the North Atlantic Oscillation (NAO), have longer-term impacts [Manuta et al. 1997; Leathers et al., 1991; Dettinger and Cayan, 1995; Mann and Park, 1994, 1996; Lall, 1995] that may be expected to impact the interdecadal variability in the risk of extreme events, including large floods and low flows. Goldenberg et al. [2001] indicate that hurricane activity in the North Atlantic shows prolonged periods of increased activity, which is consistent with the findings of Walker et al. [in review] that flood risk in the mid-Atlantic and Southeast U.S. varies on the order of decades. Eltahir [1996] attempted to relate long-term persistence identified by the Hurst phenomenon to identifiable climate signals.

Rajagopalan et al. [2000] present a summary of decadal variations in ENSO teleconnections, continental temperatures, precipitation, streamflow, and drought indices around the globe. However, they state that the cause of these variations is not clear, nor is it clear whether ENSO should be considered as an episodic behavior of the climate system or a continuous hydrologic response. They address these questions using data (global sea surface temperature (SST) anomalies, the Palmer Drought Severity Index (PDSI), and the Niño -3, NAO, and PDO indices) and methods (moving window and wavelet analyses, hypotheses testing, and correlation analysis) familiar to hydrologists. They highlighted the need to better understand the spatial structure of the tropical Pacific and global SST fields in predicting PDSI and note that the frequency structure of streamflow, lake levels, and the PDSI is a linearly or nonlinearly modulated version of the climate forcing.

Angel and Huff [1995] outline the need for a comprehensive look at seasonal rainfall frequency and soil moisture for better water resource design in the U.S. Midwest. This need is supported by Kunkel [1996] and Rodenhuis [1996] who noted high and persistent soil moisture conditions in the months leading up to the great Mississippi River flood. Baldwin and Lall [1999] demonstrated the use of graphical methods for highlighting decadal- and century-scale trends. They also evaluated trends in the seasonal cycle of the upper Mississippi River flows, precipitation, and temperature data using spectral analysis and by evaluating the relationship between high- and low-flow modes in empirical probability distributions for select months.

Finally, the weight of evidence collected over the last decade has lead to a general consensus within the scientific community that the climate of the Earth is changing, with gatherings of scientists producing statements like that embodied in the Amsterdam Declaration on Global Change. The declaration states that "the Earth system has moved well outside the range of natural variability exhibited over the last half million years at least. The nature of changes now occurring simultaneously in the Earth System, their magnitudes and rates of change are unprecedented" [The Amsterdam Declaration on Global Change, A Global Change Open Science Conference, Amsterdam, Netherlands, July 10-13, 2001].

EXAMPLES OF INCORPORATING CLIMATE INFORMATION INTO WATER SYSTEM MANAGEMENT

Sharma [2000a, b, c] presented a method for seasonal to interannual probabilistic rainfall forecasts conditional on current values of ENSO indices and sea surface temperature anomalies (SSTA). He developed a nonparametric probability distribution function (pdf) that was able to represent the bimodal shape inherent in many rainfall distributions, which conventional pdfs (shown in Figure 1) cannot accommodate. He found that better and more stable forecasts with lead times greater than one season are possible using SSTA, since they have longer memory than ENSO indices.

Pagano et al. [2001] present an example in which water managers and agencies in Arizona used quantitative indices and qualitative judgment based on previous experience to forecast the impact of the 1997-1998 El Niño event on local water resources. The most notable action was the decision by the Salt River Project (SRP) to reduce ground water pumping in anticipation of increased precipitation produced by El Niño events. They state that SRP's bold response, months ahead of actual regional impacts, saved the agency $1 million in ground water pumping costs. Pagano et al. [2001] further outline barriers to the use of climate forecasts and recommend ways to make climate forecasts more useable in the future.

CONCLUSIONS

Incorporating large-scale climate data and the effects of spatial and temporal dependence of hydroclimatological data into hydrologic analysis was once a daunting, if not impossible task. However, methods and data now exist to allow for the development of more complex and more explanatory hydrologic models. Although debates will continue about whether anthropogenic climate change or natural climate variability are the cause of the observed changes in the hydrologic data, Barros and Evans [1997] well note that this concern is largely irrelevant from the perspective of infrastructure engineering; either cause implies the need to re-engineer variability into existing regulations and guidelines, which must accommodate a broader range of extremes.

Science is still unraveling the complex relationships that define our Earth's climate, and climate models continue to improve. The challenge remains to integrate climate forecasts into a dynamic risk framework that incorporates a more predictive approach to hydrologic risk analysis. In order to identify the periods of higher hydrologic risk, researchers must continue to explore the linkages between major hydrologic events and climatic variations or watershed conditions, as well as to develop methods of risk estimation that incorporate such variation. This chapter has cited some (but far from all) of the recent research documenting the links between global-scale climatic processes and local- to regional-scale hydrology. Methods developed in the future should accommodate climatic variability by making use of the available large-scale hydroclimatological data and model predictions. As our understanding of the relationships between local and global processes continues to advance, explanatory hydrologic models that link hydroclimatic variables through observed global teleconnections should become the norm for risk-based hydrologic design.

REFERENCES

Angel, J. R. and F. A. Huff, Seasonal distribution of heavy rainfall events in the Midwest, *J. Water Res. Plann. Mgmt*, 121 (3): 110-115, 1995.

Baldwin, C. K. and U. Lall, Seasonality of streamflow, *Water Resour. Res.*, 35 (4): 1143-1154, 1999.

Barros, A. P. and J. L. Evans, Designing for climate variability, *J. Professional Issues in Engineering Education and Practice*, 123 (2): 62-65, 1997.

Bence, J. R., Analysis of short time series: correcting for autocorrelation, *Ecology*, 76 (2): 628-639, 1995.

Berryman, D., B. Bernard, D. Cluis, and Haemmerli, J., Nonparametric tests for trend detection in water quality time series, *Water Resour. Bull.*, 24 (3): 545-556, 1988.

Chu, P. -S. and Wang, J. -B., Recent climate change in the tropical western Pacific and Indian Ocean regions as detected by outgoing longwave radiation, *J.Climate*, *10*, 636-646, 1997.

Dettinger, M.D. and D. R. Cayan, Large scale atmospheric forcing of recent trends toward early snowmelt in California, *J. Climate*, 8, 606-623, 1995.

DeWalle, D. R., B. R. Swistock, T. E. Johnson and K. J. McGuire, Potential effects of climate change and urbanization on mean annual streamflow in the United States, *Water Resour. Res*, 36 (9): 2655-2664, 2000.

Douglas, E.M., R.M. Vogel and C.N. Kroll, Trends in flood and low flows across the U.S., *J. Hydrology*, 240(1-2): 90-105, 2000.

Douglas, E.M., R.M. Vogel and C.N. Kroll, Impact of streamflow persistence on hydrologic design, *J. Hydrologic Engineering*, 7 (3): 220-227, 2002.

Efron, B., Bootstrap methods: another look at the jackknife. *Ann. Stats.*, *7(1)*, 1-26, 1979.

Eltahir, E., El Nino and the natural variability in the flow of the Nile River, *Water Resour. Res*, 32 (1): 131-137, 1996.

Fernandez, B. and J.D. Salas, Occurrence interval and risk of hydrologic events, I: Mathematical formulation, *J. Hydrologic Engineering*, 4(4): 297-307, 1999*a*.

Fernandez, B. and J.D. Salas, Occurrence interval and risk of hydrologic events, II: Applications, *J. Hydrologic Engineering*, 4(4): 308-316, 1999*b*.

Genta, J.L., Perez-Iribarren, G., and C.R. Mechoso, A recent increasing trend in the streamflow of rivers in southeastern South America. *J. Climate*, 11: 2858-2862, 1998.

Goldenberg, S.B., C.W. Landsea, A.M. Mestas-Nunez, and W.M. Gray, The recent increase in Atlantic hurricane activity: Causes and implications. *Science*, 293:474-479, 2001.

Gumbel, E. J., The occurrence interval of flood flows, *Ann. Math. Stat.*, 12 (2), 163-190, 1941.

Hamed, D. H. and A. R. Rao, A modified Mann-Kendall trend test for autocorrelated data, *J. Hydrology*, 204: 182-196, 1998.

Harcum, J. B., J. C. Loftis, and R. C. Ward, Selecting trend tests for water quality series with serial correlation and missing values, *Water Resour. Bull.*, 28 (3): 469-478, 1992.

Helsel, D. R. and R. M. Hirsch, *Statistical methods in water resources*, Studies in Environmental Science 49, Elsevier Science B. V., Amsterdam, The Netherlands, 1992.

Hirsch, R. M., J. R. Slack, and R. A. Smith, Techniques of trend for monthly water quality data, *Water Resour. Res*, 18 (1): 107-121, 1982.

Hirsch, R. M. and J. R. Slack, A nonparametric trend test for seasonal data with serial dependence, *Water Resour. Res*, 20 (6): 727-732. 1984.

Hosking, J. R. M. and J. R. Wallis, *Regional Frequency Analysis: An approach based on L-moments*, Cambridge University Press, 1997.

Hurst, H, Long term storage capacity of reservoirs, *Trans. Am. Soc. Civ. Eng.,* 116, 770-808, 1951.

Jain, S. and U. Lall, Floods in a changing climate: Does the past represent the future? *Water Resour. Res,* 37(12), 3193-3206, 2001.

Kendall, M. G., *Rank correlation methods*, 3rd ed., Hafner Publishing Co., New York, 1962.

Klemes, V, The Hurst phenomenon: A Puzzle? *Water Resour. Res*, 10, 4, 675-687, August, 1974.

Knapp, H. V., Hydrologic trends in the upper Mississippi river basin, *Water Intl.,* 19, 199-206, 1994.

Knox, J. Fluvial Responses to Small Scale Climate Changes, in *Developments and Application of Geomorphology,* Springer-Verlag, Berlin, 1984.

Kunkel, K. E., A hydroclimatological assessment of the rainfall, in *The Great Flood of 1993: Causes, impacts and responses*, edited by S. A. Changnon, Westview, Boulder, CO, 1996.

Krzyscin, J. W., Detection of a trend superposed on a serially correlated time series, *J. Atmospheric and Solar-Terrestrial Physics*, 59 (1): 21-30, 1997.

Lall, U., The Great Salt Lake: A barometer of low frequency climatic variability, *Water Resour. Res,* 31, 10, 2503-2515, October, 1995.

Leathers, D, B. Yarnal and M. A. Palecki, The Pacific-North American teleconnection pattern and United States climate, Part I: Regional temperature and precipitation associations." *J. Climate*, 4 (5): 517-528, 1991.

Lettenmaier, D. P., Wood, E. F. and J. R. Wallis, Hydro-climatological trends in the continental United States, 1948-88, *J. Climate, 7*, 586-607, 1994.

Lins, H.F., Seasonal Hydrologic Variability and Relations with Climate, Ph.D. Dissertation, University of Virginia, Charlottesville, VA, 1993.

Lins, H. F. and J. R. Slack, Streamflow trends in the United States. *Geophys. Res. Let., 26*, 227-230, 1999.

Livezy, R. E. and W. Y. Chen, Statistical field significance and its determination by Monte Carlo techniques. *Mon. Weather Rev., 111*, 46-59, 1983.

Lloyd, E.H., Return period in the presence of persistence, *J. Hydrology*, 10 (3): 291-298, 1970.

Mandelbrot, B. and J. R. Wallis, Robustness o fhte rescaled range R/S in the measurement of non-cyclic long-run statistical dependence, *Water Resour. Res*, 5 (5), 1969.

Mann, M.E. and J. Park. Global scale modes of surface temperature variability on interannual to century time scales, *J. Geophys. Res.,* 99, 25819-25833, 1994.

Mann, M.E. and J. Park, Joint spatial-temporal modes of surface temperature and sea level pressure variability in the Northern Hemisphere during the last ventury, *J. Climate*, 9, 2137-2162, 1996.

Mantua N. J., S. R. Hare, Y. Zhang, J. M. Wallace, and R. C. Francis, A Pacific interdecadal climate oscillation with impacts on salmon production, *Bull. Am. Meteorol. Soc.*, 78, 1069-1079, 1997.

Matalas, N. C., Stochastic hydrology in the context of climate change, *Climatic Change*, 37 (1):89-101, 1997.

National Research Council, *Flood Risk Management and the American River Basin: An Evaluation*. National Academy Press, Washington, D.C., 1995.

Olsen, J. R., Lambert, J. H. and Y. Y. Haimes, Risk of extreme events under nonstationary conditions, *Risk Analysis, 18 (4)*, 497-510, 1998.

Olsen, J. R., Stedinger, J. R., Matalas, N. C. and E. Z. Stakhiv, Climate variability and flood frequency estimation for the upper Mississippi and lower Missouri rivers, *J. Am. Water Resour. Assoc.*, 35 (6): 1509-1523, 1999.

Pagano, T. C., H. C. Hartmann and S. Sorooshian, Using climate forecasts for water management: Arizona and the 1997-1998 El Nino, *J. Am. Water Resour. Assoc.*, 37 (5): 1139-1153, 2001.

Piechota, T. and J. Dracup. Drought and regional hydrologic variation in the United States with El Niño-Southern Oscillation, *Water Resour. Res,* 32 (5): 1359-1373, 1996.

Potter, K. W., Evidence of nonstationarity as a physical explanation of the Hurst phenomenon, *Water Resour. Res*, 12 (5), October 1976.

Potter, K. W., Hydrological impacts of changing land management practices in a moderate-sized agricultural catchment, *Water Resour. Res*, 27 (5):845-855, 1991.

Rajagopalan, B. and U. Lall, Interannual variability in western US precipitation, *J. Hydrology,* 210: 51-67, 1998.

Rajogopalan, B., E. Cook, U. Lall and B. K. Ray, Spatiotemporal variability of ENSO and SST teleconnections to summer drought over the United States during the twentieth century, *J. Climate*, 13 (24): 4244-4255, 2000.

Rodenhuis, D. R., The weather that led to the flood, in *The Great Flood of 1993: Causes, impacts and responses*, edited by S. A. Changnon, Westview, Boulder, CO, 1996.

Ropelewski, C. and M. Halpert, North American precipitation and temperature patterns associated with the El Nino-Southern Oscillation (ENSO), *Mon. Weather Rev.*, 114:

2352-2362, 1996.

Salas, J. D., Analysis and modeling of hydrologic time series, Chapter 19 in *Handbook of Hydrology*, D. R. Maidment, Ed., McGraw-Hill, Inc., 1993.

Sen, Z., Simple risk calculations in dependent hydrological series, *Hydro. Sci.*, 44 (6): 871-878, 1999.

Shabbar, A. Bonsal, B. and Khandekar, M., Canadian precipitation patterns associated with the southern oscillation. *J. Climate, 10*, 3016-3027, 1997.

Sharma, A, Seasonal to interannual rainfall probabilistic forecasts for improved water supply management: 1 – a strategy for system predictor identification, *J. Hydrology*, 239: 234-239, 2000*a*.

Sharma, A, Seasonal to interannual rainfall probabilistic forecasts for improved water supply management: 2 – Predictor identification of quarterly rainfall using ocean-atmosphere information, *J. Hydrology*, 239: 240-248, 2000*b*.

Sharma, A, Seasonal to interannual rainfall probabilistic forecasts for improved water supply management: 3 – a nonparametric probabilistic forecast model, *J. Hydrology*, 239: 249-258, 2000*c*.

Sun, H. and D. Furbish, Annual precipitation and river discharges in Florida in response to El Nino and La Nina sea surface temperature anomalies, *J. Hydrology*, 199: 74-87, 1997.

Suppiah, R. and Hennessy, K., Trends in total rainfall, heavy rain events and number of dry days in Australia, 1910-1990. *Intl. J. Climatology, 10*, 1141-1164, 1998.

Stedinger, J.R., R. M. Vogel, and E. Foufoula-Georgiou, The Frequency Analysis of Extreme Events, Chapter 18 in *Handbook of Hydrology*, D. R. Maidment, Ed., McGraw-Hill, Inc., 1993.

Thomas, H. A., Frequency of minor floods, *J. Boston Soc. Civ. Eng.*, 35 (1), 425-442, 1948.

United States Geological Survey (USGS), Peak annual flows at station 11446500 of American River, www.ugsg.gov, 1999.

Vogel, R.M., Reliability indices for water supply systems *J. Water Res. Plann. Mgmt*, 113 (4): 563-579, 1987.

Vogel, R. M. and C. N. Kroll, Low-flow frequency analysis using probability plot correlation coefficients, *J. Water Res. Plann. Mgmt., 115(3)*, 338-357, 1989.

Vogel, Richard, Yushiou Tsai, and James F. Limbrunner, The regional persistence and variability of annual streamfiow in the United States, *Water Resour. Res*, 34, 12, 3445-3459, 1998.

Von Storch, H., Misuses of statistical analysis in climate research, in *Analysis of Climate Variability: Applications of statistical techniques*, edited by H. von Storch and A. Navarra, Springer-Verlag, Berlin, Germany, 1995.

Walker, F. Russell, Statistical Analysis of Hydrologic and Environmental Data. Ph.D. Dissertation, Cornell Univ., Ithaca, NY, June, 1999.

Walker, F.R., and J.R. Stedinger, Long-term variability in the arrival rate of flood events as evidenced by flood clustering, *EOS Trans. AGU*, 2000 Spring Meeting, 81(19), S200, May 9, 2000.

Walker, F.R., J.R. Stedinger, and E. S. Martins, Flood clustering as evidence of variability in flood arrival rates, *Water Resour. Res*, in review.

Westmacott, J.R. and D.H. Burn, Climate change effects on the hydrologic regime within the Churchill-Nelson River basin. *J. Hydrology*, 202: 263-279, 1997.

Wilks, D. S, Statistical significance of long-range "Optimal Climate Normal" temperature and precipitation forecasts. *J. Climate, 9*, 827-839, 1996.

F. Russell Walker, Jr.; e-mail: rw31@cornell.edu

Ellen Marie Douglas, Water Systems Analysis Group, Complex Systems Research Center, Institute for the Study of Earth, Oceans and Space, University of New Hampshire, Durham, New Hampshire 03824; Ellen.Douglas@unh.edu

14

Advanced Hydrologic Predictions for Improving Water Management

Holly C. Hartmann, Allen Bradley, and Alan Hamlet

INTRODUCTION

Droughts, floods, and increasing demands on available water supplies consistently create concern, and even crises, for water resources management. As options for infrastructural solutions to water problems become constrained, the focus of water management must increasingly shift to make better use of existing resources, even in the face of extreme natural variability. Forecasts of weather, climate, and hydrologic conditions have long promised to enhance the ability of resource managers to accommodate the vagaries of nature. Short-term forecasts, covering minutes to days, potentially inform reaction and response decisions (e.g., to flash floods), while long-term forecasts, covering monthly to seasonal time scales, potentially enable proactive planning and adaptive responses (e.g., to seasonal water supply shortages). Although exact accounting is difficult, potential values associated with appropriate use of accurate hydrometeorologic predictions generally range from the millions to the billions of dollars [e.g., National Hydrologic Warning Council, 2002]; there are also nonmonetary values associated with more efficient, equitable, and environmentally sustainable decisions related to water resources.

Governments have made large investments to improve weather, climate, and hydrologic predictions over the past decades through satellites, in situ measuring networks, supercomputers, and research programs. However, there has been broad disappointment in the extent to which improvements in hydroclimatic science from large-scale research programs have affected resource management policies and practices [Pielke, 1995, 2001; National Research Council, 1998a,b, 1999a,b,c; Kates et al., 2001]. Several national and international programs have explicitly identified as an important objective ensuring that improved data products, conceptual models, and predictions are useful to the water resources man-

Water: Science, Policy, and Management
Water Resources Monograph 16

10.1029/016WM17

agement community [Endreny et al., this volume]. Many reasons exist for the slow adoption of advanced predictions in water management, including lack of familiarity with forecast products, disconnect between the forecasted variables and those relevant to decision makers, skepticism about forecast quality, and institutional impediments [Changnon, 1990; Pulwarty and Redmond, 1997; Pagano et al., 2001, 2002; Jacobs and Pulwarty, this volume]. However, the fact remains that many new technologies exist for generating advanced hydrometeorologic predictions useful for water management.

The onus is not simply on the water management community to become more adaptable, however. Reflecting the linkages between natural variability, predictive technologies, and water management decisions, an end-to-end prediction system extends from data through large-scale predictions, regional forecasts, forecast evaluation, impacts assessment, applications, and evaluation of applications [Miles et al., 2000]. With this perspective, more effective application of advanced forecasts can be seen to require coordinated efforts among the research, operational forecasting, and water management communities. In this chapter, we attempt to foster more extensive, informed, and interconnected application of hydrometeorologic predictions in water resources management by addressing (1) recently developed and evolving techniques for improving forecasts, (2) the multi-dimensional nature and implications of forecast uncertainty, and (3) efforts to integrate atmospheric and hydrologic predictive systems with decision-making processes.

ADVANCED HYDROLOGIC FORECASTING TECHNIQUES

In an operational setting, forecast hydrologists may have many watersheds and stream locations for which to generate and issue predictions, and for flash flood watches and warnings many forecasts must be issued in a short period of time. For example, within the United States, the National Weather Service (NWS) provides streamflow forecasts on a routine, continuous basis for about 4000 locations encompassing diverse hydrologic regimes and time scales ranging from minutes to months [Fread, 1999]. Figure 1 illustrates various components important to hydrologic forecasting. While each component is amenable to improvement, in practice, technological advances can be difficult to implement in operational systems. For the practical application of any new predictive technology, the stream of data inputs for the hydrologic models must be dependable, without downtime or large data gaps, and data processing, model simulation, and creation of forecast products must be fast and efficient. This section addresses three areas where technological advances exist, but have not found ready application in water resources management applications. They include new hydrologic models, model inputs, and the use of ensembles (i.e., collections of individual time series of model predictions) to produce probabilistic forecasts.

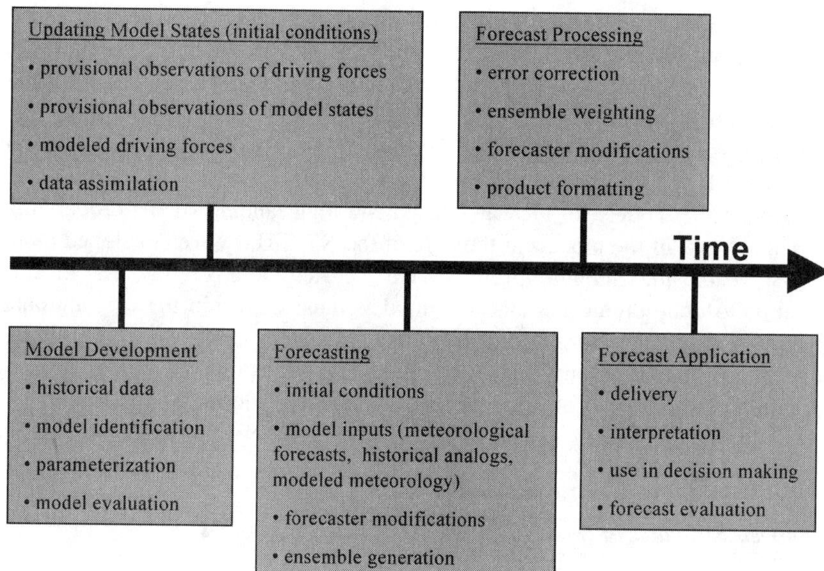

Figure 1.

Modeling Advances

At the core of any forecasting system is predictive capacity, whether by a simple statistical relationship or a complex dynamic numerical model. Advances in hydrologic modeling have been notable, especially those associated with the proper identification of a predictive model and its parameters [e.g., Duan et al., 2002] and the development of models that consider the spatially distributed characteristics of watersheds rather than treating entire basins as a single point [Grayson and Bloschl, 2000]. However, the incorporation of new models in operational hydrologic forecasting has been frustratingly slow [Hartmann et al., 2002a]. From the NWS perspective [Fread, 1999], a central challenge to improving hydrologic forecasts has simply been maintaining the reliability of the NWS River Forecast System (NWSRFS), which consists of over 400,000 lines of computer code.

The contrast between meteorological and hydrologic forecasting within the NWS [Hartmann et al., 2002a] is useful to examine, because it reveals how the rate of adoption of emerging modeling technologies can depend on institutional philosophies. The state of meteorological forecasting resulted from a distinct shift in NWS institutional philosophy [Mittelstadt, 1997], and is characterized by rapid evolution, with new predictive models moving relatively quickly from research to experimental to operational status. Previously, NWS meteorological forecasting models were limited to those passing development and evaluation thresholds. Forecast models were used unchanged until major scientific and technological advancements were incorporated and evaluated. Subsequently, however, model changes were incorporated as soon as they passed initial testing and operational adjustments (e.g., data handling) could be implemented.

In contrast, the state of hydrologic forecasting is fairly characterized as evolving slowly, based on institutional preferences for uniformity in operations (i.e., using the same models for all watersheds) and longstanding standard operating procedures. All of the models at the core of the NWSRFS were developed more than 30 years ago, although upgrades have occurred more recently. The institutional philosophy predicates that new models must fit within the existing data management infrastructure and demonstrate improved performance in an independent operational setting over several years. Those requirements have frustrated members of the hydrologic research community who have developed new models that they thought should be incorporated into NWS operational hydrologic forecast systems.

Improved Estimates of Initial Conditions and Driving Forces

As seen in Figure 1, both initial conditions and inputs used to drive predictive models are required for hydrologic forecasting. The role of initial conditions varies depending on the time scale of the forecast and the extremity of both the initial watershed states and subsequent meteorological conditions. As atmospheric science and the quality of meteorological forecasts have improved, hydrologists have increasingly been interested in the use of weather and climate forecasts for driving hydrologic models and predictions.

Initial conditions. Although meteorological uncertainty may be high for the periods addressed by streamflow forecasts, accurate estimates of the state of watershed conditions prior to the forecast period are important because they are used to initialize hydrologic model states, with significant consequences for forecast results. However, they can be difficult to measure, especially when streamflow forecasts must be made quickly, as in the case of flash flood forecasts. One option is to continuously update watershed states by running the hydrologic models continuously, using inputs from recent meteorological observations and/or atmospheric models. Regardless of the source of inputs, Westrick et al. [2002] found it essential to obtain observational estimates of initial conditions to keep streamflow forecasts realistic; storm-by-storm corrections of model biases determined over extended simulation periods were insufficient.

Where streamflows may be largely comprised of snowmelt runoff, quality estimates of snow conditions are important. The importance of reducing errors in the timing and magnitude of snowmelt runoff are especially acute in regions where a large percentage of annual water supplies derive from snowmelt runoff, snowmelt impacts are highly non-linear with increasing deviation from long-term average supplies, and reservoir storage is smaller than interannual variation of water supplies. However, resources for on-site monitoring of snow conditions have diminished rather than grown, relative to the increasing costs of errors in hydrologic forecasts [Davis and Pangburn, 1999]. Research activities of the NWS National Office of Hydrology Remote Sensing Center (NOHRSC) have long been directed at improving estimates of snowpack conditions through aerial and satellite remote sensing [Carroll, 1985]. However, the cost of aerial flights prohibits routine use [T. Carroll, NOHRSC, personal communication, 1999], while satellite estimates have qualitative limitations (e.g., not considering fractional snow coverage over large regions) and have not found broad use operationally, except on the Canadian prairies where snow water volumes are based on passive microwave satellite data [Walker and Goodison, 1993].

Recent experimental end-to-end forecasts of streamflow produced in a simulated operational setting [Wood et al., 2001] highlighted the critical role of quality estimates of spring and summer soil moisture used to initialize hydrologic model states for the eastern U.S. Land data assimilation systems (LDAS) offer various approaches for optimal adjustment of model states, such as soil moisture, based on observed conditions. Ongoing LDAS research is focused on mixing model and instrumental measurements of land surface variables in complex dynamic numerical models developed by the atmospheric science community [e.g., Cosgrove et al., 2002].

Using atmospheric models to drive hydrologic models. Streamflow and other hydrologic variables are intimately responsive to atmospheric factors, especially precipitation, that drive a watershed's hydrologic behavior. However, obtaining quality precipitation estimates is a formidable challenge, especially in the western U.S. where orographic effects produce large spatial variability and there is a scarcity of real-time precipitation gauge data and poor radar coverage. In principal, outputs from atmospheric models could serve as surrogates for observations, as well as providing forecasts of meteorologic variables that can be used to drive hydrologic models. One issue in integrating atmospheric model output into hydrologic models for small watersheds (<1000 km^2) is that the spatial resolution of atmospheric models is lower than the resolution of hydrologic models. For example, quantitative precipitation forecasts (QPFs) produced by some atmospheric models may cover several thousand square kilometers, but the hydrologic models used for predicting daily streamflows require precipitation to be downscaled to precipitation fields for watersheds covering only tens or hundreds of square kilometers.

One approach to produce output consistent with the hydrologic model needs involves nested atmospheric models, whereby outputs from large scale but coarse resolution models are used as boundary conditions for models operating over smaller extent with higher resolution. For example, Westrick and Mass [2001] used outputs from a nested mesoscale atmospheric model [MM5; Dudhia et al., 1999] over the Puget Sound drainage basin, in the state of Washington, to drive the Distributed Hydrology Soil and Vegetation Model [DHSVM; Wigmosta et al., 1994, 2002], a spatially-distributed hydrologic model. However, the quality of information from local-scale atmospheric models is highly sensitive to the boundary conditions imposed by larger-scale models [Anthes et al., 1989]. Alternatively, Clark et al. [1999] used a subset of atmospheric variables (e.g., total column precipitable water) in a series of multiple regression equations to predict regional and local variations in the meteorologic variables required by hydrologic models (i.e., precipitation and temperature), with a separate set of equations for each forecast lead-time.

The error characteristics of atmospheric model products (e.g., bias in precipitation and air temperature) can have significant effects on subsequent streamflow forecasts. Bias corrections require knowledge of the climatologies (i.e., long-term distributions) of both modeled and observed variables. The premise is that atmospheric model forecasts may have a useful signal if interpreted relative to the model climatology rather than the observed climatology. Hamlet and Lettenmaier [2000] found that simple adjustments of climate model outputs (i.e., shifting the mean and rescaling the variance) were insufficient to counter the complex patterns of bias in climate forecasts. A more detailed approach, which transforms the climate forecast distribution according to the observed climatological distribution, performed much better in test evaluations in the Columbia River Basin and a tributary of the Ohio River Basin [Wood et al., 2002].

However, as noted earlier, operational atmospheric models undergo frequent evolution. Any forecast-corrective schemes dependent on reanalysis of atmospheric model error characteristics will soon be outdated. Clark et al. [2001] found that, in the context of streamflow prediction, it was better to use outdated error characteristics and a stable numerical weather prediction (NWP) model rather than an evolving state-of-the-art NWP model with unknown error characteristics. They used a circa-1998 version of the National Centers for Environmental Prediction (NCEP) operational NWP model to develop statistical relationships between model outputs (e.g., maximum precipitable water) and observations of precipitation and air temperature, which served as hydrologic model inputs and were thus downscaled and free of systematic biases.

Artificial neural networks. The use of atmospheric models for input to hydrologic models requires extensive data processing and computational resources, raising questions about their advantage over simpler, faster streamflow forecasting approaches, especially considering that forecasters must issue products for many locations at once during periods when the forecasts are most important (i.e., when flooding potential is greatest).

Artificial neural networks (ANNs) offer a means for quick generation of quantitative precipitation forecasts used to drive hydrologic models for flash flood forecasting. ANNs are similar to regression models in that they provide system responses based on a set of predictors, but they can more effectively accommodate highly complex relationships. An ANN model is composed of simple processing units (neurons) arranged in layers. Each unit transforms inputs received into a single output that subsequently provides input to other units; a net input is computed as a weighted sum of inputs. ANN flexibility comes from being able to specify multiple layers of neurons with nonlinear transform functions and alternative methods for computing the net input. ANN calibration iteratively adjusts neuron weights to optimize the specified objective function (e.g., minimization of root mean squared error between ANN output and observations).

Barros et al. [1999] found ANN QPFs, based on non-linear relationships between atmospheric model output and rainfall and radiosonde data, significantly improved forecast skill at locations in Pennsylvania compared to (a) nesting of atmospheric models or (b) downscaling of atmospheric model output based on relationships between precipitation processes and the scaling properties of regional precipitation. Incorporation of the evolving structure and frequency of intense weather systems, derived from satellite-derived storm properties (e.g., storm life time, area, eccentricity, and track), further improved ANN QPFs and their utility for flood forecasting [Kim and Barros, 2001].

Incorporating climate variability. Great strides have been made in understanding and predicting interannual climate phenomena such as the El Nino-Southern Oscillation (ENSO). This improved understanding has resulted in long-lead (up to about a year) climate forecast capabilities that can be exploited in streamflow forecasting. Techniques have been developed to directly incorporate variable climate states into probabilistic streamflow forecast models based on linear discriminant analysis (LDA) with various ENSO indicators, e.g., the Southern Oscillation Index (SOI), Wright sea surface temperatures (SSTs) [Piechota and Dracup, 1999; Piechota et al., 2001]. The best forecasts using SOI and Wright SSTs were obtained for spring through summer runoff, predicted with lead-times ranging from three to seven months. However, the salient variables depend on the specific application. For example, Gutierrez and Dracup [2001] found the multivariate ENSO index (MEI), SOI, and Nino 4 SST anomalies to work best for basins in Colombia.

Use of Ensembles

In recent years, there has been increasing interest in moving from forecasts of categories (e.g., flood/no flood) or continuous variables (e.g., a peak streamflow of 2000 ft^3/s) to probabilistic forecasts [Krzysztofowicz, 2001]. In its strategic plan, 'Vision 2005,' the NWS explicitly commits to "provide weather, water, and climate forecasts in probabilistic terms by 2005" [NWS, 1999]. The principal motivation is that probabilistic forecasts enable quantitative estimation of the inevitable uncertainties associated with weather and climate systems, which are inherently chaotic [Hansen et al., 1997]. From a decision maker's perspective, probabilistic forecasts are more informative because they explicitly communicate uncertainty, and more useful because they can be directly incorporated into risk-based calculations (e.g., expected consequences).

Probabilistic forecasts can be created by overlaying a single prediction with a normal distribution of estimation error determined at the time of calibration [Garen, 1992]. However, to account for future meteorologic uncertainty, new developments have focused on ensembles, whereby multiple possible futures (each termed an ensemble trace) are generated; statistical analysis of the ensemble distribution then provides the basis for a probabilistic forecast.

Chaotic systems, like the atmosphere, are highly sensitive to initial conditions. Thus, in atmospheric models, ensembles are typically generated by running a single dynamic numerical model several times, each time using a slightly different set of initial conditions (e.g., SSTs). Different meteorological forecasting institutions use different methods for adjusting the initial conditions [e.g., Molteni et al., 1996; Toth and Kalnay, 1993, 1997; Houtekamer et al., 1996]. The distribution of initial values affects the ensemble distribution, although the impact depends on model characteristics, especially time scale [Hamill et al., 2000].

Hydrologic forecast ensembles, on the other hand, use several meteorological sequences to produce several hydrologic predictions, keeping constant the hydrologic model's initial conditions (e.g., snowpack or soil moisture). Early implementations of ensemble streamflow prediction (ESP) used historical meteorological sequences to represent future possibilities, and treated the ensemble traces as a random sample, whereby each trace was independent and equally likely to recur [Day, 1985]. For large river basins, use of historical meteorological records sidesteps the problem of adequately representing regional processes through large-scale models, but only to the extent that the historical data represent modern processes (i.e., while it is appropriate for representing meteorological processes, it is not appropriate for hydrologic processes because important watershed and river characteristics are non-stationary). However, some meteorological sequences may be more or less likely during certain periods. Hamlet and Lettenmaier [1999] implemented an ESP system for the U.S. Pacific Northwest using conditional sampling, reflecting the conditional hydrologic response of the Columbia River Basin to the ENSO state. They used historic meteorological data, gridded at a one-degree spatial resolution, to drive the two-layer variable infiltration capacity (VIC) macroscale hydrology model [Wigmosta et al., 1994, 2002] over the entire Columbia River Basin above The Dalles, Oregon. Ensemble traces, consisting of precipitation and air temperature time series covering the season important for seasonal water supplies (October-July), were associated with one of three predefined SOI categories in the historical record. Given a forecast of the ENSO climate signal for the coming water year, the meteorological traces were then used as "forecasts" to drive the VIC model based on the initial soil and snow conditions as of the forecast date.

Alternatively, meteorological sequences could be selected based on consistency with weather and climate forecasts. Subjective selection based on matching characteristics of NWS monthly and seasonal forecasts [Croley and Hartmann, 1987; Smith et al., 1992] proved too cumbersome to implement operationally for multiple basins after the NWS climate outlooks were extended to include 13 different seasons and lead-times. Several researchers [Perica, 1998; Perica et al., 1999; Smith et al., 1992; Croley and Lee, 1993; Croley, 1996, 1997] have developed procedures for efficiently considering multiple meteorological forecasts by restructuring the set of possible future scenarios. Croley [1997] uses all possible meteorological sequences as hydrologic model input, then biases the resulting hydrologic ensemble through differential weighting of the traces, to match the meteorological forecast probabilities. Croley [2000] provides software for combining meteorological forecasts that specify multiple event probabilities (e.g., with variable probabilities for several categories as in the NWS seasonal climate outlooks) and most-probable event outlooks (e.g., seasonal climate outlooks issued by the Canadian Meteorological Centre).

Enhancements to ESP techniques include 'post-processors' that adjust ESP outputs according to quantification of other sources of forecast uncertainty. Perica and Schaake [2000] tested three models describing forecast error, adjusting the ESP forecasts in a transformed standardized Gaussian space to conform to monthly climatological streamflow characteristics. Krzysztofowicz and Herr [2001] developed a Bayesian forecast post-processor that adjusts forecasts of streamflow probability distributions using conditional biases determined over a forecast evaluation period. Hashino et al. [2002] found that each of three methods for correcting for probabilistic forecast bias method affected different attributes of forecast performance, but that quantile mapping produced forecasts with the strongest probability statements. Quantile mapping substitutes the observed flow with the same nonexceedance probability as the simulated flow for each ensemble trace [Hashino et al., 2002]. Other sources of uncertainty exist but are not monitored explicitly, including uncertainties associated with estimation of initial conditions, identification of a realistic hydrologic model, and specification of model parameters. Thus, ESP forecasts still have a tendency to underestimate true forecast uncertainty, while also being unable to statistically reproduce climatological probabilities.

An ongoing question concerns how to combine results from a number of different models, which may themselves consist of an ensemble of individual forecast members, or be of completely different character (i.e., statistical versus dynamical models). At present, NWS atmospheric forecasts are combined subjectively, based on the knowledge of forecasters about the strengths and weaknesses of individual techniques, dominant processes and unusual conditions affecting different regions, and results of recent research. Potential exists to improve probabilistic hydrologic forecasts by considering a traditional single-value forecast as one ensemble member, or using it to make relative adjustments to individual probabilistic forecast ensemble members.

COMMUNICATING FORECAST UNCERTAINTY

Because most all resource management decisions require some sort of hydroclimatic forecast, implicitly or explicitly, forecasts constitute an important link between science and society. Each time a prediction is made, science must address and communicate the strengths and limitations of current understanding. Each time a decision is made, managers must confront their understanding of scientific information and forecast products. Further, each prediction and decision provides opportunities for interaction between scientists and decision makers, and for making clear the importance of investments in scientific research. Perceptions of poor forecast quality are a significant barrier to more effective use of hydroclimatic forecasts [Changnon, 1990; Pagano et al., 2001, 2002; Rayner et al., 2001]. This section presents recent examples of assessing hydroclimatic forecast performance in ways that have meaning and applicability for decision makers.

Evaluations of meteorological forecasts have a long history (Clayton [1889] is an early example). However, information on forecast performance has rarely been available to and framed for decision makers, beyond reviews of specific events [Brooks et al., 1997; Hartmann et al., 2002a], although the situation is improving for seasonal climate forecasts [e.g., Wilks, 2000; Hartmann et al., 2002b; Wilks and Godfrey, 2002]. While hydrologic forecasts are reviewed annually by the issuing agencies in the U.S. [Hartmann et al., 2002a], comprehensive reviews are lacking; for example, Shafer and Huddleston [1985] represents the most recent review of operational water supply forecasts for the U.S. West.

There are a myriad of criteria for evaluating forecast quality.[1] To paraphrase Tolstoy in *Anna Karenina*, while all perfect forecasts resemble one another, imperfect forecasts are imperfect in their own unique ways. Murphy [1993] describes nine complementary ways in which forecasts can be imperfect. Many other measures have been used or proposed [Ward and Folland, 1991; Krzysztofowicz, 1992; Winkler, 1994; Wilks, 1995; Mason and Graham, 1999]. Additionally, the values of various performance criteria are not necessarily informative for decision makers [Hoffrage et al., 2000]; criteria differ in their upper and lower bounds, their desired value, and the magnitude of change that indicates improvement. The diversity of forecast performance measures, with each possessing different characteristics and interpretations, presents real potential for confusing forecast users. Which aspects of quality are most important to users depend on the specifics of each decision making situation. Further, decision makers possess a range of abilities to access, interpret, and apply forecast evaluation results.

Hydrologic forecast evaluation has traditionally used standard summary statistics (e.g., mean errors, bias, correlation) [e.g., Shafer and Huddleston, 1985]. New approaches are being applied, based on meteorological forecast evaluation techniques, by focusing on important portions of both forecast and observation distributions and considering conditional and marginal distributions. The linear error in probability space (LEPS) [Wilks, 1995] recognizes that common events should be easier to forecast well, while expectations of skill should be relaxed for rare events. The LEPS score calculates the difference in the probability of occurrence between the forecasts and observations with respect to the scale of the climatological cumulative distribution function, rather than comparing the magnitude of the difference between specific forecasts and observations as in traditional error analyses (e.g., mean absolute error). Using LEPS analysis, Schwein [2002] showed that the NWS flow and stage crest forecasts for the unprecedented flooding in April 1997 of the Red River of the North, in North Dakota and Minnesota, had high skill even though absolute errors were large relative to historical forecast errors associated with more typical conditions for the same locations.

Hartmann et al. [2002b] used a continuum of evaluation criteria to enable users to increase the sophistication of their understanding about probabilistic forecasts, uncertainty, and implications for decision making. Forecast performance criteria based on 'hitting' or 'missing' associated observations (e.g., probability of detection, false alarm rate) [Wilks, 1995] offers users conceptually easy entry into discussions about forecast quality for forecasts associated with categories (e.g., flood/no flood). They are relatively simple to compute and communicate and can be related to specific user concerns. However, they unfairly penalize probabilistic forecasts by neglecting differences between weak and strong confidence statements. The Brier score [Wilks, 1995] considers the strength of probability statements for two-category forecasts (e.g., flood/no flood), while the Ranked Probability Score (RPS) [Wilks, 1995] considers the distribution of forecast probabilities among multiple observation categories (e.g., floods with different return intervals) and is more appropriate for users interested in the full range of conditions. Recent evaluations of NWS CPC seasonal precipitation outlooks using the RPS [Hartmann et al., 2002b] generally showed the highest skill (relative to climatological probabilities) during the winter, in Florida, the Southwest, and Pacific Northwest; the outlooks were best with short lead times (1-4 months), but skill existed even at moderate leads (5-8 months). Temperature forecasts showed skill, more during the winter than summer, at even the longest lead times (13 months), partly due to secular warming trends. Evaluating ESP seasonal water supply hindcasts (forecasts made retrospectively in a simulated operational setting) from the NWSRFS [Hydrologic Research Laboratory, 1998] for headwater locations in the Colorado River Basin, Franz (2002) determined that even forecasts issued January 1 (covering various periods extending as late as July 31) showed RPS skill; Upper Basin forecasts issued June 1 were nearly 50% better than using climatological probabilities.

The most comprehensive assessments of forecast performance are produced by examining various combinations of the conditional and marginal distributions of paired time series of forecasts and observations [Murphy and Winkler, 1987, 1992]; each combination provides a different perspective on forecast performance (e.g., reliability, discrimination, resolution, refinement). The advantage of evaluating conditional distributions is the ability to identify specific situations whereby forecasts perform particularly well or poorly, which is especially important for hydrologic forecasts that may be good near historical median values but poor for extreme values where quality matters most. Wilks [2000], Hartmann et al. [2002b], and Wilks and Godfrey [2002] provide distributions-oriented evaluations of recent seasonal climate outlooks, while Franz [2002] examined ESP hindcasts of seasonal water supply outlooks for Colorado Basin headwater locations. The evaluations by Hartmann et al. [2002b] illustrate the variable implications of forecast performance for water managers of the Upper versus Lower Colorado River Basins. Poor climate forecast skill within the Upper Basin, where upper elevation snowpacks are the source for almost all streamflows, reinforce the vital importance of high-quality estimates of existing snowpack conditions (e.g., coverage, water content). Within the Lower Basin, snowpack is less extensive and reliable (nearly absent some years), making flow forecasts more dependent on rainfall and thus less predictable. However, significant climate forecast skill for this region during the winter and spring offers potential for improving Lower Basin streamflow predictions.

"Discrimination" [Murphy and Winkler, 1992] identifies the extent to which forecast probabilities differ between cases where the actual observations turned out to be different. Forecasts associated with the eventual observation of high flow conditions should place most of their probability distribution in the upper range of flows, and forecasts associated with subsequent low flow observations should have most of their probability distribution in the lower range of flows. Late-season water supply outlooks examined by Franz [2002] had almost perfect discrimination (Figure 2) for cases where supplies turned out to be high (in the upper 30% of the historical distribution). Forecasts issued as early as January could not discriminate between prospects for high or mid-range flows, but they did indicate that low flows were less likely. "Reliability" measures how well forecast probabilities correspond with their associated relative frequencies of "correct" observations (i.e., the conditional bias of forecast probabilities). For example, perfect reliability would indicate that over many forecasts, high flows actually occurred in 60% of the cases for which forecasts had specified a 60% chance of high flows happening. Franz [2002] found the highest reliability for forecasts issued in March (Figure 2). Forecast "resolution" increased throughout the season, meaning that extreme forecast probabilities were specified more frequently as the season progressed. Hashino et al. [2002] demonstrated that these distributions-oriented criteria comprise a sound framework for comparing approaches for correcting probabilistic forecast bias.

Long-term forecasts pose special difficulties for evaluation. Their limited sample sizes compromise even the most mathematically rigorous analyses, and spatial and temporal autocorrelation reduces effective sample sizes further. Sample

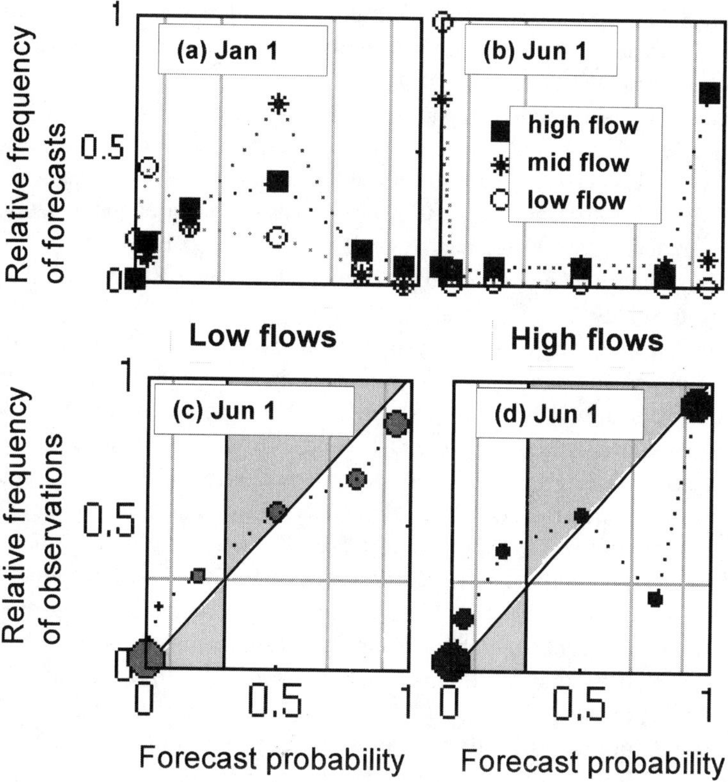

Figure 2.

size problems are exacerbated when forecast time series are disaggregated by climatic state (e.g., by ENSO state); Valdes et al. [1998] found that error characteristics of operational regional long-term hydrologic prediction models varied according to the ENSO state at the time of the prediction. Hindcast evaluation is appropriate, but only to the extent that the actual forecast process is objective, without real-time adjustments based on the expert knowledge of the forecasters. Monte Carlo simulation experiments [e.g., Bradley et al., 2002] can quantify the sensitivity and effect of small sample sizes on confidence limits of forecast quality attribute values. Further, parametric modeling of the forecast process offers a consistent framework to explore the structure, skill, and uncertainty of probabilistic hydrologic forecasts [Bradley et al., 2002]. Bayesian evaluations [e.g., Krzysztofowicz, 1983, 1991; Krzysztofowicz and Davis, 1983] that use parametric modeling can be applied to adjust forecasts based on past performance, although they have not found broad operational use. Limitations include difficulty in explaining Bayesian approaches in terms users can readily understand, but may be addressed by extending the framework of Hartmann et al. [2002b] to include Bayesian analysis and forecast adjustments as a natural progression of distributions-oriented evaluations.

Questions remain about comparing traditional deterministic forecasts with probabilistic forecasts. Official seasonal water supply outlooks issued by the NWS and NRCS have included 'reasonable minimum' and 'reasonable maximum' values derived from "jackknife" error analysis of regression equations at the time of calibration. They can be used as proxy probabilistic forecasts for comparison with hindcast evaluation of probabilistic forecasts generated using advanced techniques.

Finally, it is important to recognize that qualitative aspects of forecast products can be as important as any quantitative attribute in affecting how users interpret, apply, and ultimately judge them [Nicholls, 1999]. Important issues that deserve joint study by forecasters, social scientists, and forecast users include formatting of products, providing historical and recent contexts for forecasts, and dealing with total uncertainty (i.e., lack of predictability) [Hartmann et al., 2002b; Davis and Pangburn, 1999]. The probability-of-exceedance format used by the NWS with their seasonal climate outlooks [Barnston et al., 2000] contains more contextual information than any previous products (e.g., cumulative climatological probabilities and recent conditions). What should forecasters do when their techniques lack skill, produce conflicting results, or are unable to reduce uncertainty (e.g., from lack of data or signal)? Because they have different implications for decision making, a designation of "complete forecast uncertainty" or "no forecast confidence" is more appropriate than simply indicating climatological or equal-probability forecasts and makes clear the need for continued research investments.

INTEGRATING PREDICTIVE SYSTEMS AND DECISION MAKING PROCESSES

Recent advances in modeling and predictive capabilities naturally lead to speculation that hydroclimatic forecasts can be used to improve the operation of water resource systems. In the U.S., the Pacific Northwest, California, and the Southwest are strong candidates for the use of long-lead forecasts because ENSO and PDO signals are particularly strong in these regions and each region's water supplies are closely tied to accumulation of winter snowfall, amplifying the impacts of climatic variability. However, while water resource management has common themes, methods, and regulations, the agencies involved are as diverse as the watersheds with which they are concerned, encompassing water delivery, reservoir regulation for water supply and flood control, and emergency response, among others. Each agency has a specific context within which they must evaluate the risks and benefits of incorporating new technology, including advanced forecast systems, into their operations. This section considers the prospects for accelerating the integration of advanced predictive systems and decision making processes within the water resources sector.

Changnon [2000], Rayner et al. [2001], and Pagano et al. [2002] found that improved climate prediction capabilities are initially incorporated into water management decisions informally, using subjective, ad hoc procedures on the initiative of individual water managers. While improvised, those decisions are not necessarily insignificant. For example, the Salt River Project, among the largest water management agencies in the Colorado River Basin and primary supplier to the Phoenix metropolitan area, decided in August 1997 to substitute groundwater withdrawals with reservoir releases, expecting increased surface runoff during a wet winter related to El Nino. With that decision, they risked losses exceeding $4 million in an attempt to realize benefits of $1 million [Pagano et al., 2002]. Because these informal processes are based in part on confidence in the predictions, overconfidence in forecasts can be even more problematic than lack of confidence, as a single incorrect forecast that provokes costly shifts in operations can devastate user confidence in subsequent forecasts [e.g., Glantz, 1982].

In this context, ongoing relationships among researchers, operational forecasters, and decision makers are essential to ensure appropriate interpretation and application of evolving forecast products. Considering the diversity of water management circumstances, the continually evolving nature of predictive capabilities, the variety of forecast performance requirements and criteria, and workloads facing the research and operational forecasting communities, providing personalized support for individual water managers on a broad scale is impractical. An alternative approach is to provide accessible, understandable, and easily used decision support tools that enable individual decision makers to perform their own data exploration and analyses, customized to reflect their specific situations and concerns, with sufficient expert guidance to transform data and information into practical knowledge. Customizable dynamic decision support tools, implemented using Internet technology and accessing diverse data sources, offer a way forward, although their development is in nascent stages [Sorooshian et al., 2002]. Experience in the Climate Assessment Project for the Southwest (CLIMAS) [see Gamble et al., this volume] suggests that a decision support tool focused on forecast evaluation can help decision makers in their strategic thinking about essential forecast attributes; requisite performance thresholds; relationships among forecast quality, utility, and value; and the potential utility and value of forecast improvements.

Alternatively, end-to-end decision support tools that embody unique resource management circumstances enable formal, and more objective, linkages between meteorological, hydrologic, and institutional processes. Typically, these end-to-end tools are developed for organizations making decisions with high impact (e.g., state or national agencies) or high economic value (e.g., hydropower production), and which possess the technical and managerial abilities to efficiently exploit research advances. When linked to socioeconomic models incorporating detailed information about the choices open to decisionmakers and their tolerance for risk, these end-to-end tools enable explicit acknowledgment of the impacts of scientific and technological research advances. Decision makers that traditionally rely on statistical analysis of historical data can be reluctant to shift operations based only on theoretical advantages, but these tendencies can be countered by using the end-to-end tools in design studies [Lee, 1999, Davis and Pangburn, 1999].

Scott et al. [2000] considered the potential impact of seasonal hydroclimatic forecasts to reduce vulnerability of irrigated agriculture to low water-availability conditions, using the Yakima River Basin in the state of Washington as an example. They considered management regimes linked to different types of hydroclimatic forecast systems, including using ENSO and PDO cycles to classify climate states for contingent management actions, using climate indices (e.g., SOI, PDO) to forecast snowpack and water supplies, and using NWS CPC climate forecasts to predict snowpack and water supplies. The potential benefits of hydroclimatic forecasts were particularly strong for junior water rights holders.

Hamlet et al. [2000, 2002] evaluated the implications of using hydroclimatic forecasts for the operation of the Columbia River hydropower system. Current rules for managing the 20-reservoir system focus on hedging against drought by storing water in summer and fall until operational water supply forecasts that rely primarily on snowpack measurements become available the first of January, representing a six-month lead-time. Over a roughly 50-year evaluation period, retrospective ESP streamflow forecasts with 12-month lead-times were generated using the VIC hydrologic model [Wigmosta et al., 1994, 2002], perfect categorical forecasts of the ENSO state (warm, neutral, cool), and heuristic methods of predicting decadal-scale changes in PDO and initial soil conditions [Hamlet et al., 1999]. New rule curves based on the long-lead streamflow forecasts were constructed to respond to autumn streamflow variability by making more water available for energy production in anticipation of wet years, and less for dry years. By transferring hydropower production from spring (after snowmelt), when energy demand and prices are low, to the late summer and fall (before snowfall), when demand and prices are higher, the system produced annual benefits exceeding $150 million on average, without compromising other operational objectives (e.g., maintenance of minimum river flows). Additional economic benefits were constrained by the limited historical record that could be used to represent forecast climate conditions.

For the Folsom Reservoir in California, Carpenter and Georgakakos [2001] and Yao and Georgakakos [2001] linked an ensemble streamflow prediction model based on the operational NWSRFS [Bae and Georgakakos, 1994] and a reservoir control scheme that embodied actual system characteristics [Georgakakos et al., 1998] to dynamically fulfill multiple objectives, including minimizing reservoir spillage and flood damage, maximizing hydropower revenue, mitigating drought impacts, and accommodating hydroclimatic forecast uncertainty, multiple planning and management horizons, and decision maker preferences. Over the 1981-1987 evaluation period, using streamflow forecasts over a 60-day planning horizon, their system increased hydropower revenues by one million dollars while reducing flood damages by over five billion dollars.

CONCLUSIONS

Scientific and technological advances have long promised substantial benefits for water resources management, although the extent to which large-scale hydroclimatic research programs have actually affected resource management policies and practices generally has been disappointing. Advances in hydrologic modeling are notable, with many sufficiently developed to be implemented operationally (e.g., ensemble prediction, use of climate forecasts). Rapid changes in forecast systems enable the most advanced science to be embodied within each forecast, meaning that decision makers are receiving, with each forecast, the best science that forecasters believe is applicable to a particular situation. Alternatively, given limited resources, less frequent changes to a forecast system provide opportunities for developing detailed understanding of forecast biases and other aspects of real-world forecast performance. Decision makers would benefit if operational forecasting institutions would explicitly identify their philosophy for incorporating new advances.

Because uncertainty exists in all phases of the forecast process, forecast systems designed to support risk-based decision making need to explicitly quantify and communicate uncertainties, from the entire forecast system and from each component source, including model parameterization and initialization, meteorological forecast uncertainty at the multiple spatial and temporal scales at which they are issued, adjustment of meteorological forecasts (e.g., though downscaling) to make them usable for hydrologic models, and implementation of ensemble techniques. Organizations that make forecasts publicly available should also provide publicly available evaluations of forecast performance. Toward that end, forecasters and the research community should jointly establish archives of operational products and ancillary information. Performance evaluations should be accessible to users when they acquire forecasts, and provide information about forecast performance for the periods and lead times relevant to the users' situations, using multiple criteria that reflect user sensitivities to different forecast qualities.

Hydroclimatic research advances alone are insufficient for more extensive use of hydroclimatic information for better decision and societal outcomes. Decision makers must adapt their management to effectively exploit evolving predictability. Better familiarity with forecast products is needed, especially regarding proper interpretation and realistic understanding of forecast uncertainty. Demonstration studies of end-to-end predictive and decision making systems make clear that achieving maximum benefits from improved long-lead predictive capabilities requires replacing extant rules with dynamic decision schemes that explicitly account for forecast uncertainty. Management inertia and the lack of a centralized authority to coordinate diverse water demands can pose formidable barriers, however, to developing new decision procedures [e.g., Miles et al., 2000; Jacobs and Pulwarty, this volume]. Programs to foster demonstration projects linking hydrologic research and forecasts with water resource management [Endreny et al., this volume] promise to advance decision making in select settings, while providing insight with broader applicability throughout the water resources sector.

Increasing the utility of hydroclimatic research is ultimately an interactive, iterative process involving researchers, forecasters, and decision makers over extended periods and multiple hydroclimatic events. Focusing on incremental improvements provides opportunities for sustained interactions with the water resources sector as new products become available, and for more effective enumeration of the value of scientific advancements as reflected by improved predictability and impacts on resource management decisions. Progress in hydroclimatic research and water resource management is neither easy nor assured. However, working together offers opportunities for progress unlikely to be realized independently.

REFERENCES

Anthes, R. A., E.Y. Hsie, S. Low-Nam, and T.W. Bettge. Estimation of skill and uncertainty in regional numerical models, *Quart. J. Roy. Meteor. Soc., 115*, 763-806, 1989.

Bae, D.-H., and K.P. Georgakakos, Climatic variability of soil water in the American Midwest, I, Hydrologic modeling, *J. Hydrol., 162*, 355-377, 1994.

Barnston, A. G., Y. He, and D. A. Unger, A forecast product that maximizes utility for state-of-the-art seasonal climate prediction, *Bull. Amer. Meteor. Soc., 81,* 1271-1279, 2000.

Barros, A. P., R. J. Kuligowski, R. Bindlash, and O. Yildiz, High-resolution QPF in regions of complex terrain – an intercomparison of dynamical and post-processing downscaling approaches, *EOS Transactions AGU, 80,* Fall Meeting Supplement, Abstract H51C-11, F456, 1999.

Bradley, A. A., T. Hashino, and S. S. Schwartz, Distributions-oriented verification of ensemble predictions for small data samples, *EOS Transactions AGU, 83*, Spring Meeting Supplement, Abstract H42E-02, S184, 2002.

Brooks, H. E., A. Witt, and M. D. Eilts, Verification of public weather forecasts available via the media. *Bull. Amer. Meteor. Soc., 78,* 2167-78, 1997.

Carpenter, T. M., and K P. Georgakakos, Assessment of Folsom Lake response to historical and potential future climate scenarios, I, Forecasting. *J. Hydrol., 249,* 148-175, 2001.

Carroll, T., Snow surveying, in *Yearbook of Science and Technology*, pp. 386-388, McGraw-Hill, New York, N.Y., 1985.

Changnon, S. A., The dilemma of climatic and hydrologic forecasting for the Great Lakes, in *Proceedings of The Great Lakes Water Level Forecast and Statistics Symposium*, edited by H.C. Hartmann and M.J. Donahue, pp. 13-25, Great Lakes Commission, Ann

Arbor, Mi, 1990.

Changnon, S.A., *El Nino 1997-1998: The Climate Event of the Century,* Oxford University Press, New York, N.Y., 2000.

Cosgrove, B., D. Lohmann, K. Mitchell, P. Houser, E. Wood, A. Robock, J. Schaake, C. Marshall, J. Sheffield, L. Luo, Q. Duan, D. Lettenmaier, R. Pinker, D. Tarpley, W. Higgins, and J. Meng, Realtime and retrospective forcing in the North American Land Data Assimilation Systems (NLDAS) project, *J. Geophys. Res., Atmospheres*, in press, 2002.

Clark, M. P., L. E. Hay, G. J. McCabe, G. H. Leavesley, and R. L. Wilby, Towards the use of atmospheric forecasts in hydrologic models, I, Forecast drift and scale dependencies, *EOS Transactions AGU, 80,* Fall Meeting Supplement, Abstract H32G-10, F406-407, 1999.

Clark, M. P., L. E. Hay, and J. S. Whitaker, Development of operational hydrologic forecasting capabilities, *EOS Transactions AGU, 82,* Fall Meeting Supplement, Abstract H31D-0267, F421, 2001.

Clayton, H. H., Verification of weather forecasts, *Amer. Meteor. J., 6,* 211-219, 1889.

Croley, T. E., II, Using NOAA's new climate outlooks in operational hydrology, *J. Hydrol. Engin., 1,* 93-102, 1996.

Croley, T. E. II, Mixing probabilistic meteorology outlooks in operational hydrology, *J. Hydrol. Engin., 2,* 161-168, 1997.

Croley, T. E. II, *Using Meteorology Probability Forecasts in Operational Hydrology,* American Society of Civil Engineers Press, Reston, Va., 2000.

Croley, T. E., II, and Hartmann, H.C., Near real-time forecasting of large lake supplies, *J. Water Resour. Planning and Manag., 113,* 810-823, 1987.

Croley, T. E., II, and D. H. Lee, Evaluation of Great Lakes net basin supply forecasts, *Water Resour. Bull., 29,* 267-282, 1993.

Davis, R. E. and T. Pangburn, Development of new snow products for operational water control and management in the Kings River Basin, California. *EOS Transactions AGU, 81,* Spring Meeting Supplement, Abstract H22D-07, S110, 1999.

Day, G.N., Extended streamflow forecasting using NWSRFS. *J. Water Resour. Planning and Manag., 111,* 157-170, 1985.

Duan, Q., H. V. Gupta, S. Sorooshian, A. N. Rousseau, and R. Turcotte, *Calibration of Watershed Models,* American Geophysical Union, Washington, D. C., 2002.

Dudhia J., D. Gill, Y. R. Guo, D. Hansen, K. Manning, and W. Wang, *PSU/NCAR Mesoscale Modeling System Tutorial Class Notes and Users' Guide,* Universities Corporation for Atmospheric Research, Boulder, Co., 1999.

Franz, K., Evaluation of NWS ensemble streamflow prediction (ESP) water-supply forecasts. M.S. thesis, University of Arizona, Tucson, 2002.

Fread, D.L., Moving hydrologic research into NWS forecasting operations, *EOS Transactions AGU, 80,* Spring Meeting Supplement, Abstract H22D-05, S109-S110, 1999.

Garen, D.C., Improved techniques in regression-based streamflow volume forecasting, *J. Water Resour. Planning and Manag., 118,* 654-670, 1992.

Georgakakos, A. P., H. Yao, M. G. Mullusky, and K. P. Georgakakos, Impacts of climate variability on the operational forecast and management of the Upper Des Moines River Basin, *Water Resour. Res., 34,* 799-821, 1998.

Glantz, M.H., Consequences and responsibilities in drought forecasting- the case of

Yakima, 1977, *Water Resour. Res., 18,* 3-13, 1982.

Grayson, R., and G. Bloschl, *Spatial Patterns in Catchment Hydrology: Observations and Modellig,* Cambridge University Press, Cambridge, U. K., 2000.

Gutierrez, F., and J. A. Dracup, An analysis of the feasibility of long-range streamflow forecasting for Columbia using El Niño- Southern Oscillation indicators, *J. Hydrol., 246,* 181-196, 2001.

Hamill, T. M., C. Snyder, and R. E. Morss, A comparison of probabilistic forecasts from bred, singular vector, and perturbed observation ensembles, *Mon. Wea. Rev., 128,* 1835-1851, 2000.

Hamlet, A. F., and D. P. Lettenmaier, Columbia River streamflow forecasting based on ENSO and PDO climate signals, *J. Water Resour. Planning and Manag., 125,* 333-341, 1999.

Hamlet A. F., and D. P. Lettenmaier, Long-range climate forecasting and its use for water management in the Pacific Northwest region of North America. *J. Hydroinformatics, 2,* 163-182, 2000.

Hamlet, A. P., D. Huppert, and D. P. Lettenmaier, Economic value of long-lead streamflow forecasts for Columbia River hydropower, *J. Water Resour. Planning and Manag., 128,* 91-101, 2002.

Hansen, J. and Coauthors, Forcings and chaos in interannual to decadal climate change, *J. Geophy. Res., 42,* 25679-25720, 1997.

Hartmann, H. C., R. Bales, and S. Sorooshian, Weather, climate, and hydrologic forecasting for the U.S. Southwest: a survey, *Climate Research, 21,* 239-258, 2002a.

Hartmann, H. C., T. C. Pagano, R. Bales, and S. Sorooshian, Confidence builders: evaluating seasonal climate forecasts from user perspectives, *Bull. Amer. Meteor. Soc., 83,* 683-698, 2002b.

Hashino, T., A. A. Bradley, and S. S. Schwartz, Assessment of bias-correction methods for probabilistic forecasts of monthly streamflow volumes, Proc. Mississippi River Climate and Hydrology Conference, NWS Office of Global Programs, Silver Spring, Md., 200, 2002.

Hoffrage, U., S. Lindsey, R. Hertwig, and G. Gigerenzer, Communicating statistical information, *Science, 290,* 2261-2262, 2000.

Houtekamer, P. L., L. Lefaivre, and J. Derome, The RPN ensemble prediction system, Proc. ECMWF Seminar on Predictability, II, European Centre for Medium-range Weather Forecasts, Reading, U. K., 121-146, 1996.

Hydrologic Research Laboratory, *National Weather Service River Forecast System (NWS-RFS) User's Manual*, National Weather Service Office of Hydrology, Silver Spring, Md., 1998.

Kates, R. W., W. C. Clark, R. Corell, J. M. Hall, C. C. Jaeger, I. Lowe, J. J. McCarthy, H. J. Schellnhuber, B. Bolin, N. M. Dickson, S. Faucheux, G. C. Gallopin, A. Grubler, B. Huntlyey, J. Jager, N. S. Jodha, R. E. Kasperson, A. Mabogunjie, P. Matson, H. Mooney, B. Moore, T. O'Riordan, and U. Svedin, Sustainability science, *Science, 292,* 641-642, 2001.

Kim, G., and A. P. Barros, Quantitative flood forecasting using multisensor data and neural networks, *J. Hydrol., 246,* 45-62, 2000.

Krzysztofowicz, R., Why should a forecaster and a decision makers use Bayes' theorem? *Water Resourc. Res., 19,* 327-336, 1983.

Krzysztofowicz R., and D. R. Davis, A methodology for evaluation of flood forecast-response systems, *Water Resourc. Res., 19,* 1423-1454, 1983.

Krzysztofowicz, R., Bayesian analyses of seasonal runoff forecasts, *Stochastic Hydrol. and Hydraulics, 5,* 295-322, 1991.

Krzysztofowicz, R., Bayesian correlation score: a utilitarian measure of forecast skill, *Mon. Wea. Rev., 120,* 208-219, 1992.

Krzysztofowicz, R., The case for probabilistic forecasting in hydrology, *J. Hydrol., 249,* 2-9, 2001.

Krzysztofowicz, R., and H. D. Herr, Hydrologic uncertainty processor for probabilistic river stage forecasting: precipitation-dependent model, *J. Hydrol., 249,* 46-68, 2001.

Lee, D. H., Institutional and technical barriers to implementing risk-based water resources management: a case study, *J. Water Resour. Planning and Manag., 125,* 186-193, 1999.

Mason, S. J., and N. E. Graham, Conditional probabilities, relative operating characteristics, and relative operating levels, *Wea. and Forecasting, 14,* 713-725, 1999.

Miles, E. L., A. K. Snover, A. F. Hamlet, B. Callahan, and D. Fluharty, Pacific northwest regional assessment: the impacts of climate variability and change on the water resources of the Columbia river basin, *J. Amer. Water Resour. Assoc., 36,* 399-420, 2000.

Mittelstadt, J., Current and upcoming changes to the Eta models, *Western Regional Technical Attachment No 97-06,* National Weather Service Forecast Office, Salt Lake City, Ut., 1997.

Molteni F., R. Buizza, T. N. Palmer, and T. Petroliagis, The ECMWF ensemble prediction system: methodology and evaluation. *Quart. J. Roy. Meteor. Soc., 122,* 73-119, 1996.

Murphy, A. H., What is a good forecast? An essay on the nature of goodness in weather forecasting, *Wea. and Forecasting, 8,* 281-293, 1993.

Murphy, A. H., and R. L. Winkler, A general framework for forecast verification, *Mon. Wea. Rev., 115,* 1330-1338, 1987.

Murphy, A. H., and R.L. Winkler, Diagnostic verification of probability forecasts, *Int. J. Forecasting, 7,* 435-455, 1992.

National Hydrologic Warning Council, *Use and Benefits of the National Weather Service River and Flood Forecasts,* National Weather Service Office of Hydrologic Development, Silver Spring, Md., 2002.

National Research Council, *GCIP: A Review of Progress and Opportunities,* National Academy Press, Washington, D. C., 1998a.

National Research Council, *Hydrologic Sciences: Taking Stock and Looking Ahead,* National Academy Press, Washington, D. C., 1998b.

National Research Council, *A Vision for the National Weather Service: Road Map for the Future,* National Academy Press, Washington, D. C., 1999a.

National Research Council, *Making Climate Forecasts Matter,* National Academy Press, Washington, D. C., 1999b.

National Research Council, *Hydrologic Science Priorities for the U.S. Global Change Research Program: An Initial Assessment,* National Academy Press, Washington, D. C., 1999c.

National Weather Service, *Strategic Plan for Weather, Water, and Climate Services; 2000-2005,* Department of Commerce, Washington, D. C., 1999.

Nicholls, N., Cognitive illusions, heuristics, and climate prediction, *Bull. Amer. Meteor. Soc., 80,* 1385-1398, 1999.

Pagano, T. C., H. C. Hartmann, and S. Sorooshian, Using climate forecasts for water management: Arizona and the 1997-98 El Nino, *J. Amer. Water Resour. Assoc., 37,* 1139-

1153, 2001.

Pagano, T. C., H. C. Hartmann, and S. Sorooshian, Use of climate forecasts for water management in Arizona: a case study of the 1997-98 El Niño, *Clim. Res., 21,* 59-269, 2002.

Perica S., Integration of meteorological forecasts/climate outlooks into an ensemble streamflow prediction system, *Preprints, 14th Conference on Probability and Statistics in the Atmospheric Sciences*, American Meteorological Society, Boston, Ma., 130-133, 1998.

Perica, S., J. Schaake, and D. J. Seo, National Weather Service River Forecast System (NWSRFS) operational procedures for using short and long range precipitation forecasts as input to ensemble streamflow prediction (ESP), *Preprints, 14th Conference on Hydrology*, American Meteorological Society, Boston, Ma., 1999.

Perica, S., and J. C. Schaake, Hydrologic uncertainty post-processor for ensemble streamflow prediction, *EOS Transactions AGU, 81,* Spring Meeting Supplement, Abstract H31D-10, S201, 2000.

Piechota, T. C., and J. A. Dracup, Long range streamflow forecasting using El Niño-Southern Oscillation indicators, *J. Hydrol. Engineer., 4,* 144-151, 1999.

Piechota, T. C., F. H. S. Chiew, J. A. Dracup, and T. A. McMahon, Development of an exceedance probability streamflow forecast using the El Niño-Southern Oscillation, *J. Hydrol. Engineer., 4,* 20-28, 2001.

Pielke, R. A., Jr., Usable information for policy: an appraisal of the U.S. global change research program, *Policy Sciences, 38,* 39-77, 1995.

Pielke, R. A., Jr., The development of the U.S. global change research program: 1987 to 1994. Policy Case Study, National Center for Atmospheric Research, Boulder, Co, 2001.

Pulwarty, R. S., and K. T. Redmond, Climate and salmon restoration in the Columbia River basin: the role and usability of seasonal forecasts. *Bull. Amer. Meteor. Soc., 78,* 381-397, 1997.

Rayner, S., D. Lach, H. Ingram, and M. Houck, Why water resource managers don't use climate forecasts, International Research Institute on Climate Prediction, Palisades, N. Y., 2001.

Schwein, N. O., A methodology for determining river forecasting skill using monthly cumulative distribution functions of mean daily flow, *Preprints, 16th Conference on Probability and Statistics in the Atmospheric Sciences,* American Meteorological Society, Boston, Ma., 40-47, 2002.

Scott, M. J., L. W. Vail, J. A. Jaksch, and K. K. Anderson, Considerations for management of irrigation water with climate variability, *EOS Transactions AGU, 81,*Spring Meeting Supplement, Abstract H32B-06, S205, 2000.

Shafer, B. A., and J. M. Huddleston, Analysis of seasonal volume streamflow forecast errors in the western United States, in *A Critical Assessment of Forecasting in Western Water Resources Management,* edited by J. Cassidy and D. Lettenmaier, pp. 117-126, American Water Resources Association, Bethesda, Md., 1985.

Smith, J. A., G. N. Day, and M. D. Kane, Nonparametric framework for long-range streamflow forecasting, *J. Amer. Water Resour. Assoc., 118,* 82-92, 1992.

Sorooshian, S., S. Ram, B. Imam, R. Bales, and H. Gupta, An integrated information system for multi-purpose user-adaptive access to land surface hydrologic and water resources data, Report NAG5-8503, University of Arizona, Tucson, Az., 2002.

Toth, Z., and E. Kalnay, Ensemble forecasting at NMC: the generation of perturbations, *Bull. Amer. Meteor. Soc., 74,* 2317-2330, 1993.

Toth, Z. and E. Kalnay,. Ensemble forecasting at NCEP and the breeding method, *Mon. Wea. Rev., 12,* 3297-3319, 1997.

Valdes, J. B., D. Entekhabi, and Z. Liu, Merging and error analysis of regional hydrometeorology forecasts conditioned on climate forecasts, *EOS Transactions AGU, 79,* Spring Meeting Supplement, Abstract H51E-05, S146, 1998.

Walker, A. E., and B. E. Goodison, Discrimination of wet snow cover using passive microwaver satellite data, *Annals of Glaciology, 17,* 307-311, 1993.

Ward, M. N., and C.K. Folland, Prediction of seasonal rainfall in the north Nordeste of Brazil using eigenvectors of sea-surface temperatures, *Int. J. Climatol., 11,* 711-743, 1991.

Westrick, K. J., and C. F. Mass, An evaluation of a high resolution hydrometeorological modeling system for prediction of a cool-season flood event in a coastal mountainous watershed, *J. Hydrometeorol., 2,* 161-180, 2001.

Westrick, K. J., P. Storck, and C. F. Mass, Description and evaluation of a hydrometeorological forecast system for mountainous watersheds, *Wea.and Forecasting, 17,* 250-262, 2002.

Wigmosta, M. S., L. W. Vail, and D. P. Lettenmaier, A distributed hydrology-soil-vegetation model for complex terrain, *Water Resour. Res., 30,* 1665-1679, 1994.

Wigmosta, M. S., B. Nijssen, P. Storck, and D. P. Lettenmaier, The distributed hydrology soil vegetation model, in *Mathematical Models of Small Watershed Hydrology and Applications*, edited by V. P. Singh and D.K. Frevert, pp. 7-42, Water Resource Publications, Littleton, Co., 2002.

Wilks, D. S., *Statistical Methods in the Atmospheric Sciences,* Academic Press, San Diego, Ca., 1995.

Wilks, D. S., Diagnostic verification of the Climate Prediction Center long-lead outlooks, 1995-1998, *J. Clim., 13,* 2389-2403, 2000.

Wilks, D. S., and C. M. Godfrey, Diagnostic verification of the IRI net assessment forecasts, 1997-2000, *Preprints, 16th Conference on Probability and Statistics in the Atmospheric Sciences,* American Meteorological Society, Boston, Ma., 31-36, 2002.

Winkler, R. L., Evaluating probabilities: asymmetric scoring rules, *Manag. Sciences, 40,* 1395-1405, 1994.

Wood, A., A. Hamlet, D. P. Lettenmaier, and A. Kumar, Experimental real-time seasonal hydrologic forecasting for the Columbia River Basin, *Proc., 26th Annual Climate Diagnostics and Prediction Workshop*, National Weather Service, PB92-167378, National Technical Information Service, Springfield, VA, 2001.

Wood, E., E. P. Maurer, A. Kumar, and D. P. Lettenmaier, Long-range experimental hydrologic forecasting for the eastern United States, *J. Geophys. Res., D20*, 4429-4423, 2002.

Yao, H., and A. Georgakakos, Assessment of Folsom Lake response to historical and potential future climate scenarios, 2, reservoir management, *J. Hydrol., 249*, 176-196, 2001.

Holly C. Hartmann, P.O. Box 210011, Harshabarger Bldg. 11, Department of Hydrology and Water Resources, The University of Arizona, Tucson, Arizona, 85721.

Allen Bradley, 404 Hydraulics Laboratory, Institute of Hydraulic Research, University of Iowa, Iowa City, Iowa 52242.

Alan Hamlet, 170 Wilcox Hall, Department of Civil and Environmental Engineeting, University of Washington, Seattle, Washington 98195.

Introduction To Section 4

This section focuses on the integration of science into the many forms of public and private decision-making related to policy development for water management. Examples range from traditional applied research to capacity building in developing countries. Despite the best efforts of many scientists and policy makers, integrating science and policy has been hit-or-miss in most applications, perhaps because successful integration depends on a conjunction of factors that are not well understood. Be that as it may, case studies provide examples of successful integration of science into policy and management from which to draw lessons for future efforts. As the science and policy dialogue increases, more and better tools for integrating science with policy and management will be developed on an accelerating schedule. Already, the interface of science and policy is yielding insights into the necessary conditions for successful integration.

Some principal aspects of successful science-policy dialogue outlined in this section deal with problem identification, the use of assessments, and a new paradigm for bringing together scientists with policy makers and managers. Problem identification often grows out of policy debates. Scientists can play an important role as fact finders in these debates; a role that is frequently successful, especially when policy makers are committed to giving weight to results of scientific studies. Equally important is the sometime necessity for scientists to market their information and tools to decision makers. There is no substitute, however, for direct reciprocal communication between scientists and policy makers. Another way to ensure science influences policy is to adopt a new paradigm that places scientists/engineers, policymakers/managers, and communities/interest groups on equal footing. As indicated here, at both the national and international levels this paradigm has been used to produce more successful integration of science and policy for problems that are spatially distributed and variable, such as non-point source pollution and the tailoring of climate information to local water management problems.

Water: Science, Policy, and Management
Water Resources Monograph 16

10.1029/16WM18

15

Integration of Science and Policy During the Evolution of South Carolina's Drought Program

Hope P. Mizzell and Venkat Lakshmi

INTRODUCTION

Climatological records show that South Carolina has been plagued with recurrent drought conditions for nearly a century. Periods of dry weather have occurred in each decade since 1818 [National Water Summary 1988-1989 Hydrologic Events and Floods and Droughts, 1991]. The earliest records of drought indicate that some streams in South Carolina went dry in 1818, and fish in smaller streams died from lack of water in 1848. The most damaging droughts in recent history occurred in 1954, 1986, and 1998-2002. Less severe droughts were reported in 1988, 1990, 1993, and 1995.

Droughts have had severe adverse impacts on the people and economy of South Carolina. Drought impacts are diverse, causing a ripple effect through the economy [Wilhite, 1993]. This was made especially clear during the drought of 1998-2002 that impacted many sectors, including agriculture, forestry, tourism, power generation, public water supply, and fisheries. The economic repercussions associated with the 1998-2002 drought will likely surpass any other drought in South Carolina's history. During the past 50 years, droughts have caused the second highest economic loss in South Carolina of any natural hazard [Mizzell, 2001]. Only Hurricane Hugo caused more damage to the economy.

Drought is only one aspect of the water-supply concerns facing the State, however. Water use in South Carolina has increased by 50 percent during the past century [Castro and Hu, 1997]. The water resources of South Carolina are under increasing stress because of the many uses they serve, including hydroelectric power generation, nuclear-plant cooling, industrial cooling and process water,

Water: Science, Policy, and Management
Water Resources Monograph 16

10.1029/016WM19

drinking water, water-based recreation, and disposal of sewage and non-point source pollution. With today's increasing demands, impacts from future drought may be much greater than those witnessed in the past. Mostscientists agree that vulnerability to drought is increasing, primarily due to increasing pressure from population growth on limited water resources [Wilhite, 1993].

Drought differs from other natural hazards because the effects accumulate slowly over a considerable time and may linger for years after the event has ended [Wilhite et. al., 2000]. Through mitigation and preparedness, however, the impacts of drought can be reduced. It is important to develop a plan to deal with drought periods in a timely, systematic manner [Wilhite et. al., 2000]. Developing such a plan is the first step toward reducing societal vulnerability to drought [Wilhite, 1993]. Scientists and policy makers must understand the characteristics and complexity of drought in order to establish viable assessment and response strategies. South Carolina's response strategies include a drought early-warning system [Glantz and Degefu, 1990], committee designation based on river basins, and trigger mechanisms [Hrezo et. al., 1986] for demand management [Little and Moreau, 1991].

South Carolina has one of the oldest drought response programs in the United States [Hayes and Knutson, 2001]. South Carolina began examining drought impacts and occurrences in 1978 while most of the United States was experiencing severe drought conditions [Rouse et al., 1985]. Since then there have been several plans and laws established to monitor, manage, and conserve the State's water resources in the best interest of all South Carolinians. South Carolina recognized the need for formalizing the drought plan by passing legislation in 1985. South Carolina is unique in dealing with drought management through legislation and its associated regulations. Passing and amending the law has required a cooperation between scientists and policy makers. The purpose of this paper is to discuss the integration of science associated with drought indices, climatological data, and hydrological variables with policy during the evolution of South Carolina's drought program.

BACKGROUND

History of South Carolina Drought Management

A study conducted in 1978 by the South Carolina Water Resources Commission revealed that the State had no drought provisions except during emergencies. Subsequently, the South Carolina Water Resources Commission established a Drought Advisory Committee in 1983, which formulated the first comprehensive drought response plan. The following year the legislation was introduced in the General Assembly, and the Drought Response Act was passed on April 29, 1985. The regulations to implement the Act were approved on May 17, 1986. This planning effort was the fifth drought plan established by a state

[Hayes and Knutson, 2000]. According to the National Drought Mitigation Center, as of December 2000 only 37 states had developed, or were in the process of developing, a plan to respond to drought.

The Drought Response Act established procedures for sustaining the availability of water during periods of drought. It established six drought management areas based on climatic divisions and a local drought response committee within each management area, and established a state drought coordinator position. According to the National Drought Mitigation Center, the appointment of a permanent state drought coordinator is the only such position within the United States and is a unique asset to South Carolina.

In 1998, the South Carolina Department of Natural Resources adopted a State Water Plan that contains guidelines for the efficient, equitable, and environmentally responsible management of the State's water resources. The Drought Response Act was amended in 2000 to implement guidelines set forth in the State Water Plan, i.e. to adjust drought management areas to correspond with the State's four major river basins, restructure local drought committees, and clarify existing procedures so as to identify and address water shortages.

State Drought Plans

South Carolina's drought response is different from that of other states, owing primarily to the creation and existence of the Drought Response Act. Like many other states, South Carolina has a Drought Response Plan and a State Water Plan. While the plans consist of detailed actions and responses, these are only recommen dations and not actually enforceable laws. Through experience, South Carolina decision-makers learned that when dealing with an issue as controversial as water use and restrictions it is imperative to have a law and associated legislation with mandated actions. In 1985, South Carolina's first drought law was adopted. It was amended in 2000. The disadvantage of South Carolina's drought law is that it addresses a limited number of actions. During the original creation of the law and the amendment process, organizations and agencies lobbied against sections of the act that were detrimental to their operations. For instance, agriculture-related groups lobbied that agriculture should be considered an essential water user and therefore should not be subject to water-use curtailment. Their argument was based on people's need for food to survive, thus making agriculture essential. The legislature finally agreed that agriculture could not be excluded from the jurisdiction of the drought law, since each and every sector could argue that it is an essential user of water - such as the need for power generation facilities. Ideally, the drought law or additional legislation would address ways to assist these groups if water-use curtailment is mandatory; for instance, creation of a program that pays farmers not to irrigate or that establishes stronger incentives for private investment in water conservation.

According to the National Drought Mitigation Center, thirty states in the United States have drought contingency plans in place. The difference in state drought

Table 1. Comparison of South Carolina's Drought Act (including supporting regulations) and Other-State Plans[a]

Title	Lead Agency	Year Developed	Committee/Task Forces	Primary Impacts Addressed
South Carolina Drought Response Act	Dept.of Natural Resources	Created 1985 Amended 2000	State and local representatives from: Natural Resources, Health and Environmental Control, Emergency Preparedness, Agriculture, Forestry, Counties, Municipalities, Private Water Suppliers, Public Service Districts, Domestic Users, Industry, Regional Councils of Government, Commissions of Public Works, Special-Purpose Districts, Power Generation Facilities, Conservation Districts	Public water supplies
Colorado Drought Response Plan	Colorado Div. of Disaster Emergency Services	Created 1981 Amended 1990	Municipal, Water, Wildfire Protection, Agricultural, Industry, Tourism, Wildlife Economic Impacts, Energy Loss, Health	Agriculture, wildfire protection, municipal supply, commerce, tourism, and fish/wildlife preservation
New York State Drought Plan	Dept. of Environmental Conservation	Created 1982 Amended 1988	Task Force includes invited participants with five voting members - State Emergency Management Office, Dept. of Health, Dept. of Environmental Conservation, Dept. of Agriculture and Markets, Canal Authority	Public water supplies

Table 1 (continued). Comparison of South Carolina's Drought Act (including supporting regulations) and Other-State Plans[a]

Title	Lead Agency	Year Developed	Committee/Task Forces	Primary Impacts Addressed
Idaho Drought Plan	Dept.of Water Resources	Created 1990 Amended 1995	Subcommittees on Water Data, Public Information; Agricultural; Energy; Economic; Municipal, Industrial, and Water Quality; Fish, Wildlife, Recreation and Environment	Agriculture, municipal supplies and water quality, recreation and tourism, fish and wildlife, energy, economics
North Carolina Drought Assessment and Response Plan	Div.of Water Resources	Created 2001	State Agency representatives: Division of Water Resources, Environmental Health, Water Quality, Soil and Water Conservation, Emergency Management, Community Assistance, Utilities Commission, State Extension Service, Department of Agriculture, State Climate Office Task Forces include: Agriculture, Economic Impact, Energy Loss, Health, Water Sources	Agriculture, economic impact, energy loss, health, water sources
Missouri Drought Response Plan	Dept.of Natural Resources	Created 1995 Amending 2002	Agriculture, Natural Resources and Environmental, Recreation, Water Supplies and Wastewater, Health, Social, Economic, Post Drought Evaluation	Multipurpose, with emphasis on drinking water and agriculture

Table 1 (continued). Comparison of South Carolina's Drought Act (including supporting regulations) and Other-State Plans[a]

Monitoring		
State	**Conditions/Indices Monitored**	**Triggers**
South Carolina Drought Response Act	Climatic variables, streamflow, drinking-water supplies, surface and groundwater, Palmer Drought Index, Crop Moisture Index, Standard Precipitation Index, Keetch-Byram Drought Index, US Drought Monitor, short- and long-range National Weather Service forecasts	Incipient Drought: PDSI of -0.50 to -1.49, CMI of 0.00 to -1.49, SPI of 0.00 to -0.99, KBDI of 300 to 399, USDM of D0, Avg. daily streamflow is 111%-120% of minimum flow for 2 consecutive weeks, Static water level in aquifer is between 11 feet and 20 feet above trigger level for 2 consecutive months Moderate drought alert phase: PDSI of -1.50 to -2.99, CMI of -1.50 to -2.99, SPI of -1.00 to -1.49, KBDI of 400 to 499, USDM of D1, Avg. daily streamflow is 101%-110% of minimum flow for 2 consecutive weeks, Static water level in aquifer is between 1 feet and 10 feet above trigger level for 2 consecutive months Severe drought alert phase: PDSI of -3.00 to -3.99; CMI of -3.00 to -3.99; SPI of -1.50 to -1.99; KBDI of 500 to 699; USDM of D2; Avg. daily streamflow between minimum flow and 90% of minimum for 2 consecutive weeks; Static water level in aquifer is between trigger level and 10 feet below for two consecutive months; Extreme drought alert phase: PDSI of -4.00 and below; CMI reaches or falls below -4.00; SPI reaches or falls below -2.00, KBDI reaches or exceeds 700; USDM of D3 or higher; Avg. daily streamflow is less than 90% of minimum for 2 consecutive weeks; Static water level in aquifer is more than 10 feet below trigger level for two consecutive months. Indication by one index alone does not mandate a declaration

plans often depends on the type of water allocation system (i.e. prior appropriation, riparian, modified riparian), the state's physical environment, size of urban and rural population, or the state and local political conditions (Hrezo et al., 1986). Similar to South Carolina, many states designate a water or natural resources department as the lead agency. Najarian (2000) in an analysis of all

Table 1 (continued). Comparison of South Carolina's Drought Act (including supporting regulations) and Other-State Plans[a]

State	Conditions/Indices Monitored	Triggers
Colorado Drought Response Plan	Snowpack, soil moisture, reservoir levels, streamflow, precipitation, temperature, Standard Precipitation Index (SPI), Palmer Index, Surface Water Supply Index (SWSI)	Palmer or SWSI 0 to plus: Normal Conditions Palmer or SWSI 0 to -1.0: Normal Conditions Palmer or SWSI -1.0 to -2.0: Phase 1 Palmer or SWSI -2.0: Phase 2 Declared by Task Force: Phase 3
New York State Drought Plan	Climatological data, reservoir/lake storage, streamflow, groundwater levels, Palmer Drought Index, State Drought Index	Unique Palmer and State Drought Index values for eight drought regions Four stages include: Drought Watch, Drought Warning, Drought Emergency, and Drought Disaster
Idaho Drought Plan	Climatological data, snow surveys, stream-flow, Palmer Drought Index, Surface Water Supply Index	Triggers established by Water Supply Committee as drought develops

state drought plans, identified five sectors and programs addressed in most state plans: water availability, municipal water, water shortage and conservation, agricultural industry, and public information/education. South Carolina's drought act addresses all of these sectors. It does not address other sectors identified by Najarian (2000), such as fish and wildlife, health, commerce and tourism/economy, wildlife protection/public lands, energy, and social services (counseling). Najarian (2000) found that only Washington, New Mexico, and Montana address nearly all sectors in their drought plans.

Another common feature among state drought plans is the designation of a monitoring system and committee. Only five states (North Dakota, Nebraska,

Table 1 (continued). Comparison of South Carolina's Drought Act (including supporting regulations) and Other-State Plans[a]

State	Conditions/Indices Monitored	Triggers
North Carolina Drought Assess-ment and Response Plan	Streamflow, groundwater and reservoir/lake levels, water supply, rainfall, agriculture weather and crop reports, climate forecasts, current Keetch-Byram Drought Index, US Drought Monitor, Palmer Index, Climate Prediction Center reports	Uses the U.S. Drought Monitor and/or Palmer Drought Severity Index to trigger a sequence of response actions
Missouri Drought Response Plan	Water monitoring data and weather data, Palmer Drought Index	Palmer >= -1.0: Phase 1 (Advisory Phase) Palmer -1.0 to -2.0: Phase 2 (Drought Alert) Palmer -2.0 to -4.0: Phase 3 (Conservation Phase) Palmer <= -4.0: Phase 4 (Possible Local Rationing Phase)

[a]Based on evaluation of each state drought plan, state drought plan analysis by Najarian (2000) and personnel communication with Warren Lavery (New York State Department of Environmental Conservation), Setve McIntosh (Missouri Department of Natural Resources), and Woody Yonts (North Carolina Division of Water Resources).

Iowa, Michigan, Indiana) do not specify in their plan the data to be monitored or indices used to monitor the onset of drought episodes (Najarian, 2000). Most states, like South Carolina, specify the data monitored, such as precipitation, streamflow, reservoir levels, groundwater, soil moisture, and temperature. Fourteen states, including South Carolina, have a triggering mechanism in their plan (Najarian,2000). Establishing criteria for declaring drought emergencies and triggering various mitigation and response activities are essential components of state drought plans (Wilhite, 1991). South Carolina's neighboring state, North Carolina, has a drought response plan, Georgia and Virginia are currently developing a plan, and Florida delegates planning to local authorities. Table 1 gives a comparison of South Carolina's Drought Act and various plans (Colorado, New York, Idaho, North Carolina, and Missouri) across the United States. Compared to the other state plans listed in Table 1, South Carolina's act and supporting regulation specifies the most trigger mechanisms, including the threshold values for each drought declaration. According to Table 1, the primary impact addressed by the New York Drought Plan and South Carolina's Drought Act is public water supply, whereas the plans for Colorado, Idaho, North Carolina, and Missouri specifically address multiple impacts such as agriculture and energy.

South Carolina's Water Resources

The amount of water available in South Carolina's rivers, lakes, and aquifers varies in time and space. Replenishment of water in the State is primarily from precipitation. Annually, South Carolina receives about 48 inches of water as precipitation and about 8 inches as streamflow from adjacent states. The greatest loss of water, about 34 inches results from evapotranspiration. The remaining water, about 22 inches, is available for use [South Carolina State Water Plan, 1998].

South Carolina water use was approximately 43,440 MGD (million gallons per day) in 2000. Hydroelectric-power use was 36,170 MGD, thermoelectric-power use (fossil and coal) was 5,750 MGD, and all other uses combined were 1,520 MGD. These other uses are industry (37 percent), public supply (36 percent), crop irrigation (17 percent), golf-course irrigation (6 percent), and domestic use (4 percent).

South Carolina had more than 3,000 active water-supply systems (including small systems such as subdivisions, mobile home parks, hospitals, etc.) in 2000, with 100 of these providing over 97 percent of all water pumped for public supply. Lakes and streams provide about three-fourths of the public supply use and the rest comes from wells and springs [Newcome, 2000].

Water use in South Carolina during the past century has increased by 50 percent. Water withdrawals, excluding hydroelectric and thermoelectric power, are projected to increase from 1,522 MGD in 2000 to 2,290 MGD in 2045 [Castro and Hu, 1997]. Table 2 shows the projected total demand for each use type in 2045. During the drought of 1999-2002, many water suppliers faced water shortages and instituted mandatory reduction in use by their customers.

State Water Plan and Drought Response Act

The South Carolina Department of Natural Resources (SCDNR) has examined the water issues facing South Carolina, including the problems of water shortage and the effects of long-term withdrawal and use of water. The South Carolina Drought Response Act was first established in 1985 and designated SCDNR as the primary agency to monitor drought conditions, or potential for drought, throughout the State and to coordinate the State's response. The Act applied to all water suppliers and to all ground and surface water resources of the State except private ponds. It required that all public water suppliers (municipalities, counties, public service districts, and commissions of public works) adopt a Drought Response Ordinance or Plan and file that Ordinance or Plan with

Table 2. South Carolina Total Water Use for 2000 and Projected Demand for 2045

Use Type	Total Water Use (million gallons per day)	
	2000	2045[a]
Industry	566	1,000
Public Supply	542	750
Irrigation	253	300
Golf Course	97	160
Domestic	64	80
Total	1,522	2,290

[a] Source: Castro and Hu, 1997

SCDNR. The Act established six Drought Management Areas—Northwest, North-Central, Northeast, West- Central, Central, and Southern. Drought Response Committees were established for each Drought Management Area, comprising both state and local interests.

The Act established four levels of drought; Incipient, Moderate, Severe, and Extreme, requiring different actions with respect to conservation. The drought phase was determined by the Drought Response Committee on the basis of the Palmer Drought Severity Index. The Drought Response Committee also has the authority to evaluate drought conditions in drought management areas to determine if a need exists for action beyond the scope of local government that may involve mandatory water-use curtailment.

The 1998 State Water Plan presents goals and objectives for sustaining the availability of water for current and future use. In an effort to implement the State Water Plan recommendations, SCDNR proposed changes to the South Carolina Drought Response Act of 1985. There was also support for amending parts of the Drought Response Act to include the use of improved monitoring techniques.

In 1999, SCDNR submitted to the General Assembly several amendments to the South Carolina Drought Response Act. The proposed changes were to adjust drought management areas to correspond with the State's four major river basins rather than six climate divisions, restructure local drought committees, use multiple indices for drought declarations, and clarify existing procedures. There was opposition to the amendments by the South Carolina Department of Health and Environmental Control (SCDHEC), South Carolina Chapter of the American Water Works Association (SCAWWA), and industrial representatives.

SCDHEC is responsible for the enforcement of federal and state environmental laws and regulations, and for issuing permits, licenses and certifications for

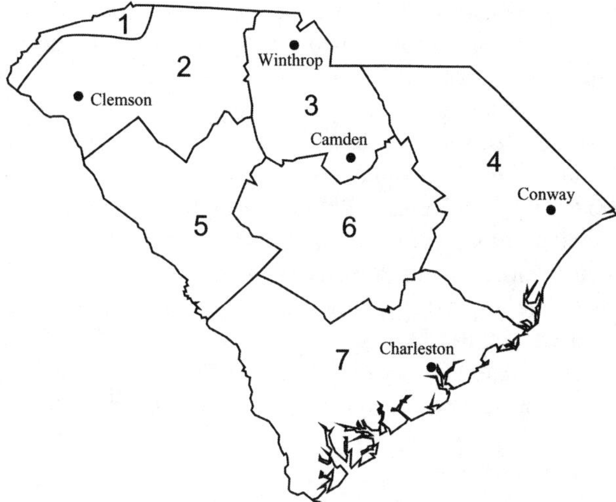

Figure 1. Location of stations and climate divisions used for trend analysis.

activities which may affect the environment. Their primary concern was the possible conflict between their Groundwater Use and Reporting Act and the Drought Response Act. They wanted to exclude the use of declining water levels in confined aquifers as an indicator of drought. Their argument was that decline in water levels in confined aquifers is a result of pumping and long-term groundwater withdrawals not drought. SCAWWA represents public water supply systems in the State. Their concern was with the overall authority of the original drought response act. Businesses may lose revenue when a drought declaration is made by the State on a regional basis, placing mandatory water restrictions on users and water system operators who have stored water for such shortfalls.

INTEGRATION OF SCIENCE AND POLICY

Climatological Trends

The first study analyzed historical climate trends for five South Carolina stations since 1900. The South Carolina station data used for the analysis were extracted from the National Climatic Data Center's United States Historical Climatology Network (USHCN). USHCN is a high-quality data set of monthly averaged maximum, minimum, and mean temperature and total monthly precipitation developed to assist in the detection of regional climate change. The stations selected for the South Carolina study were Clemson, Winthrop, Camden, Charleston, and Conway. These stations had less than one percent missing data and were geographically distributed across the State (Figure 1).

The second study analyzed Palmer Drought Severity Index (PDSI) values for 1900-1999 for the seven climate divisions in South Carolina. The data were analyzed as a time series and monthly. Since the PDSI values are autocorrelated, the residuals were computed for the analysis.

A Kendall Tau Statistical Test was used to determine whether there was a trend in monthly and seasonal precipitation and temperature data and in the PDSI residuals 1900-1999. The Kendall Tau measures the strength of the relationship between precipitation, temperature, and PDSI over time (years). Tau is a rank-based procedure and is therefore resistant to the effect of a small number of unusual values. The test statistic S measures the monotonic dependence of precipitation, temperature, or PDSI (y) on time (x). The estimate of the trend slope for precipitation, temperature, or PDSI (y) over time (x) can be computed as the median of all slopes between data pairs [Helsel and Hirsch, 1992].

Multiple Drought Indices

One major recommended change to the Drought Act of 1985 was the use of multiple drought indices for the drought declarations rather than sole dependence on the PDSI as the primary trigger mechanism. The PDSI was selected as the primary trigger mechanism in the Drought Response Act of 1985 because of its national acceptance as a reliable drought index and its availability; however, there are limitations and assumptions in the calculation [Alley, 1984].

The PDSI calculation is based on precipitation and temperature data and the local available water content of the soil. Since the backbone of the method is a water- balance computation, PDSI computes potential soil-water recharge, potential evaporation loss, and potential runoff [Alley, 1984]. The objective of the PDSI is to provide measurements of moisture conditions that are standardized so that comparisons can be made between locations and between months [Palmer, 1965].

South Carolina Drought Committee members and policy makers realized the limitations of using the PDSI as the State's primary trigger mechanism during the drought of 1998 because the PDSI was showing that the State was experiencing moist to very moist conditions in late June when actually the State was experienc ing a drought-induced agricultural disaster. The PDSI showed this value because of the built-in computational lag picking up the very wet winter and spring. The job of the State's climatologists and hydrologists was to test and introduce new drought and water-monitoring indices to justify the need to consult more than one index before making a decision, realizing that none of the major indices are applicable in all circumstances [Hayes, 1996, Heim, 2002].

Interaction Between Groundwater and Surface Water

Even though there is little debate that surface water is hydraulically connected with groundwater, this interaction is often ignored in water management policies [Winter et al., 1998]. The interaction between groundwater and surface water is ignored because the connection is difficult to observe and measure and, therefore, even more difficult to consider in policy. Groundwater interacts with all forms of surface water, such as streams, lakes, and wetlands, but the relationship is difficult to visualize.

In South Carolina, an average of 40 percent of the streamflow is contributed by groundwater [Winter et. al. , 1998]. An estimated 1,100 inches of water are stored in aquifers in South Carolina's Coastal Plain. Groundwater stored in the Piedmont region is controlled by the location and size of fractures in the bedrock and the thickness of the overlying weathered portion of the bedrock [South Carolina Water Plan, 1998]. Lakes and streams provide about three-fourths of the public supply use and the rest comes from groundwater. Wells supply most rural residents with drinking water. The cities of Sumter and Florence have the largest public systems supplied only by wells [Newcome, 2000].

South Carolina's water resources are managed and protected by numerous laws, such as the Safe Drinking Water Act, Water Use Reporting and Coordination Act, Groundwater Use Act, Interbasin Transfer of Water Act, State Recreational Waters Act, and the Drought Response Act (over 15 additional laws not listed). The SCDHEC is responsible for the management of most of these water-resource related laws. The SCDNR is responsible for the management of the Drought Response Act and other fishing and recreational based laws and regulations. There has been considerable debate between the agencies on the consideration of groundwater and surface water as a single resource and use of groundwater levels as an indicator of drought. In an effort to clarify the connection between surface water and groundwater, data were analyzed from an extensive statewide surface water and groundwater monitoring network.

RESULTS

The SCDNR was able to answer the questions and concerns of the South Carolina General Assembly and other interested groups by justifying the amendments to the drought policy with scientific documentation. Science was not able to resolve all discrepancies, owing, in some cases, to conflicting results and in others to political barriers. There was an overall mutual interest among the stakeholders, policy makers, and scientists in utilizing science-based management. In some cases the science helped the stakeholders recognize the ineffectiveness of strategies based on protecting turf [Clark et al., 1998]. The use of multiple indices for drought declarations, including indices of groundwater levels, was included in the Drought Act of 2000. The amendments to the Drought Act would not have been approved by the South Carolina General Assembly (owing to the controver sial nature of droughts and water rights) without the provided scientific documentation coupled with the State's ongoing severe drought.

Table 3a. South Carolina Precipitation Analysis 1900-1999

Kendall's rank correlation tau

	Clemson		Winthrop		Camden		Charleston		Conway	
	p	slope	p	slope	p	slope	p	slope	p	slope
January	0.258	0.008	0.072	0.015	0.000	0.022	0.050	0.012	0.000	0.026
February	0.759	-0.002	0.732	-0.002	0.910	0.001	0.150	-0.009	0.786	-0.002
March	0.017	0.019	0.069	0.013	0.098	0.011	0.385	0.006	0.040	0.012
April	0.174	0.009	0.640	0.003	0.662	0.003	0.555	0.003	0.133	0.009
May	0.052	0.013	0.636	-0.004	0.503	0.003	0.342	-0.005	0.024	0.014
June	0.432	-0.006	0.428	-0.006	0.307	-0.008	0.953	0..00	0.988	-0.001
July	0.282	-0.009	0.112	-0.013	0.668	-0.003	0.252	-0.012	0.314	0.010
August	0.973	0.000	0.207	-0.010	0.602	0.005	0.924	-0.001	0.373	0.008
September	0.313	0.008	0.103	0.013	0.176	0.011	0.104	0.020	0.029	0.020
October	0.033	0.018	0.101	0.011	0.128	0.009	0.854	-0.001	0.310	0.006
November	0.001	0.021	0.003	0.017	0.007	0.014	0.967	0.000	0.005	0.011
December	0.088	-0.016	0.122	-0.010	0.891	-0.001	0.960	0.000	0.333	0.005
Winter	0.180	-0.023	0.791	-0.004	0.110	0.020	0.922	-0.001	0.016	0.029
Spring	0.001	0.044	0.415	0.010	0.278	0.012	0.732	0.004	0.000	0.039
Summer	0.274	-0.018	0.043	-0.029	0.623	-0.007	0.833	-0.006	0.066	0.025
Fall	0.003	0.044	0.001	0.043	0.004	0.040	0.087	0.028	0.005	0.040
Annual	0.110	0.054	0.530	0.017	0.008	0.077	0.280	0.031	0.000	0.155

Gray indicates slope is statistically significant, using Kendall Tau test.

Trend Analysis

The purpose of the first study was to determine whether climatological trends have changed over the past century, causing drought events to become more frequent. The results were inconsistent both spatially and temporally. The only consistent pattern among all stations was an increase in Fall precipitation (~0.04) (Table 3a). The other monthly and seasonal results were highly variable. No station for any given month showed a statistically significant decreasing trend. Conway depicted the most months with an increasing trend in precipitation and a 0.155 inch increase in annual precipitation. The local regression model showing the annual precipitation trend for Conway, 1900-1999, is displayed in Figure 2. No station exhibited a trend in June, July, and August precipitation. Clemson data indicated four months, March, May, October, November, with an increasing precipitation trend. Only one month displayed a significant trend for Winthrop and Charleston. Winthrop displayed a positive trend in November, and the Charleston data indicated a positive trend in January. Camden data showed a positive trend for both January and November.

The temperature trend data were also highly variable, with no consistent trend temporally or geographically (Table 3b). The monthly trends that were

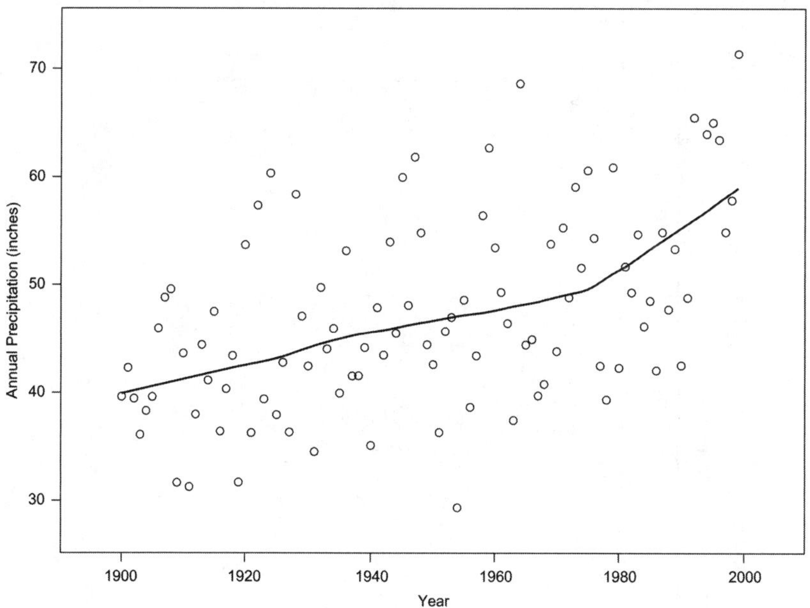

Figure 2. Local regression model showing annual precipitation trend for Conway, S.C., 1900-1999.

Table 3b. South Carolina Temperature Analysis 1900-1999
Kendall's rank correlation tau

	Clemson		Winthrop		Camden		Charleston		Conway	
	p	slope	p	slope	p	slope	p	slope	p	slope
January	0.013	0.032	0.218	0.019	0.692	0.006	0.486	0.001	0.166	-0.022
February	0.877	0.002	0.741	-0.005	0.171	-0.018	0.383	-0.011	0.280	0.016
March	0.738	0.003	0.448	0.007	0.993	0.000	0.307	0.008	0.619	-0.007
April	0.025	-0.02	0.110	-0.014	0.029	-0.020	0.347	-0.006	0.027	0.020
May	0.824	0.001	0.606	-0.004	0.565	-0.004	0.967	0.000	0.497	0.005
June	0.003	0.018	0.699	0.002	0.092	0.011	0.007	0.014	0.066	0.015
July	0.005	0.018	0.514	0.004	0.507	0.004	0.938	0.000	0.000	0.022
August	0.066	-0.017	0.232	-0.011	0.326	0.010	0.502	-0.006	0.005	0.015
September	0.935	-0.001	0.381	-0.009	0.632	-0.006	0.668	0.003	0.662	-0.003
October	0.055	0.022	0.145	0.015	0.257	0.011	0.013	0.024	0.730	0.004
November	0.008	0.030	0.341	0.012	0.507	-0.009	0.418	0.009	0.005	0.033
December	0.835	0.003	0.133	-0.025	0.009	-0.043	0.094	-0.026	0.206	0.017
Winter	0.399	0.009	0.683	-0.004	0.083	-0.018	0.439	-0.009	0.688	0.005
Spring	0.477	-0.003	0.362	-0.006	0.090	-0.008	0.991	0.000	0.394	0.005
Summer	0.191	0.006	0.621	-0.003	0.761	0.001	0.365	0.003	0.000	0.018
Fall	0.047	0.014	0.670	0.003	0.677	-0.004	0.180	0.010	0.227	0.007
Annual	0.162	0.005	0.455	-0.003	0.189	-0.006	0.976	0.000	0.155	0.007

Gray indicates slope is statistically significant, using Kendall Tau test.

Table 4a. South Carolina Palmer Drought Serverity Index (residuals)
Monthly Time Series Trend Analysis
Kendall Rank Correlation Tau

Climate Division	P-Value	Slope
1	0.4234	+0.0130
2	0.5212	+0.0011
3	0.9282	+0.0002
4	0.6497	-0.0008
5	0.6136	-0.0089
6	0.5711	-0.0104
7	0.4086	-0.0145

significant for Clemson, Charleston, and Conway were primarily positive, but the Camden data indicated a decreasing temperature trend. Clemson exhibited a decreasing trend in April, but January, June, July, and November temperature increased. Winthrop showed no trend for any month or season.

Owing to the spatial, geographical, and temporal variability, it is difficult to conclude that there is a significant trend in either precipitation or temperature, except during the Fall, since 1900. A conclusion could be made that Fall precipitation has increased statewide, since all stations indicated a significant positive trend. Since only one season (summer) for one station (Winthrop) indicated a decreasing precipitation tend, the precipitation data do not support more frequent drought events over the past century.

The PDSI trend analysis further supported this conclusion that drought events have not become more frequent. Both the time series and monthly trend analysis concluded that there was no trend for any climate division. The trend analysis tests were performed on the residual values due to the autocorrelation in the PDSI values. Table 4a displays the probability value and the slope for the 1900-1999 time series analysis for each climate division. Table 4b displays the monthly probability value for each climate division (as indicated in Figure 1). The null hypothesis was accepted for all tests, indicating that there is no significant trend in the PDSI residuals since 1900.

Table 4b. South Carolina Palmer Drought Severity Index (Residuals) Trend Analysis 1900 - 1999
Kendall's rank correlation tau

	CD 1	CD 2	CD 3	CD 4	CD 5	CD 6	CD 7
	p	p	p	p	p	p	p
January	0.963	0.885	0.834	0.791	0.876	0.963	0.989
February	0.915	0.787	0.980	0.572	0.708	0.696	0.576
March	0.851	0.833	0.989	0.680	0.859	0.741	0.680
April	0.989	0.949	0.994	0.721	0.994	0.762	0.576
May	0.821	0.842	0.933	0.911	0.795	0.894	0.902
June	0.704	0.821	0.980	0.758	0.817	0.855	0.980
July	0.838	0.799	0.976	0.894	0.894	0.827	0.989
August	0.766	0.812	0.928	0.976	0.941	0.774	0.967
September	0.774	0.758	0.825	0.894	0.977	0.989	0.863
October	0.783	0.817	0.937	0.842	0.812	0.723	0.783
November	0.859	0.855	0.989	0.911	0.816	0.906	0.745
December	0.967	0.932	0.941	0.876	0.889	0.963	0.894

No statistically significant trend for any month, using Kendall Tau test.

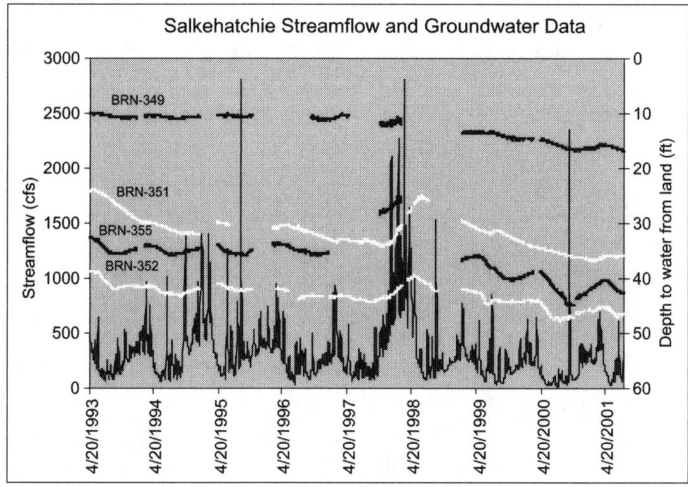

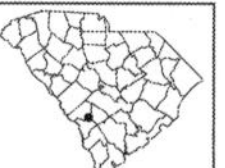

Figure 3. Salkehatchie River streamflow and groundwater levels in Barnwell County, S.C., April 20, 1993-August 1, 2001.
BRN-351 Aquifer depth 95 feet; screen placed 80 to 90 feet
BRN-352 Aquifer depth 293 feet; screen placed 278 to 288 feet
BRN-355 Aquifer depth 701 feet; screen placed 686 to 696 feet
BRN-349 Aquifer depth 1045 feet; screen placed 1030 to 1040 feet

Multiple Drought Indices

Over the past decade, water supply planners and drought experts have gained more confidence and access to different drought indices. Mostuse more than one index in making decisions, since no index is superior in all circumstances. Access to multiple indices that provide sector-specific drought data prompted the amendment of the Drought Act. The primary proposed indices to be used in the drought declaration process, in addition to the PDSI [Palmer, 1968], include the Crop Moisture Index [Palmer, 1968], the Standard Precipitation Index [McKee et. al., 1993, Guttman, 1999], the Keetch-Byram Drought Index [Keetch and Byram, 1968], the National Drought Monitor [Svoboda et. al, 2002], two-week average streamflow, and aquifer levels [South Carolina State Water Plan, 1998]. Each of these indices provides more sector-specific drought data, which provides a more accurate depiction of the type of drought (i.e. agricultural or hydrological), the severity, and the needed response. Table 5 shows the various drought declaration levels based on the different drought indices. For example, on May 27, 2000, for Climate Division 5, the drought status based on the Crop Moisture Index was incipient (first drought stage), based on the PDSI it was moderate (second drought stage), based on the three month Standard Precipitation Index, the Keetch Byram Drought Index, and the National Drought Monitor the status was severe (third stage), and based on the streamflow the status was extreme (fourth and final drought stage). There was no unanimous agreement in any climate division. The Drought Act states that each declaration may be made if any of the indices indicate the drought status; however, indication by one index alone does not mandate a declaration. There was no opposition to using multiple indices for the drought declaration process after documentation such as displayed in Table 5 was provided.

Surface Water and Groundwater—A Single Resource

The SC DNR supports the theory that surface water and groundwater are a single resource and should be treated as such in policies such as the Drought Response Act, as opposed to the SCDHEC's viewpoint. Their primary concern was the possible conflict between their Groundwater Use and Reporting Act and the Drought Response Act. The SCDNR used surface water and groundwater data from an extensive statewide surface water and groundwater monitoring network to justify their position that water availability in streams, lakes, and aquifers (confined and unconfined) must be considered in the drought declarations, and any conflict between laws would have to be considered by the Drought Response Committee. Water levels in the confined aquifers may be a result of water withdrawals and/or long-term drought (less recharge or replacement) or a combination of both. During long-term and persistent drought, the water level status in the confined aquifers is a very important indicator of the drought severity. Figure 3 shows the streamflow and groundwater levels in Barnwell County, S.C., on the Salkehatchie River. The groundwater data are from a well cluster with depth-to-water measurements for screen intervals of 80-90 feet, 278-288 feet, 686-696 feet, and 1030-1040 feet. The water levels in the aquifers are being affected by a combination of drought (1998- 2001) and pumping by the City of Barnwell and rural homes. It is impossible to separate the effects of drought and pumping, but it is important to document and consider the impact of both on the water resources of that area. The well measurements indicate that since the drought began in 1998 there has been a 3- to 10-foot drop in the groundwater levels. Prior to 1998 the groundwater levels varied seasonally, but no continual decline could be detected. Figure 4a displays the streamflow, groundwater, and precipitation data in Aiken County on the South Fork Edisto River. This is a shallow well with screen at 82-92 feet. The variability in the groundwater levels at this well is primarily from drought. Because the well is relatively shallow, the groundwater level responds quickly to precipitation events or extended periods of no precipitation. This figure shows declining groundwater levels due to the drought of 1998-2002. The well has had a 3-foot drop in water level since June 1998. The groundwater level has reached its lowest point since measurements began at this site in 1993. Figure 4b shows the

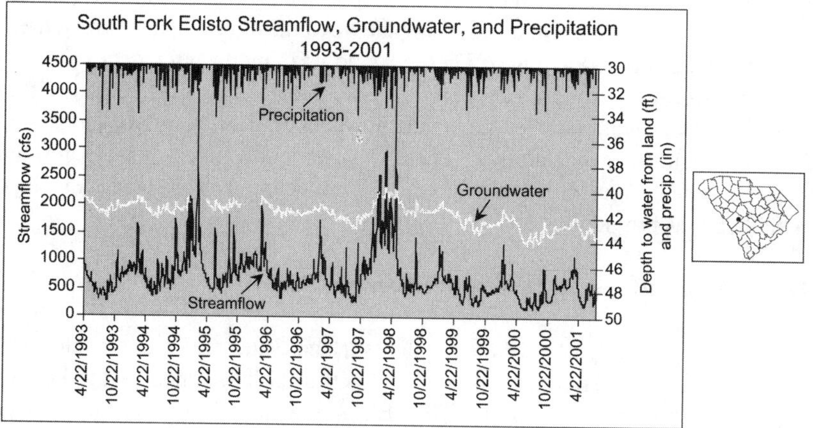

Figure 4a. South Fork Edisto River streamflow, groundwater levels and daily precipitation[a] at Aiken, S.C., April 22, 1993-August 1, 2001. Groundwater well located on South Fork Edisto River with screen at 82 - 92 feet.

[a]In order to display the precipitation values on the secondary axis, an arbitrary value of 30 was added to each daily precipitation amount.

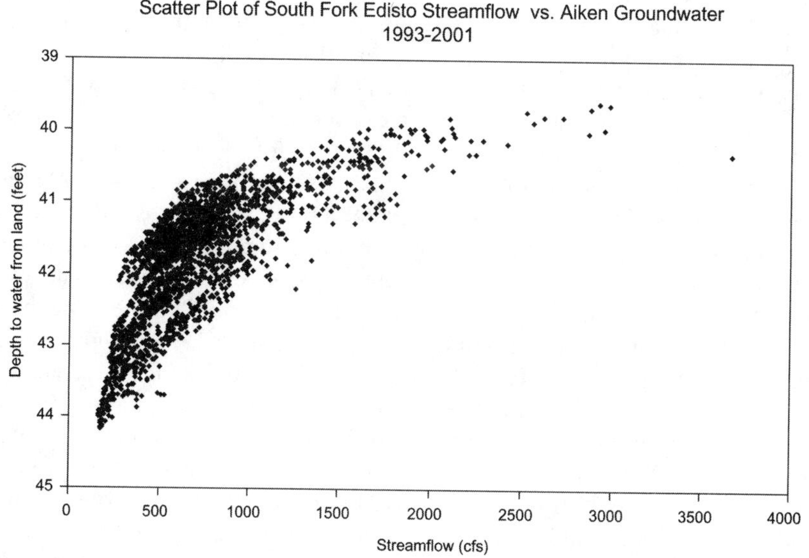

Figure 4b. Scatter plot of daily South Fork Edisto Streamflow (cfs) and Aiken groundwater (feet), April 22, 1993 - August 1, 2001.

the Keetch Byram Drought Index, and the National Drought Monitor the status was severe (third stage), and based on the streamflow the status was extreme (fourth and final drought stage). There was no unanimous agreement in any climate division. The Drought Act states that each declaration may be made if any of the indices indicate the drought status; however, indication by one index alone does not mandate a declaration. There was no opposition to using multiple indices for the drought declaration process after documentation such as displayed in Table 5 was provided.

Surface Water and Groundwater—A Single Resource

The SC DNR supports the theory that surface water and groundwater are a single resource and should be treated as such in policies such as the Drought Response Act, as opposed to the SCDHEC's viewpoint. Their primary concern was the possible conflict between their Groundwater Use and Reporting Act and the Drought Response Act. The SCDNR used surface water and groundwater data from an extensive statewide surface water and groundwater monitoring network to justify their position that water availability in streams, lakes, and aquifers (confined and unconfined) must be considered in the drought declarations, and any conflict between laws would have to be considered by the Drought Response Committee. Water levels in the confined aquifers may be a result of water withdrawals and/or long-term drought (less recharge or replacement) or a combination of both. During long-term and persistent drought, the water level status in the confined aquifers is a very important indicator of the drought severity. Figure 3 shows the streamflow and groundwater levels in Barnwell County, S.C., on the Salkehatchie River. The groundwater data are from a well cluster with depth-to-water measurements for screen intervals of 80-90 feet, 278-288 feet, 686-696 feet, and 1030-1040 feet. The water levels in the aquifers are being affected by a combination of drought (1998- 2001) and pumping by the City of Barnwell and rural homes. It is impossible to separate the effects of drought and pumping, but it is important to document and consider the impact of both on the water resources of that area. The well measurements indicate that since the drought began in 1998 there has been a 3- to 10-foot drop in the groundwater levels. Prior to 1998 the groundwater levels varied seasonally, but no continual decline could be detected. Figure 4a displays the streamflow, groundwater, and precipitation data in Aiken County on the South Fork Edisto River. This is a shallow well with screen at 82-92 feet. The variability in the groundwater levels at this well is primarily from drought. Because the well is relatively shallow, the groundwater level responds quickly to precipitation events or extended periods of no precipitation. This figure shows declining groundwater levels due to the drought of 1998-2002. The well has had a 3-foot drop in water level since June 1998. The groundwater level has reached its lowest point since meas-

Table 5. Comparison of Drought Indices

Date	CD[a]	PDSI[b]	CMI[c]	3 Month SPI[d]	KBDI[e]	DM[f]	Streamflow (cfs)[g]
May 27, 2000	1	-2.65 Moderate	-0.31 Incipient	N/A No Drought	450 Moderate	D1 Moderate	76.25 Extreme
	2	-2.77 Moderate	-0.79 Incipient	-0.99/0.99 Incipient	450 Moderate	D2 Severe	276.19 Extreme
	3	-2.08 Moderate	-0.54 Incipient	-1.00/-1.49 Moderate	450 Moderate	D1 Moderate	14.65 Extreme
	4	-0.59 Moderate	-0.20 Incipient	-0.99/0.99 Incipient	350 Incipient	D0 Incipient	39.85 Extreme
	5	-2.46 Moderate	-1.41 Incipient	-1.50/-1.99 Severe	600 Severe	D2 Severe	14.65 Extreme
	6	-1.85 Moderate	-0.91 Incipient	-0.99/0.99 Incipient	550 Severe	D1 Moderate	8.99 Extreme
	7	-1.55 Moderate	-1.05 Incipient	-0.99/0.99 Incipient	550 Severe	D1 Moderate	244.79 Extreme

Table 5 (continued). Comparison of Drought Indices

Date	CD[a]	PDSI[b]	CMI[c]	3 Month SPI[d]	KBDI[e]	DM[f]	Streamflow (cfs)[g]
January 6, 2001	1	-2.38 Moderate	0.34 No Drought	-1.00/-1.49 Moderate	200 No Drought	D3 Extreme	58.7 Extreme
	2	-2.52 Moderate	0.04 No Drought	-1.00/-1.49 Moderate	250 No Drought	D2 Severe	193.11 Extreme
	3	0.21 No Drought	-0.01 Incipient	-1.50/-1.99 Severe	350 Incipient	D1 Moderate	25.55 Extreme
	4	1.16 No Drought	-0.01 Incipient	-1.00/-1.49 Moderate	250 No Drought	– No Drought	200.73 Extreme
	5	0.71 No Drought	-0.01 Incipient	-1.00/-1.49 Moderate	250 No Drought	D1 Moderate	30.70 Extreme
	6	0.52 No Drought	-0.02 Incipient	-1.00/-1.49 Moderate	250 No Drought	D0 Incipient	25.19 Extreme
	7	1.17 No Drought	0.57 No Drought	-1.00/-1.49 Moderate	150 No Drought	– No Drought	348.57 Extreme

[a] Climate Division [b] PDSI = Palmer Drought Severity Index [c] CMI = Crop Moisture Index
[d] 3 Month SPI = 3 Month Standard Precipitation Index [e] KBDI = Keetch Byram Drought Index [f] DM = Drought Monitor
[g] Streamflow (cfs) = Two week average daily streamflow in cubic feet per second

urements began at this site in 1993. Figure 4b shows the close interaction between groundwater and surface water (r^2=0.74). In general, the higher the groundwater levels the greater the streamflow. Figure 5 displays groundwater data for a well in Georgetown County with a screen at 490-740 feet. The steady decline in the water level is primarily due to pumping by the Georgetown Rural Water District and not to drought. This region of the State has received normal to above-normal rainfall since 1998 - yet the well indicates the greatest decline in groundwater. It is this situation that concerns the SCDHEC, because this area would fall under the Capacity Use Area designation by their Groundwater Use Act rather than the Drought Response Act. In situations where excessive groundwater withdrawal presents potential adverse effects to the natural resources, or poses a threat to public health or safety, or where conditions pose a significant threat to the long-term integrity of a groundwater source, including saltwater intrusion, the SCDHEC board may designate a capacity use area. The Groundwater Use and Reporting Program issues Ground Water Use Permits to all ground water systems located in a designated capacity use area designed to withdraw and use groundwater equal to or greater than three million gallons in any month. [Groundwater Use and Reporting Act, Section 49-5-10, et seq., Code of Laws of South Carolina 1976, as amended]. Therefore, the SCDHEC and the SCDNR reached an agreement to clarify the drought act with a statement that a decline in aquifer water levels due to withdrawals not associated with drought should not be used for declaration of drought-alert phases.

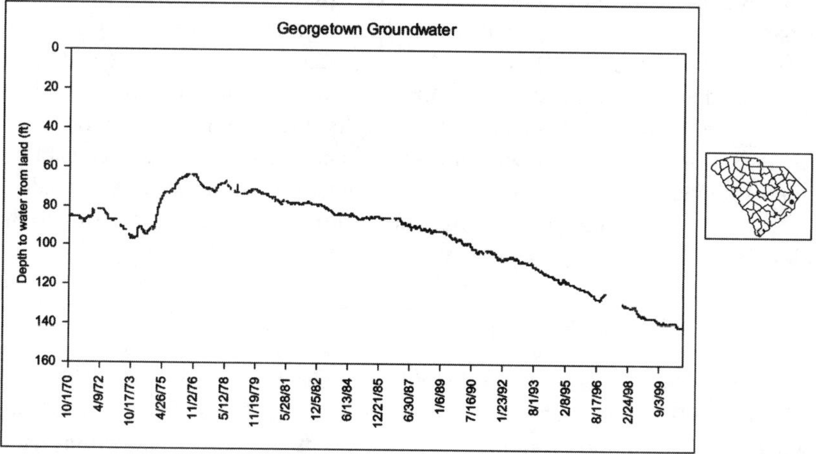

Figure 5. Groundwater levels at Georgetown, S.C., October 1, 1970-September 30, 2000, with screen at 490-740 feet.

Amendments and Compromises

Compromises were made between the SCDNR and other interested groups that did not require scientific justification by either stakeholder. A compromise was reached with the South Carolina Chapter of the American Water Works Associa tion to the effect that during an incipient, moderate, or severe drought, public water systems with aquifer storage and recovery facilities or water stored in managed watershed impoundments would be excluded from mandatory water use restric tions; however, during the final stage of drought (extreme) they may be required to implement water use curtailment. As a business, some of the water systems invest large amounts of money preparing for times of shortage. The compromise would help to ensure that their mitigation actions will not be negated, thereby allowing them to collect revenue by distributing water.

There were other unopposed amendments based on recommendations from the SCDNR and/or other stakeholders - such as the designation of new drought management areas that will allow for more effective drought management. The new drought management areas are based on river basin (Savannah, Santee, Pee Dee, and ACE) and geopolitical boundaries instead of climate divisions. An amendment was added to increase the local representation on the drought response committee to include power- generation facilities, special-purpose districts, and Soil and Water Conservation Districts. These members are in addition to the other local committee members (agriculture, industry, public service districts, Commis sion of Public Works, domestic user, regional council of governments, private water supplier, county, and municipality) and five state agency committee members (Department of Natural Resources, Department of Health and Environmental Control, Department of Agriculture, Forestry Commission, and Emergency Preparedness Division). The specification was added, however, that there may not be more than two members on a local committee from the same county, and the local committee members are appointed by the Governor with the advice and consent of the Senate rather than recommendation from the legislative delegation. A section was also added so that any party affected by a drought declaration can appeal that action to the Administrative Law Judge Division.

CONCLUSIONS AND FUTURE DIRECTIONS

This paper summarizes, for a limited geographical region (South Carolina) and a limited time period (1900-2002), the climatological, hydrological and policy trends related to droughts. Whereas, scientific analysis alone may yield multiple answers, it is nevertheless important to use a scientific basis for policy decisions involving water resource management. The complex interaction between science and politics has shaped South Carolina's drought policy.

South Carolina is the only state with a law directly relating to drought response. The interaction between science and policy was successful because of the open communication between stakeholders, policy makers, and the scientists. There were many obstacles to overcome especially with each stakeholder and scientist wanting to protect his/her "turf" [Clark et al., 1998]. Science was not able to resolve all discrepancies, owing in some cases to conflicting results and in other cases to political barriers. The trend analysis did not prove that climate trends have changed over the past century, thereby causing drought events to become more frequent; however, use of the multiple drought indices analysis and the streamflow and groundwater data were instrumental in justifying compromise on the drought act amendments.

As we look to the future of drought response in South Carolina, the drought policy will continue to evolve. As new theories and facts emerge, policy makers will be faced again with formulating changes. There is a growing need to synergize the various drought indices to achieve unanimity in decisions. There is also a gap between theory and observations. Observations are mostly point-based, whereas the theories outlined in this paper and elsewhere are universal and spatially distributed. Distributed land-atmosphere modeling, coupled with remote sensing and geographic information systems, may help to bridge the gaps between theory, observations, modeling and, most importantly, policy decisions in the conservation of water resources.

Acknowledgements. The authors thank Dr. Michael Helfert and Mr. Milton Brown for their support and encouragement. The authors would also like to express their appreciation for the editorial guidance provided by Mr. Roy Newcome and Ms. Ann Nolte.

REFERENCES

Alley, W., The Palmer Drought Severity Index: limitations and assumptions, *Journal of Climate and Applied Meteorology, 23*, 1100-1109, 1984.

Castro, J. E., and J. Hu, Distribution and rate of water use in South Carolina, Water Resources Report 18, South Carolina Department of Natural Resources, 1997.

Clark, R.N., and E.E. Meidinger, Integrating Science and Policy in Natural Resource Management: Lessons and Opportunities from North America, United States Department of Agricultural Technical Report PNW-GTR-441, 1998.

Hayes, M.J., Drought Indices, National Drought Mitigation Center, Lincoln, Nebraska, 1996.

Hayes, M.J., and C. Knutson, South Carolina Drought Mitigation and Response Assessment:1998-2000, Drought, Natural Hazards Research and Applications Informa tion Center Report, 2001.

Heim, R.R., A Review of Twentieth-Century Drought Indices Used in the United States, *Bulletin of the American Meteorological Society*, 83(8), 1149-1165.

Helsel, D.R., R. M. Hirsch, Statistical Methods in Water Resources, Elsevier B.V., The Netherlands, 1992.

Hrezo, M.S., P. G. Bridgeman, W.R.Walker, Integrating Drought Planning into Water

Resources Management, *Natural Resources Journal*, 26, 142-167, 1986.

Hrezo, M.S., P.G. Bridgeman, W.R. Walker, Managing Droughts Through Trigger Mechanisms, *Journal of American Water Works, 46-51*, 1986.

Glantz, M., W. Degefu, Drought Issues of the 1990s, Proceedings of the Second World Climate Conference, 253-263, 1990.

Guttman, N.B., Accepting the Standardized Precipitation Index: A calculation algorithm. *Journal of the American Water Resources Association, 35(2)*, 311-322, 1999.

Guttman, N.B., Comparing the Palmer Drought Index and the Standardized Precipitation Index. *Journal of the American Water Resources Association, 34(1)*, 113-121, 1998.

Keetch J.J., G.M. Byram, A Drought Index for Forest Fire Control, United States Forest Service Research Paper SE-38, November 1968.

Little, K., and D. Moreau, Estimating the Effects of Conservation on Demand During Droughts, *Journal of American Water Works Association, 48-54*, 1991.

McKee, T.B., N.J. Doesken, and J. Kleist, The relationship of drought frequency and duration ot time scales. Eighth Conference on Applied Climatology, American Meteorological Society, Jan 17-23, 1993, Anaheim CA, 179-186, 1993.

Mizzell, H., 1998-2001 Drought Impacts, South Carolina Drought Program Report, 2001.

Najarian, P.N., An Analysis of State Drought Plans: A Model Drought Plan Proposal, Masters Thesis, University of Nebraska, Lincoln, Nebraska, 2000.

National Water Summary 1988-1989 - Hydrologic Events and Floods and Droughts, 2375, United States Geological Survey, United States Government Printing Office, Denver, Colorodo, 1991.

Newcome, R., The 100 Largest Public Water Supplies in South Carolina, Water Resources Report 21, South Carolina Department of Natural Resources, 2000.

Palmer, W. C., Keeping track of crop moisture conditions, nationwide: the new Crop Moisture Index, *Weatherwise,* 21:156-161, 1968.

Palmer, W.C., Meteorological Drought. Research Paper No. 45, U.S. Department of Commerce Weather Bureau, Washington, D.C., 1965.

Rouse, R. R., M. Perry, J. Purvis, and A. Zupan, South Carolina Drought Response Plan, AR-18, South Carolina Water Resources Commission, 1985.

South Carolina Code of Laws 1976, as amended, Title 49, Chapter 5, Groundwater Use and Reporting Act, 2000.

South Carolina State Water Plan, South Carolina Department of Natural Resources, 1998.

Svoboda, M., D. LeComte, M. Hayes, R. Heim, K. Gleason, J. Angel, B. Rippey, R. Tinker, M. Palecki, D. Stooksbury, D. Miskus, S. Stephens, The Drought Monitor, *Bulletin of the American Meteorological Society*, 83(8), 1149-1165.

Wilhite, D.A., The Enigma of Drought, in *Drought Assessment, Management, and Planning, Theory and Case Studies*, edited by D. A. Wilhite, pp. 3-15, Kluwer Academic Publishers, Nowell, M.A., 1993.

Wilhite, D.A., Planning for Drought: A Methodolody, in *Drought Assessment, Management, and Planning, Theory and Case Studies*, edited by D. A. Wilhite, pp. 87-108, Kluwer Academic Publishers, Nowell, M.A., 1993.

Wilhite, D. A., State actions to mitigate drought: Lessons learned, *Journal of the American Water Resources Association*, 33(5), 961-968, 1997.

Wilhite, D. A., M. Hayes, C. Knutson, and K. Smith, Planning for Drought: Moving From Crisis to Risk Management, *J. of the American Water Resources Association*, 36, 697-710, 2000.

Wilhite, D.A., A Methodology for Drought Preparedness, *Natural Hazards,* 13, 229-252, Kluwer Academic Publishers, the Netherlands, 1996.

Wilhite, D.A., Drought Planning: A Process for State Government, Water Resources Bulletin, 27(1), American Water Resources Association, 1991.

Winter, T., J. Harvey, L. Franke, W. Alley, Ground Water and Surface Water A Single Resource, Circular 1139, United States Geological Survey, Denver Colorado, 1998.

Hope P. Mizzell, South Carolina Department of Natural Resources, South Carolina State Climatology Office, 1201 Main Street, Suite 1100, Columbia, SC 20901; mizzell@water.dnr.state.sc.us

Venkataraman Lakshmi, University of South Carolina at Columbia, Department of Geological Sciences, 700 Sumter Street, Columbia, SC 209208; vlaksmi@geol.sc.edu

16

Assessing the Impact of Climate Variability and Change on Regional Water Resources: The Implications for Stakeholders

Janet L. Gamble, John Furlow, Amy K. Snover, Alan F. Hamlet, Barbara J. Morehouse, Holly Hartmann, and Thomas Pagano

INTRODUCTION

The nation's water supplies—those that provide clean and plentiful drinking water, those that meet the needs of industry, agriculture, and power production, and those that support transportation and recreation—are vulnerable to multiple and interacting factors, including alterations in the hydrologic cycle related to climate variability and change, changes in land use, in population growth, and in the policies that affect water supply and demand. These factors are of particular interest to those charged with managing water supplies and others whose decisions influence or are influenced by the availability and quality of water in the U.S. Some believe (Schilling and Stakhiv 1998) that the impacts of climate variability and change on water supplies will be adequately handled via existing management practices. They contend that water resource managers make decisions based on a belief that the impacts of climate change on water supplies are within the normal range of weather variability, concluding that "approaches for effectively dealing with climate change are little different than the approaches already available to manage existing variability" [Gleick, 2000]. As other demands on water resources increase, however, research suggests that fluctuations in climate will challenge the capacity of current management techniques and planning horizons, thus requiring new approaches to effectively manage and plan for reliable water supplies.

They may be right, in total or in part. But, important questions remain: If water resource managers and other decision makers do not incorporate near-term climate variability and longer-term climate change in their plans and decisions, is

Water: Science, Policy, and Management
Water Resources Monograph 16

10.1029/016WM20

this because of assumptions they have made about the adequacy of existing management strategies or does it derive from some other factor, such as:

- a failure in information flows (to and from scientists and stakeholders),
- a failure in the skill or scale of forecasts that precludes their use for planning or management purposes,
- a lack of adaptive capacity in the institutions that control water resources, or is it symptomatic of decision gridlock within watersheds controlled across multiple jurisdictions?

In the case studies that follow, we explore the experiences of three regional water resource assessment teams, their attempts to communicate climate information to stakeholders, and their efforts to incorporate stakeholder concerns into the research, modeling, and assessment processes. Not surprisingly, these case studies suggest a complex picture, one that may begin to diverge from that described by Schilling and Stakhiv. It appears that the impacts of climate variability in the near term and climate change over the longer term have sparked the interest of stakeholders. Managers are beginning to express an openness to climate information, particularly when the scale of relevant models (both geographic and temporal) and their skill allow for watershed-level applications. And, while resource managers continue to rely on existing management practices, an increasing number of managers are considering climate information, at least informally, in their management decisions. The evidence now suggests that active stakeholder participation in the assessment process is essential for developing useful climate-related products and ultimately for increasing the adaptive capacity of managed systems.

Regional Water Assessments: The Context

Each of the regional assessments presented here contributed to the larger U.S. National Assessment. The Great Lakes Regional Assessment (GLRA) was one of 19 regional assessments conducted for the First National Assessment of the Potential Consequences of Climate Variability and Change [NAST, 2000] under the auspices of the U.S. Global Change Research Program (USGCRP). The Environmental Protection Agency, a member agency of the USGCRP, funded the GLRA. The Climate Impact Group's (CIG) assessment of Pacific Northwest watersheds, funded by the National Oceanic and Atmospheric Administration (NOAA) Office of Global Programs, was well underway when the National Assessment began, but it too contributed to the National Assessment. The third case study—the CLIMAS project (the Climate Assessment for the Southwest)—is a long-term effort supported by NOAA. While CLIMAS was less tightly integrated with the National Assessment than the other regional efforts, the CLIMAS team shared findings with the Southwest Regional Assessment.

The National Assessment was undertaken to satisfy the requirements of the Global Change Research Act of 1990 [PL 101-606]. The Act calls for periodic assessments of the effects of global change in the U.S. and requires that an assessment be conducted that "analyzes the effects of global change on the natural environment, agriculture, energy production and use, land and water resources, transportation, human health and welfare, human social systems, and biological diversity [PL 101-606 Sec. 106]." To make the assessment tractable, the nation was divided by region and sector. Nineteen regional assessments and five sectoral assessments were undertaken (including assessments of water resources, coastal issues, forests, agriculture, and human health).

Each regional or sectoral assessment began with a workshop. In these workshops, stakeholders (including planners, resource managers, local officials and other decision makers) met with scientists to clarify information needs and to coordinate their respective roles in the assessment process (e.g., with data or tools or applied expertise). Early involvement of the stakeholder community in the assessment ensured that the selection of topics, the development of assessment strategies, and the communication of findings were informed by the end-user's perspective. The regional case studies that follow detail the process of stakeholder involvement and suggest a future for research and assessment projects characterized by sustained collaboration between scientists and decision makers. Certainly, a heightened awareness among scientists of the needs and requirements of the stakeholder community is evident in each of the case studies.

REGIONAL WATER RESOURCE ASSESSMENTS: INCORPORATING THE STAKEHOLDER'S PERSPECTIVE

The Great Lakes Regional Assessment (GLRA)

At the outset of the Great Lakes Regional Assessment, the assessment team learned from stakeholders in the region that water resources would be the dominant assessment issue. The Great Lakes influence everything: regional politics, recreation, the economy – even the weather. The challenge to the assessment team was to discern which areas were of greatest interest to stakeholders, and then to understand how climate change and variability, both directly and as mediated through changes in the Lakes, would affect those areas of concern. Stakeholder interests tended to be expressed in terms of economics or quality of life. The assessment team felt that part of its role was to provide practical scientific information to stakeholders to help them make connections between climate and the Lakes and other aspects of life in the region. The team understood that GCMs do not provide information that is directly useful to stakeholders; a projection of a change of a few inches of precipitation or a few degrees temperature would not address questions such as, "Will my company be able to ship goods on the

Lakes?" "Will my power plant have to redesign its cooling system?" Furthermore, temperatures rise more between 6 AM and noon on a typical day than they are expected to rise, on average, by the end of the century. Resource managers have noted that we manage for that variability, and ask if climate change is of such magnitude that its effects will extend beyond the variability we observe and plan for.

Given the wide range of possible impacts and the need to develop practical information for stakeholders, the assessment team sought to narrow its focus by encouraging the stakeholders who might benefit from the assessment to help in the selection of assessment topics. To accomplish this, the assessment team organized a stakeholder workshop. Drawing on information from several previous studies, including *Adapting to Climate Change and Variability in the Great Lakes-St. Lawrence Basin* (Mortsch, et al., 1998), and the binational study *The Great Lakes: An Environmental Atlas and Resource Book* (USEPA and Environment Canada, 1995), the team informed stakeholders of current conditions and discussed how climate change might affect several economically and socially important sectors. Next, they asked the stakeholders to identify particular areas of concern within the region. Over one hundred representatives of industry, academia, government, and environmental and other interest groups attended that first workshop [Sousounis, 1998]. They expressed a wide range of concerns, from the scientific (such as the need for new general circulation models) to the political and legal (such as, who has the right to the waters of the Great Lakes and how may increased competition for fresh water be managed?). For each area, stakeholders considered current conditions and stresses, possible impacts of climate change, research and information needs, and adaptation options.

The stakeholders identified issues of concern and worked with the assessment team to determine the information they would need to facilitate adaptation to climate change. The assessment team used those identified needs to guide their investigation of how climate change might impact stakeholder interests. This process—translating broad interests into specific scientific questions to be analyzed and answered and translated back into language and tools that stakeholders can use—is complicated but vital to the effectiveness of the assessment.

From the outset, the researchers and stakeholders agreed that water resources bind together most aspects of life in the region. "Understanding the impacts on water resources is the linchpin to understanding impacts of climate change on other sectors. The water in the Great Lakes—St. Lawrence Basin serves as a resource for sustaining human life, ecology, agriculture, trade, energy generation, and recreation, to name a few [Sousounis and Albercook, 1998]." Some issues of concern, such as agriculture, were affected by multiple stresses. In other cases, the same stress affected multiple areas of interest—for instance, shipping, recreational boating, and water utilities were all affected by lake levels. The stakeholders at the workshop agreed with the assessors that there were five general areas of highest concern: water resources, water ecology, land ecology, agriculture, and quality of life (which includes health, the rural landscape and lifestyle, and recreation and tourism).

Following the initial workshop, the assessment's steering committee and principal investigator identified researchers from academic and research institutions in the region to conduct studies of the sectors identified by the workshop participants. The assessment team faced an organizational challenge: should they organize the assessment around the endpoints identified by the stakeholders, or should they organize it around the scientific disciplines of the assessors which would require further integration to address specific stakeholder questions. Each approach has its advantages and disadvantages; organizing around teams tasked with addressing specific stakeholder issues risks duplication of efforts in certain disciplines. Organizing along different disciplines risks a failure to integrate those disciplines in a way that is useful to the stakeholders. In the end, the Great Lakes team took a hybrid approach, identifying the disciplines needed to address stakeholder questions and integrating cross-disciplinary findings in the assessment report to match stakeholder needs. The Great Lakes Overview Report, available at http://www.epa.gov/globalresearch, demonstrates the hybrid approach taken by the assessment team. In cases where multiple stakeholders are affected by a common event, such as a change in lake levels, the approach is to address the common impact. For stakeholders with unique concerns about multiple stresses, the approach is to address the sector of concern directly. For example, the concerns of farmers are addressed by looking at changes in precipitation, temperature, CO_2, as well as market and government forces in the context of agriculture. A new round of workshops is underway. Again, an emphasis is being placed on eliciting stakeholders' information needs, both near and long-term.

Stakeholders' near-term interests and information needs challenge the long-term focus of GCM output. One of the most important tasks facing the assessment team is that of translating stakeholder interests and time lines into issues that the scientists can successfully address. Stakeholder skepticism continues as another significant hurdle. The following examples demonstrate the range of information requested by stakeholder groups and the difficulties in matching those requests with standard modeling and assessment output.

Transportation managers. Transportation managers, from airport authorities to city street managers, face a wide range of weather-related problems. How much salt will be needed in a given winter? How many days may runways be out of service due to heavy snows? Snow and rainfall in the region are influenced by what is known as the "lake effect." Cold winter air moves down from Canada, crosses the lakes and picks up warmth and moisture. This moisture is then deposited downwind. Cities on the downwind shores of the Lakes receive significantly more snows than those on the upwind sides. In order to answer the questions of transportation managers, the assessment team looked at how changing temperatures and changes in precipitation may alter lake-effect snow patterns.

By examining historical records for circumstances that were favorable to heavy lake- effect snow, and then looking at climate models for the occurrence of similar conditions, the assessment team estimated that lake-effect snowfall may be reduced due to warmer air temperatures, though lake-effect rains may increase, providing transportation managers with a different set conditions to address.

The commercial shipping and boating industry. When the initial stakeholder workshop was held in 1998, representatives of shipping and boating interests were not particularly concerned about lake water supplies; water levels in the Great Lakes were at near-record highs [Sousounis, 2000]. The chance that lake levels could drop—a possible future suggested by the assessment—seemed remote. In March of 2001, when the Great Lakes assessment team held a follow-up workshop focusing on the impacts to shipping and boating that the near-record low lake levels have had over the past two years, the stakeholders' interest increased. Though still unconvinced that the recent drop in lake levels could be an indication of the way things could be, these stakeholders are taking a greater interest in how climate projections could help them plan for the future. Of greatest concern is the possibility that locks and shallow channels may become too shallow for big ships to pass through, forcing them to offload freight. Marina owners already face shallow slips for boaters that make it difficult for them to launch their crafts into the lakes. Off-loading freight and losing rents on marina space translates into immediate revenue losses. To reduce the draft of a 1,000 foot ship by one inch, its load must be reduced by 270 tons [Ryan, 2001]. It is estimated that marinas in the Great Lakes lost over $11 million in 1999 due to the impacts of low water levels [Mahoney et. al., 2000].

The assessment team's projections of lake levels are based on what the climate models say about changes in temperature and precipitation; the balance between lake evaporation and basin precipitation determines water levels. Warmer winters mean a later freeze-over for the lakes, allowing more winter evaporation and longer shipping seasons. According to the U.K.'s Hadley model, evaporation will be offset by more rain and snowfall. But, utilizing the Canadian model, the other model used in the National Assessment and which produces results more consistent with earlier GCMs, scientists project declines of lake levels of up to two feet on Lake Superior and up to five feet on Lake Michigan [Lofgren, Quinn, et.al., 2000]. As confidence improves in the models' projections of the timing of ice formation and breakup and of lake water levels, policy makers will be able to make decisions to address these issues. The models will have to be more reliable or current conditions will have to be more severe or unrelenting, for the shipping sector to consider adaptation measures like lengthening the shipping season, redesigning ships to have a shallower draft, redesigning locks and channels to accommodate the new ships, or dredging shallow channels.

Farmers. Agriculture in the Great Lakes Region is largely dependent on the weather. Agricultural practices are guided by an understanding of climate, but individual crop success is dependent on the weather. Farmers base their decisions on information about the length of the growing season and the timing and availability of water. This is especially true in the Great Lakes, where most crops depend on rainfall rather than irrigation [Andresen et.al., 2000]. Agricultural decision making will be influenced by a combination of inter-related factors, including climate change, government regulations, and economics. A decrease in growing season precipitation could be offset by increased irrigation, if adequate supplies of ground and surface water are available, and if regional rules allow for an expansion of irrigation. Warmer temperatures could allow for a longer growing season, but farmers will have to be confident that the timing of the first and last frost has truly shifted in order to take advantage of the opportunity.

Climate change will have direct impacts on temperature, precipitation, and CO_2 concentrations, which will influence growing season, soil moisture, irrigation needs, and plant growth. The assessment team looked at how changes in these factors, as simulated by the Hadley and Canadian models, would impact production of several crops [Andresen et.al., 2000]. The outlook for agriculture is good. Due to CO_2 enrichment and a warmer, wetter growing season, modeled yields of soybeans increase by 276% under the conditions described by the Hadley model, while maize increases by 373% under the Canadian model by 2099 (assuming crops occupy the same geographical area as they do today). The models also suggest a decrease in short-term droughts, reducing the need for irrigation [Andresen et.al., 2000]. The climate information may not have short-term value for the farmers, but it may inform longer-term decision making. The temperature projections suggest that the growing season may be lengthened, allowing for changes in crops or double cropping, thus increasing economic yield per acre. Farmers will probably only embrace such changes as time passes and experience supports the model projections. Changes in the population and distribution of pests and implications for pest control are impacts of climate change that will require further investigation [Winkler and Andresen, 2000].

Water utilities. There are more than 40 drinking water plants that draw water from Lake Michigan [Johnson and Soucie, 2001]. These plant managers face problems due to changes in lake water level as well as temperature. High waters and heavy surf can flood a plant. Low waters reduce this threat, but extremely low water levels can affect the ability to pump water. For instance, if there is no water over the intake point, the intake must be moved to deeper water; and if water levels are low over the intake, pumping costs may increase [Johnson and Soucie, 2001]. As decisions are made concerning where to site an intake point, and whether to move one, the same information on lake levels that would help the shipping industry would be useful to water utility operators. Drinking water plants are designed to handle wide variations in the quantity of source water associated with variable lake levels. Water levels can vary across the lakes due to winds and other storm conditions—water on one shore may be five to ten feet higher than on the opposite shore. That kind of weather-related displacement, on top of longer-term trends in water levels, can be significant. Temperature and turbidity are monitored and can change quickly. So while plant operators expect that they will be able to provide clean drinking water, the costs of doing so may vary depending on changes in the quality of source water. Changes in source water quality can affect the taste and smell of finished water. Drinking water managers have reported changes in the use, and associated costs, of carbon to control taste and odors. Information on the probability of high-turbidity events and algal blooms and research on alternative cost-effective approaches for controlling those problems will be helpful for planning [Johnson and Soucie, 2000].

Electric utilities. Electric utilities around the Great Lakes rely on lake water for the cooling cycle of electricity generation. The efficiency of plants can be impacted by several factors relating to lake water [Michaud, 2001]. First, the temperature of the water used for cooling is important. Although plants are designed with temperature fluctuations in mind, variations do affect the efficiency of the steam cycle. A sustained rise in lake water temperature of even a few degrees represents a significant net loss in the efficiency of all plants using the Lakes for cooling, on the order of hundreds of megawatts of power. This lost capacity would have to be recouped through other types of generation, assuming demand is constant or increasing. (It is important to note that the impact of venting of cooling waters back into the Lake creates a very small thermal footprint. The amount of hot water vented into the Lakes is so small relative to the total volume of water in the Lakes that the impact is minimal. This is not true on small rivers, where thermal pollution is a problem. However, there are only a few power plants built on small rivers, and few are likely to be impacted [Michaud, 2001]).

Another concern for power plant operators is change in lake levels. If water levels fall enough that intakes take in air, or cavitate, there are significant effects on plant efficiency. Air affects the transfer of heat. Cavitation, though unlikely where monitoring at the intake is ongoing, would require immediate attention [Michaud, 2001]. Most of the nuclear plants around Lake Michigan were built in the 1960s, the time when lake levels reached record lows. However, if current lake levels fall much lower, action may have to be taken at some plants, and the remediation can be expensive.

The Great Lakes Regional Assessment team has begun an iterative process of engagement between stakeholders and assessors. In the initial workshop, the team identified areas of general concern to stakeholders, and began to model the possible long-term impacts of climate change on those areas. In the second round of workshops the focus is on specific sectors and impacts; a greater understanding is being developed of the information stakeholders need to influence decisions. In the future, continued collaboration between assessors and stakeholders will be needed to better understand decision processes and to learn how climate information can support those processes. The decision to change behavior, particularly in cases where there is economic risk, should be taken seriously.

Projections of climate change that are supported by observation may provide stakeholders with the confidence to make decisions that take advantage of developing opportunities or that mitigate potential damages. A clear understanding of how decisions are made and how credible information can influence those decisions is critical. Cooperation and trust between decision makers and assessors is essential to our growing understanding of climate change.

The Climate Impacts Group (CIG) Assessment of the Pacific Northwest

The Climate Impacts Group (CIG) at the University of Washington was created in 1995 based on a contract with the Office of Global Programs (OGP) of the National Oceanic and Atmospheric Administration (NOAA). CIG assesses the consequences of climate variability and climate change for the natural and human systems associated with four "sectors" of the Pacific Northwest[1,2] (PNW): hydrology and water resources, forest ecosystems, marine and aquatic ecosystems, and coastal activities. The assessment team studies past climate variations, their impacts on the natural and human systems of the region, and the institutional context surrounding human management of and dependence on natural systems, in order to characterize regional climate-related vulnerability. These lessons from the past are then used to guide an assessment of the likely regional consequences of anthropogenic (human-caused) climate change. In addition to its research activities, CIG strives to enhance the resilience of PNW natural resources management to changes in climate by providing regional managers with the information and tools required to better incorporate an understanding of climate and its impacts into resource planning and management practices. In this section, the focus is on the development of CIG's strategy for assessment of climate impacts on hydrology and water resources and the resulting interface between CIG's research and regional water resource management processes.[3]

CIG began to examine PNW water resources management in the context of climate variability and climate forecasting in 1996. At that time, relatively little was known about the ability of water resources management agencies to utilize the improved water resource forecast information likely to arise from advances in large-scale climate forecasting capabilities. This was a crucial piece of information for the climate forecasting community, because it would indicate whether new kinds of resource forecasts could be utilized directly by the management community, or whether education, outreach, or other preliminary tasks would be

[1] For the purposes of the assessment, the Pacific Northwest is defined as the states of Washington, Oregon and Idaho and all of the Columbia River Basin.

[2] The CIG began the first of the U.S. Global Change Research Program's Regional Integrated Scientific Assessment (RISA) Program's evaluations of the impacts of climate variability and climate change on specific regions of the United States.

[3] For more information about CIG's other work, see CIG [1999] and Miles et al. [2001].

required to facilitate this transfer of information. To address this question, researchers from CIG [Callahan et al., 1999] interviewed natural resource managers to determine the degree to which they incorporated climate forecast information into their operational decision-making processes, and to determine their capability to make use of seasonal to interannual climate forecasts. In interviews with managers from twenty-eight organizations involved in PNW water resource management, CIG found that neither climate forecasts nor the associated resource forecasts played a prominent role in operational decision making. Many managers used forecasts informally as background information, but rarely used them during actual decision-making. On the whole, the management community knew little about the predictability of climate variations or associated water resources impacts and, as was later determined, even less about the potential impacts of anthropogenic climate change. These results confirmed earlier findings of Pulwarty and Redmond [1997].[4]

This assessment of the management community's ability to adopt new forecast information played a significant role in CIG's strategic planning. First, it was clear that simply developing pilot water resource forecasting methods in an academic setting would not produce the desired outcome, i.e., their use by water managers. In addition to the development of improved forecasting methods, a well- coordinated outreach effort would be required to (1) introduce the water management community to the potential role of interannual climate forecasts in water resources management, and (2) to facilitate the transfer of information from the research context to one of practical water management applications. Second, because there was little understanding of (or contextual framework for) the potential implications of climate change, it was clear that linkages should be made between the water managers' general understanding of climate variations in the historic record and the implications of climate change. The most promising approach for engaging managers in the issue of planning for climate change was to link the projected impacts to past climate variations that had challenged management. Finally, a knowledge base sufficient to support these two goals—a concerted outreach effort and an understanding of the linkages between climate variability and change—needed to be developed; it did not exist at the time.

CIG's strategy for building the knowledge base required to facilitate functional connections to the water management community was refined over time, and ultimately comprised the following:

---

4 This is not to say that the academic community lacked an understanding of the influence of climate on water resources. Research dating from the 1980s indicated the general nature of the hydrologic sensitivities of the western United States to global warming [Gleick, 1987a,b; Lettenmaier and Gan, 1990; Nash and Gleick, 1991; Lettenmaier and Sheer, 1991]. As the interviews by Pulwarty and Redmond (1997) and Callahan et al. (1991) showed, however, this information had not found its way into the hands of water resource managers in the PNW.

- The identification and quantification of the relationships between global climate phenomena (e.g., El Niño/Southern Oscillation (ENSO), Decadal Oscillation (PDO), anthropogenic climate change)[5] and regional hydrologic processes based on an analysis of the historical record [Mantua et al., 1997; Leung et al., 1999; Hamlet and Lettenmaier, 1999a,b; Mote et al., 1999; Hamlet and Lettenmaier, 2000; Mantua and Hare, 2001; Mantua and Mote, 2001; Mote, 2001; Hamlet et al., 2001b].
- A determination of the impacts of these hydrologic variations on PNW water resources systems, i.e., the users and management of these water resources (again using the historical record) [Mote et al., 1999; Hamlet and Lettenmaier, 1999a,b; Miles et al., 2000; Hamlet et al., 2001b].
- The development and demonstration of general hydrologic forecasting techniques based on advances in interannual climate forecasting abilities and onthe understanding of the regional hydrologic response to natural climate variations [Hamlet and Lettenmaier, 1999a; Leung et al., 1999; Huppert et al. 2001].
- The development of specific applications of hydrologic forecasts for PNW water resources management and demonstration of their feasibility and benefits [Hamlet et al., 2001a; Huppert et al., 2001].
- An analysis of the institutional[6] context of regional water resources management, including institutional sources of vulnerability to climate [Callahan et al., 1999; Gray, 1999], the characterization of barriers to the use of climate forecasts in management processes and the identification of strategies for realizing the potential value of forecasts for water resources management [Callahan et al., 1999].
- Finally, an integration of the understanding of the physical processes that determine climate impacts, and of the technical advances in climate and resource forecasting and modeling of global climate change, with a comprehensive analysis of the fundamental water resources problems facing the region and their institutional context [Miles et al., 2000; Hamlet et al., 2001b; Miles and Snover, 2001; Mote et al., 2001].

---

[5] El Niño/Southern Oscillation (ENSO) and Pacific Decadal Oscillation (PDO) are naturally occurring patterns of Pacific climate variability that include changes in sea and air temperatures, winds, and precipitation and influence climate around the world. ENSO is a tropical Pacific phenomenon with a two to seven year period. PDO has its strongest signature in the North Pacific and has a 50 to 70 year period.

[6] Institutional refers to the variety of formalized actions that underlie human social activity, including standards of behavior, formal decision rules and decision making procedures and grants of authority to prescribe policy. For instance, legal systems are institutions.

Imbedded in this overall strategy was the fundamental concept that an understanding of the implications of anthropogenic climate change must be derived primarily from an understanding of natural variability and its impacts on both natural and human systems [Miles et al., 2000; Snover and Miles, 2001].

Many of CIG's research findings that resulted from this approach helped, in turn, to shape both future research and outreach strategies. For example, because CIG's institutional analysis indicated that the regional water resources management system has a much lower capacity to respond to the threat of droughts than to the threat of floods, CIG focused much of its outreach on preparing for droughts and emphasized the drought-related impacts of anthropogenic climate change. Because management inflexibility was shown to increase the region's vulnerability to droughts, CIG has focused its research and outreach on ways to use climate information to increase flexibility. Finally, because regional vulnerability to drought is increased by a fragmented management structure, CIG has worked to engage and inform stakeholders from all user groups about using climate information.

With the continued development of the research foundation described above, CIG began a focused outreach program in 1998, working to communicate these and other findings to the regional management community. The need for such a formal program became a pparent during the highly-publicized 1997-1998 El Niño event, which so overwhelmed CIG's outreach response capability that new full-time staff were needed to cope with the regional demand for information. This experience demonstrated that environmental events can be useful triggers for stimulating interest in climate information and its use in planning and decision-making processes. CIG similarly used the 1998-1999 La Niña and the 2001 drought to engage regional interest in climate impacts.

The form of CIG outreach efforts has evolved over time, but CIG convenes annual water workshops aimed primarily at local and regional water resource managers, occasional specialty workshops and press conferences based on specific events (e.g., the 1997-1998 El Niño, CIG's PNW contribution to the U.S. National Assessment of climate impacts, and the 2001 drought), and periodic, high-profile, climate change workshops aimed at upper level policy-makers.

The annual water workshops, typically held in the fall (during the beginning of the water year) are convened in locations on both the east and west sides of the hydrologic, ecologic, economic, and cultural divide of the Cascade Mountains. The workshops provide an ongoing connection between PNW water managers and the CIG, and feature an exposition of the expected climate for the coming winter, the influence of climate variability on water resources, the predictability of those climate variations and their impacts, reports from regional managers on their use of climate information, and presentations about the possible impacts of anthropogenic climate change. CIG also uses these workshops to highlight newly developed forecasting techniques and applications. These workshops also provide crucial feedback to CIG about the specific needs of the water management community, which ultimately informs CIG's research agenda, as well as the design of future water workshops. CIG learned, for example, that regional differences in climate and in the spatial scale of watersheds need to be accommodated within the workshops. This led to the practice of holding several parallel workshops at different locations around the region each fall, with the content tailored to each location, and the recent expansion of CIG's research scope to include hydrologic modeling of the smaller-scale water systems west of the Cascade Mountains.

CIG's first climate change workshop was held in July 1997 (Snover et al. 1998), with the goal of initiating a dialogue with regional stakeholders concerning potential impacts of climate change on the PNW, important regional vulnerabilities, and strategies for adaptation. This workshop, part of the National Assessment of the Potential Consequences of Climate Variability and Change on the United States [see NAST 2000], represented the first comprehensive, regional-scale examination of the potential consequences of climate change for the PNW.

In one sense, this first workshop served the CIG in much the same way as did the initial interviews with water managers concerning short-range climate forecasting. The workshop demonstrated, for example, that the region was largely unaware of the potential impacts of climate change, and was institutionally and politically unprepared to make use of the early warning of climate change impacts voiced by government and academic scientists. As in the case of interannual climate forecasting, it was clear that continued outreach and a much more focused description of the probable impacts and potential adaptation strategies would be required to overcome these obstacles. (Regional-scale descriptions of the likely impacts of climate change were insufficient to many participants, who desired an analysis of impacts and adaptation strategies on a local or jurisdictional level.) Several iterations of annual water workshops after the 1997 climate change workshop led the CIG to understand that the target of climate change outreach efforts had to be to those that could respond to such information over the appropriate time period. Most mid-level water managers (whose jobs were to carry out existing water policy) and policy makers (whose professional lives are too short to be concerned about climate change impacts 50 years out) apparently could not fulfill this important requirement.

These lessons were incorporated in a subsequent climate change workshop. In July 2001, CIG convened a Climate Change and Water Policy Workshop for senior water resource managers, policy makers, and water users in the PNW, focusing on the twenty-year time horizon (CIG 2001). Despite the uncertainties that persist in the science of climate change and climate impacts, policy makers in 2001 recognize climate change as a potentially significant threat to water resources in the region. This represents a significant step forward from 1997, when climate change was not recognized as a serious issue affecting the region. At this workshop, decision- makers and water resource managers requested more detailed information and data, as well as continued education and outreach, concerning climate impacts—not only on water resources, but on forests, coasts, salmon, and dryland agriculture (topics that were, ironically, included in the 1997 meeting). Several upper-level managers stated that if relevant data on climate impacts were accessible, they would use it in their planning. Among the important lessons learned from this workshop were:

- In the policy arena, generating the capacity to respond effectively to near-term challenges such as drought in the PNW may be largely synonymous with generating the capacity to adapt to climate change, and may be a more politically viable approach than one characterized as an adaptation to an uncertain, long-range phenomenon.
- Outreach efforts informing policy makers about climate change will need to be ongoing, and research efforts should be expanded to encompass the specific needs of this community (e.g., readily available climate change streamflow data).

Through its research, initial interviews with water managers, and water and climate change workshops, CIG has learned about the variety of barriers impeding use of improved climate and resource forecast information in water resources management. Institutional resistance to changing operations often implies that top-down direction is required to include climate information in decision-making processes; risk averse tendencies cause many managers to avoid risking failure using a new forecast application by relying on traditional, recognized methods based on "period of record" expectations. In addition, the political realities of interests associated with the status quo can make changing operations extremely difficult. In general, encouraging the use of climate forecasts requires a careful consideration of the context into which they will be introduced. The interpretation, presentation, and communication of the forecast must consider such contextual factors as the geographic location, scales, and technical capacity of potential users of the forecast information. This is all information that can best be derived from close interactions with the users such as that achieved in CIG's outreach program.

The ongoing interactions with resource managers, planners, and decision-makers has helped CIG learn more about the purview of and specific issues of concern for regional managers and policy makers, thereby ensuring that research is directed towards regionally relevant topics and problems and actively integrated with planning and policy-making realities. In addition, these contacts have provided the research team with an in-depth knowledge and understanding of the present capacity of PNW institutions to deal with climate variability and change, and have helped to create and refine an integrated research and outreach strategy designed to overcome some of the primary obstacles to the active use of climate information and forecasts in PNW water management.

Climate Assessment for the Southwest (CLIMAS)

Purpose. The Climate Assessment for the Southwest (CLIMAS) project, which has been funded by the National Oceanic and Atmospheric Administration's Office of Global Programs (NOAA-OGP) since February 1998, seeks to identify and evaluate the impacts of climate variability and change on human and natural systems in the southwestern United States and in adjacent areas of Mexico. The links between climatic variability and water resource variability have long been recognized by residents. For much of the historical period, adaptation to climatic stresses involved development of reservoirs and, with the introduction of pumping technologies, exploitation of groundwater resources. Recognition that water could be a limiting factor in regional growth and development is reflected in the ongoing controversies, among U.S. states and between the United States and Mexico, over allocation and management of the waters of the Colorado and Rio Grande rivers. More recently, such concerns may be discerned in efforts within both Arizona and New Mexico to strengthen regulation of water supply and demand. Recent advances in global and synoptic scale climatology, new findings emerging from historical and paleo research into climate impacts on water supplies, and improvements in predictive capabilities at seasonal to interannual time scales, offer opportunities to improve water resource management. At the same time, these innovations highlight the need for further scientific research to understand the interactions between climate and hydrology, as well as the interactions between these processes and society. As discussed below, integrated assessment provides a framework for examining these interactions within the U.S. Southwest and adjacent border areas of Mexico.

The initial CLIMAS research agenda was strongly influenced by recommendations articulated at a 1997 symposium on climate impacts (Merideth et al. 1998), particularly with regard to research on climate-hydrology-water resource issues. In response, one team of CLIMAS researchers focused on the sensitivity and vulnerability of water resources to climatic variability and change. Recognizing that both Arizona and New Mexico are more than 80 percent urban, the team focused on assessing the sensitivity of urban water resources to severe drought. The research, which for reasons of practicality focused on selected urban areas in Arizona, provided an opportunity to develop a methodological framework useful for carrying out sensitivity analyses in other urban settings within the Southwest. A second team took as its research focus improving snowpack estimates used as inputs to hydrologic forecasting. This work provides a wider geographical perspective on climate-water resource interactions. A third team responded to early interactions with stakeholders by undertaking an assessment of the skill of important climate and hydrologic forecasts. This work not only provides new knowledge for managers and decision makers in the Southwest, but for the entire United States as well.

Throughout the assessment process, strong emphasis is placed on both interdisciplinary collaboration and ongoing, active interaction with stakeholders. CLIMAS research also provides insights useful in other research and outreach activities. For example, findings from CLIMAS research were incorporated into the Southwest Regional Assessment report (Southwest Assessment Group, 2000). The following narrative describes some of the primary research methods used in CLIMAS research, and some of the ways in which this research interfaces with water resource management.

Methods. The research has been framed from both social science and natural science perspectives, in an effort to advance, in a coordinated manner, an understanding of climate-water resource-societal interactions. The goal has been to provide information and insights that may be used to enhance adaptation capacity and resilience to extreme events, in the context of continued high growth and development in the region. Specific methods are outlined below.

I. Social Science Methods. At the beginning of the project, a team of researchers interviewed community members in southeastern Arizona regarding their need for and use of climate information, as well as the ways in which climatic events and conditions impact their lives and livelihoods. In parallel with this effort, researchers compiled basic contextual information about the region, including socio- demographic profile and trends, economic profile, water resource availability and use, and state of knowledge about climate in the region. This information formed the foundation for subsequent research activities with regard to climate impacts on water resources, as well as on other sectors such as ranching and small rural communities.

From 1998 through 1999, an in-depth analysis of the sensitivity of water supplies to droughts as severe as the deepest one-, five-, and ten-year droughts in the historical record was conducted. The assessment focused on the Tucson and Phoenix metropolitan areas, and on the communities of Nogales, Sierra Vista, and Benson, Arizona, and selected the year 2025 as the target assessment year. This year was selected because it is the year specified in the Arizona Groundwater Code when the Phoenix, Tucson, and Nogales areas are expected to achieve safe-yield conditions (i.e., when renewable supply is in balance with demand). Sierra Vista was included because it is the site of a significant effort to devise innovative ways to manage urban growth and military water demand (a major Army base, Fort Huachuca, is adjacent to the city) in the context of a riparian area nationally recognized for its support of significant resident and migratory bird populations. The town of Benson, a small community of less than 10,000 residents, was included in the water study as a means of integrating the water impacts analysis with the rural-community ethnographic case study being conducted concurrently under CLIMAS.

A straightforward spreadsheet-based method was used to calculate the impacts of reduced precipitation on water supply. The amount of renewable supply was reduced by the same proportion that precipitation was reduced during the relevant period of record for the NWS-defined climate division in which the study area was located. Demand was adjusted by a factor based on previous research on the relationship between climate and water demand [Woodard and Horn, 1988] and on advice from hydrologists, consulted specifically about this issue.

The results [Carter et al., 2000, Morehouse et al., 2001, Morehouse et al., 2002] indicate that five- and ten-year droughts equivalent to the most severe in 100 years of historical records would generate significant challenges for water resource managers in all five communities, and would likely require enforcement of more stringent conservation measures along with decisions about where and how to allocate available water. A related study, carried out from 1999 to 2000, discussed some of the ways in which current water law may facilitate or impede effective institutional response to severe extended drought conditions. This study [Morehouse et al., 2001] was supplemented by participation in a public process, initiated by the Governor of Arizona, in the areas of the state regulated under the Groundwater Code to achieve safe-yield, and to identify possible legislative changes to the state's Groundwater Code. Active participation in this process provided opportunities to identify potential obstacles to effective water management under conditions of climatic adversity, such as lack of appropriate infrastructure and institutional arrangements.

Results of a third study, involving personal interviews with water managers in the metropolitan areas of Tucson, Phoenix, Nogales, and Sierra Vista, will be published during 2002. The semi-structured interview protocol was designed to elicit information about managers' use of climate information, their perceived need for such information, their perceptions about climate impacts on their water systems, and the strategies they would be most and least likely to employ in response to drought impacts. This research builds on similar work done in the Southwest and in other regions [see, e.g., O'Connor et al., 2000; Lach et al., 2000; Pagano et al., forthcoming; Pulwarty and Redmond, 1997; Croley, 1996; Changnon, 1990; Glantz, 1986] and reflects some of the research needs identified in *Making Climate Forecasts Matter* [NRC, 1999].

In addition, several workshops for water managers, forecasters, and researchers have been held. These workshops have proven especially valuable for identifying research needs associated with forecasting stream flow. Forecasting surface water supplies has become ever more crucial to water management in the arid West, and this is certainly the case in Arizona, where the Colorado River is tapped not only for agriculture, but to meet a growing proportion of urban water demand. This is particularly the case in the Phoenix and Tucson metropolitan areas where efforts to achieve safe-yield rely heavily on water delivered via the Central Arizona Project (CAP) canal.

II. Natural science research methods. As noted above, stakeholders consistently demonstrate a desire for better information about water availability, including improved hydrologic forecasts with a better sense of forecast skill. In conjunction with a major NSF-funded research project, SemiArid Hydrology and Riparian Areas (SAHRA), intensive research is underway to improve knowledge of surface water and groundwater hydrology, and snowpack dynamics. Within CLIMAS, assessment of forecast skill for different time scales and geographical areas has been strongly emphasized, as has development of better methods for developing snowpack estimates needed for improved streamflow forecasting in the Southwest.

Climate and hydrologic forecast assessment. CLIMAS researchers have, since inception of the project, recognized that seasonal climate and hydrologic forecasts hold significant potential to improve water resource management. However, as revealed by stakeholders and others, forecasts often play little more than a marginal role in real-world decision making [Changnon, 1990; Sonka et al., 1992; Pulwarty and Redmond, 1997; Callahan et al., 1999; Pulwarty and Melis, 2000]. Given that contests over access to water supplies will undoubtedly intensify in the coming years, interest in understanding and using climate and hydrologic forecasts is likely to increase. For this to occur, user perceptions that seasonal forecasts are undependable must be countered with more and better information about the actual "track record" of different forecasts. Stakeholders have repeatedly said that they lack any quantitative basis for judging forecast performance. They have noted that clear and consistent communication of forecast uncertainty levels, even when uncertainty is high, can increase forecast credibility [O'Grady and Shabman, 1990]. While performance evaluation of climate forecasts appear in the scientific literature [e.g., Bettge et al., 1981; Barnett and Priesendorfer, 1987; O'Lenic, 1990; Livezey, 1990; Murphy and Huang, 1991; Wilks, 2000], the results of such studies are not easily applied to specific resource management decisions because they reflect perspectives and concerns of climate modelers and forecasters rather than decision makers.

Climate forecast techniques are evolving more rapidly than hydrologic forecasting techniques and water management procedures [Hartmann et al., 1999]. Improved climate prediction capabilities are being incorporated into water resource management informally today, but are based on subjective, ad hoc procedures devised by individual water managers based on their confidence in the predictions [Pagano et al., forthcoming]. Caution must be exercised, however, for overconfidence in forecasts may be even more problematic than lack of confidence: a single incorrect forecast that prompts costly operational changes can devastate user confidence in subsequent forecasts [Glantz, 1982]. In light of these concerns, ongoing evaluation of forecasts is a top priority within CLIMAS.

Evaluation of regression-based hydrologic forecasts. Increases in skill of ENSO- related forecasts account for much of the current success in forecasts for the southwestern United States. Re-analysis and hindcasting techniques have been used to reevaluate the skill of CPC climate forecasts, including those issued and those not issued under ENSO conditions. Seasonal regression-based water supply outlooks and probabilistic forecasts were evaluated. These regression-based outlooks are jointly issued by the National Weather Service (NWS) Colorado Basin River Forecast Center (CBRFC) and the Natural Resources Conservation Service for the upcoming season of snowmelt runoffs. The outlooks are used to regulate reservoir releases, interbasin water transfers and fulfillment of water allotments.

In evaluating water supply outlooks, a comprehensive database was used that contains historical official forecasts and reconstructed "naturalized" flows. Multiple techniques were used to evaluate forecasts, including quantitative, graphical, and qualitative. These evaluations provide a baseline for assessing improvements in predictability based on use of newer models, data, and/or forecast procedures. The focus is not only on evaluating performance statistics, but also on issues related to the interpretation and use of forecasts.

These evaluations are based on a diagnostic approach that is more customarily applied to weather and climate forecasts; this allows a focus on specific conditions important to stakeholders. These analyses show, for example, that in January, forecasts overwhelmingly communicate "normal" conditions, while forecasts of extreme conditions are made rarely, if ever. For some rivers in the Southwest, such as the Verde River in Arizona, forecast performance deteriorates from March to April. Significantly, forecasts of extremely low flows are missed even in April, even though this is the time in Arizona when forecasts should be most skillful (April-June has almost no precipitation; thus flows are almost entirely dependent on snowmelt). Good forecasts of extremely low flows are very important to many stakeholders who are dependent on surface flows, most notably junior water rights holders and riparian ecosystem managers.

Other forecast evaluation research activities. Forecast evaluation activities also encompass assessment of CPC seasonal temperature and precipitation outlooks, which have utility for managing water demand as well as for managing both surface water flows and groundwater recharge. While much of the impetus for this particular work arose from interactions with wildfire managers and ranchers, water managers are also included as key stakeholder collaborators. The focus has been, for example, on format elements (such as the use of "climatology" to designate areas where equal probabilities exist for above, below or average conditions) that confuse users and negatively affect the ease, accuracy, and reliability of forecast interpretations.

A series of skill scores are devised to assist users in evaluating the CPC forecasts. Evaluations demonstrate how analyses targeted at specific user situations can produce different assessments of forecast performance. For example, winter-season precipitation forecasts issued under synoptic-scale El Niño conditions may show relatively high skill; conversely a "no-Niño" situation may produce a forecast with weak skill. With proper skill information attached, both such forecasts may be valuable to water managers.

Beginning with a special workshop for selected water managers and decision makers, the research team embarked on an ongoing series of interactions with stakeholders aimed at presenting forecast evaluation results and obtaining feedback on the effectiveness and utility of analyses. The workshop included representatives from local and regional water providers, emergency management agencies, the local office of the National Weather Service, and the NWS CBRFC. The discussions at the workshop generated valuable feedback on format and content of various climate forecasts, and an assessment of the forecast evaluation tools being developed. Equally important, the discussions assist in identifying decision makers' thresholds for forecast "quality" (that is, specifically what sorts of standards forecasts must meet if they are to be used in making decisions).

Evaluation of gridded snow estimates. High-quality forecasts of areal snow coverage and snow pack water storage are critical to improving both regression- based and ESP water supply outlooks, and for developing third-generation hydrologic models that include highly distributed dynamical land surface- atmosphere interactions. No official snow forecasts exist, except for severe storm warnings. With the cooperation of the NWS National Operational Hydrologic Remote Sensing Center (NOHRSC) and the University of Arizona's Southwest Regional Earth Science Applications Center (RESAC) project, efforts were initiated to produce a ten-year (1990-1999) database of high-resolution (1km square) gridded snow conditions, including coverage and water equivalent. The database allows, for the first time, comprehensive evaluation of the role of snow conditions in the context of hydrologic variability and predictability. The research team proposes to extend the existing research in two areas: evaluation of the predictability of seasonal snow conditions and quantifying improvements in water supply forecasts that will be made possible by availability of advanced snow estimates.

Applications at the water resource management interface. The team's scientific research is very sharply focused on applications-oriented activities. Through sustained interactions with key water managers and decision makers in the region, as well as with the water supply forecasters who also constitute CLIMAS stakeholders, relationships have been established that are essential for introducing new climate products emerging from climate and hydrologic sciences, and for disseminating information about sensitivity and vulnerability to climate conditions generated through social science research. Among the primary research applications are the distribution of climate and water supply forecast assessments, more and better information about snowpack and snow water equivalent, factors that contribute to increased or decreased vulnerability and sensitivity to climatic impacts, and information needed to devise strategies for enhancing adaptive capacity at local and regional scales to climate variability.

Through workshops and focus groups, the project web site, professional meetings and conferences, and personal interactions, user concerns are identified even as knowledge is transferred and decision tools are made available to stakeholders. A significant endeavor to design a structured transfer process is underway that will allow scientific evaluation of the actual knowledge/information transfer process. This development process involves a review of the literature on visualizing and communicating complex scientific information, small-scale experimentation with alternative communication methods that rely on a few key stakeholder collaborators, and finally delivery of products to the larger stakeholder community via workshops and the use of sophisticated web server technologies.

CONCLUSION

The regional assessments described here have addressed many different stakeholder issues but many of the same interests: adequate potable water for drinking, water for agricultural and industrial use, water for hydropower production, water for protection of aquatic species, water for transportation and recreational use. Despite the uncertainties that persist in the study of climate variability and change and the assessment of their impacts, stakeholders have come to recognize that a variable or a changing climate represents a potentially significant threat to water resources. This recognition represents an important step forward from the mid-1990's. Since that time, stakeholders have been reminded, largely by an active media response, of the extent of variability associated with naturally occurring phenomenon (e.g., the El Niño / La Niña event in 1997-1999). They have also seen many more analyses of historical climate variability.

As capabilities for modeling and forecasting climate have improved and been more widely applied, capacities for incorporating stakeholders in the process of regional climate impacts assessment have expanded and become more routine. A number of lessons have been learned about stakeholder involvement in the assessment process, including:

- The process of translating broad stakeholder interests into specific scientific questions to be investigated, answered, and translated back into language and tools that stakeholders can use is complicated, but vital to the success of the assessment. Truly participatory stakeholder processes ultimately produce information that can be utilized in management decisions and planning.
- The importance of water to all sectors of the nation's economy and societal well- being means that impacts on water resources affect stakeholders not directly associated with the management of water resources. The interests of these stakeholder groups should be considered when conducting an assessment. For instance, followup workshops organized by the GLRA have been instrumental in involving stakeholders, such as shippers and marina owners, whose interests differ from water managers but are nevertheless essential to a fully-formed understanding of climate impacts in the Great Lakes.

- Assessment teams are likely to face an organizational challenge: should they organize the assessment around the endpoints identified by the stakeholders or should they organize along scientific disciplines? Each approach has its advantages and disadvantages; but in the end, assessment teams who organize along disciplinary lines must be prepared to invest significant energies to ensure that they are responsive to stakeholder interests that fall outside neat disciplinary boundaries.
- An especially promising approach for engaging managers on the issue of climate change is to link projected impacts of climate change to consequences of historical climate variations. This approach demonstrates that a basic understanding of the implications of climate change can be derived from an understanding of natural variability and its impacts on both natural and human systems.
- Experience has demonstrated that environmental events can be useful triggers for stimulating decision makers' interest in climate information. For instance, the Climate Impacts Group found that generating the capacity to respond effectively to drought in the Pacific Northwest was largely synonymous with generating the capacity to adapt to climate change, and appeared to be more politically viable. Similarly, assessment teams found that the well-publicized El Niño and La Niño events of the late 1990s galvanized stakeholders' interest in climate. Recognizing sensitivity to climate variability may help motivate a consideration of sensitivities to future climate change.
- Local impacts matter. Scale is important and may be determinant in whether information will be useful to stakeholders. All three assessments found that regional differences in climate and in the spatial scale of watersheds need to be accommodated. In the CIG assessment, this conclusion led to the practice of holding parallel workshops at different locations across the region each fall, with the content tailored to each location. Corresponding efforts were made to downscale climate forecasts and to improve their skill.
- Stakeholders have complained that they lack any quantitative basis for judging forecast performance or skill. They have suggested that clear and consistent communication of uncertainty levels, even when uncertainty is high, can increase forecast credibility. The CLIMAS group has been developing forecast evaluation methods in an effort to address these issues.
- A variety of methods need to be employed to build and maintain stakeholder networks. Workshops have been used successfully by each of these groups. Similarly active pursuit of stakeholders' opinions is needed; surveys have been effective tools. Clearly, ongoing support of stakeholder networks requires an intensive, multiple-strategy effort.
- Understanding the context within which managers manage and planners plan is essential. For instance, assessors should be aware of limitations imposed by multiple jurisdictional authorities, by institutional arrangements

that confound efficient adaptation or impose unrealistic time lines, or by water policies that impede effective responses to changing conditions.

From these case studies, we learn that stakeholders can be effective partners in the assessment process. Information developed through climate and impacts modeling can be applied to management decisions and resource planning. Despite the uncertainties in the science and constraints in the management and policy communities, we conclude that stakeholders have come to recognize climate variability and climate change as potentially significant factors in water resources management, ones that will require revision of existing management strategies.

Challenges remain. Finer scale climate projections, suited to the requirements of watershed-level planning, are needed. Fragmented management, political, and jurisdictional arrangements continue to inhibit effective, flexible responses. Short-term planning tends to conform to the terms of elected officers but discourages the consideration of long-range phenomena, such as climate change. Solutions to these institutional impediments are not likely to be derived unless stakeholders are convinced that the evidence of climate variability and change merits a re-thinking of present strategies. Each of these case studies seem to suggest that such "re-thinking" may have begun. In any event, it appears that the partnerships between stakeholders and scientists now being forged have the potential, if undeterred, for supporting the adaptations necessary to secure the future viability of the nation's water supplies.

Acknowledgments. We would like to thank Claudia Nierenberg, Susan Julius, and the other reviewers of this chapter for their comments and suggestions. Any errors or omissions that remain are our own. This publication is partially funded by the Joint Institute for the Study of the Atmosphere and Ocean (JISAO) under NOAA Cooperative Agreement No. NA67RJ0155, Contribution #892.

REFERENCES

Andresen, J. A., G. Alagarswamy, D. F. Stead, H. H. Cheng, and W. B. Sea, Agriculture, in *Preparing for a Changing Climate: The Potential Consequences of Climate Change and Variability, Great Lakes Overview*, edited by P. J. Sousounis and J. Bisanz , pp. 69-76, University of Michigan, Ann Arbor, MI, 2000.

Barnett, T. P. and R. W. Priesendorfer, Origins and levels of monthly and seasonal forecast skill for North American surface air temperatures determined by canonical correlation analysis, *Monthly Weather Review,* 115, 1825-50, 1987.

Benequista, N. and J. S. James, Pilot stakeholder assessment report, *Climate Assessment for the Southwest*, Institute for the Study of Planet Earth, University of Arizona, Tucson, AZ, February, 1999. http://www.ispe.arizona.edu/climas/archive.html.

Bettge, T. W., D. P. Baumhefner, and R. M. Cherwin, On the verification of seasonal climate forecasts, *Bulletin of the American Meteorological Society,* 62(12), 1654-1665, 1981.

Callahan, B., E. Miles, and D. Fluharty, Policy implications of climate forecasts for water resources management in the Pacific Northwest, *Policy Sciences,* 32, 269-293, 1999.

Carter, R. H., P. Tschakert, and B. J. Morehouse, Assessing the sensitivity of the Southwest's urban water sector to climatic variability, *CLIMAS Report Series CL1-00*, Institute for the Study of Planet Earth, University of Arizona, Tucson, AZ, 2000.

Changnon, S. A., The dilemma of climatic and hydrologic forecasting for the Great Lakes, in *The Great Lakes Water Level Forecast and Statistics Symposium*, edited by H. C. Hartmann and M. J. Donahue, pp. 13-25, Great Lakes Commission, Ann Arbor, MI, 1990.

Changnon, S. A. and D. R. Vonnhame, Use of climate predictions to decide a water management problem, *Water Resources Bulletin,* 22(4), 649-652, 1986.

Climate Impacts Group (CIG), *Impacts of Climate Variability and Change: Pacific Northwest,* JISAO/SMA Climate Impacts Group, University of Washington, Seattle, WA, 1999.

Climate Impacts Group (CIG), *Climate and Water Policy Workshop Executive Summary,* Climate and Water Policy Workshop, Stevenson, WA, July, 2001. Available from jisao.washington.edu/PNWimpacts/Publications/es.pdf.

Glantz, M. H., Politics, forecasts, and forecasting: Forecasts are the answer, but what was the question? in *Policy Aspects of Climate Forecasting,* edited by R. Krasnow, pp. 81-96, Resources for the Future, Washington, DC, 1986.

Gleick, P. H., Regional hydrologic consequences of increases in atmospheric CO_2 and other trace gasses, *Climatic Change,* 10, 137-161, 1987a.

Gleick, P. H., Climate change, hydrology and water resources, *Reviews of Geophysics,* 27, 329-344, 1987b.

Gleick, P. H. and D. B. Adams, *Water: The Potential Consequences of Climate Variability and Change for the Water Resources of the United States*, Pacific Institute, Oakland, CA, 2000.

Gray, K. N., *The impacts of drought on Yakima Valley irrigated agriculture and Seattle municipal and industrial water supply*, School of Marine Affairs, University of Washington, Seattle, WA, 1999.

Hamlet, A. F. and D. P. Lettenmaier, Columbia River streamflow forecasting based on ENSO and PDO climate signals, *Journal of the Water Resources Planning and Management,* Nov/Dec, 333-341, 1999a.

Hamlet, A. F. and D. P. Lettenmaier, Effects of climate change on hydrology and water resources in the Columbia River basin, *Journal of the American Water Resources Association,* 35(6), 1597-1623, 1999b.

Hamlet, A. F. and D. P. Lettenmaier, Long-range climate forecasting and its use for water management in the Pacific Northwest region of North America, *Journal of Hydroinformatics,* 2(3), 163-182, 2000.

Hamlet, A. F., D. Huppert, and D. P. Lettenmaier, Economic value of long-lead streamflow forecasts for Columbia River hydropower, *ASCE Journal of Water Resources Planning and Management,* in press, 2001.

Hamlet, A. F., P. W. Mote, A. K. Snover, and E. L. Miles, Climate, water cycles, and water resources management in the Pacific Northwest, in *Rhythms of Change: Climate Impacts on the Pacific Northwest,* edited by E. L. Miles, A. K. Snover, and The Climate Impacts Group. MIT Press, Boston, MA, in review, 2001.

Hartmann, H. C., R. Bales, and S. Sorooshian, Weather, climate and hydrologic forecasting for the Southwest U.S., *CLIMAS Report Series CL2-99*, Institute for the Study of

Planet Earth, University of New Mexico, Tucson, AZ, 1999.

Huppert, D., J. Kaje, A. F. Hamlet, E. L. Miles, and A. K. Snover, Using climate forecasts in natural resource management, in *Rhythms of Change: Climate Impacts on the Pacific Northwest,* edited by E. L. Miles, A. K. Snover, and The Climate Impacts Group, in review, 2001.

Johnson, R. and W. Soucie, presentation at the workshop "Climate Change and the Water Ecology of the Great Lakes: The Potential Impacts, and What We Can Do." The workshop was held in Milwaukee in June, 2001.

Lach, D., H. Ingram, and S. Rayner, Coping with climate variability: municipal water agencies in southern California, Paper presented at "Climate, Water, and Transboundary Challenges in the Americas" Symposium, Santa Barbara, CA, July 16-20, 2000.

Lettenmaier, D. P. and T. Y. Gan, An exploratory analysis of the hydrologic effects of global warming on the Sacramento-San Joaquin River Basin, California, *Water Resources Research,* 26, 69-86, 1990.

Lettenmaier, D. P. and D. P. Sheer, Climatic sensitivity of California water resources, *Journal of Water Resources Planning and Management,* ASCE, 177(1), 108-125, 1991.

Leung, L. R., A. F. Hamlet, D. P. Lettenmaier, and A. Kumar, Simulations of the ENSO hydroclimate signals in the Pacific Northwest Columbia River Basin, *Bulletin of the American Meteorological Society,* 80(11), 2313-2329, 1999.

Lofgren, B. M., F. H. Quinn, A. H. Clites, R. A. Assel, and A. J. Eberhardt, Water Resources, in *Preparing for a Changing Climate: The Potential Consequences of Climate Change and Variability, Great Lakes Overview*, edited by P. J. Sousounis and J. Bisanz, pp. 29-38, University of Michigan, Ann Arbor, MI, 2000.

Mahoney, E., C. Tzu-Ching, C. Pistis, and L. Martin, The impacts of low water on Michigan Great Lakes Marinas, A Report for the Michigan Boating Industries Association, Michigan State University, 2000.

Mantua, N. J., S. R. Hare, Y. Zhang, J. M. Wallace, and R. C. Francis, A Pacific interdecadal climate oscillation with impacts on salmon production, *Bulletin of the American Meteorological Society,* 78(6), 1069-1079, 1997.

Mantua, N. J. and S. R. Hare, The Pacific Decadal Oscillation, *Journal of Oceanography,* in press, 2001.

Mantua, N. J. and P. W. Mote, The underlying rhythms: characteristics of Pacific Northwest climate, in *Rhythms of Change: Climate Impacts on the Pacific Northwest,* edited by E. L. Miles, A. K. Snover, and The Climate Impacts Group, in review, 2001.

Merideth, R., D. Liverman, R. Bales, and M. Patterson (eds.), *Climate variability and change in the Southwest: impacts, information needs, and issues for policymaking,* Final report of the Southwest Regional Climate Change Symposium and Workshop, September 3-7, 1997, Tucson, AZ, Udall Center for Studies in Public Policy, The University of Arizona, Tucson, AZ, July 1998.

Michaud, D., comments during the workshop "Climate Change and the Water Ecology of the Great Lakes: The Potential Impacts, and What We Can Do," Milwaukee, WI, June 2001. Proceedings will be posted on the Great Lakes Regional Assessment web page in Winter, 2001. Further information via personal communication.

Miles, E. L., A. K. Snover, A. F. Hamlet, B. Callahan, and D. Fluharty, Pacific Northwest regional assessment: The impacts of climate variability and climate change on the water resources of the Columbia River Basin, *Journal of the American Water Resources*

Association, 36(2), 399-420, 2000.

Miles, E. L., A. K. Snover, and The Climate Impacts Group, in *Rhythms of Change: Climate Impacts on the Pacific Northwest,* edited by E. L. Miles, A. K. Snover, and The Climate Impacts Group, in review, 2001.

Miles, E. L. and A. K. Snover, The integrated assessment: The sensitivity, adaptability, and vulnerability of the Pacific Northwest to climate variability and change, in *Rhythms of Change: Climate Impacts on the Pacific Northwest,* edited by E. L. Miles, A. K. Snover, and The Climate Impacts Group, in review, 2001.

Morehouse, B. J., Climate impacts on urban water resources in the Southwest: the importance of context, *Journal of the American Water Resources Association,* 36(2), 265-277, 2000.

Morehouse, B. J., R. H. Carter, and T. W. Sprouse, The implications of sustained drought for transboundary water management in Nogales, Arizona and Nogales, Sonora, *Natural Resources Journal,* 40, 783-817, 2000.

Morehouse, B. J. and R. H. Carter, An examination of Arizona water law and policy from the perspective of climate impacts, *CLIMAS Report Series CL2-01*, Institute for the Study of Planet Earth, University of Arizona, Tucson, AZ, 2001.

Mortsch, L. D., S. Quon, L. Craig, B. Mills, and B. Wrenn, (editors), *Adapting to climate change and variability in the Great Lakes-St. Lawrence Basin.* Proceedings of a binational symposium, Environment Canada, University of Waterloo, Waterloo, 1998.

Mote, P., M. Holmberg, and N. Mantua. Impacts of climate change: Pacific Northwest (Summary), JISAO/SMA Climate Impacts Group, University of Washington, Seattle, WA, 1999.

Mote, P. W., Possible future climate, in *Rhythms of Change: Climate Impacts on the Pacific Northwest,* edited by E. L. Miles, A. K. Snover, and The Climate Impacts Group, in review, 2001.

Mote, P. W., et al., Climate and the water, salmon, and forests of the Pacific Northwest, submitted to *Climatic Change*, 2002.

Nash, L. L. and P. H. Gleick, Sensitivity of streamflow in the Colorado basin to climatic changes, *Journal of Hydrology,* 125, 221-241, 1991.

National Assessment Synthesis Team (NAST), *Climate Change Impacts on the United States: The potential consequences of climate variability and change,* U.S. Global Change Research Program, Washington, DC, 2000.

O'Connor, R. E., B. Yarnal, R. Neff, R. Bord, N. Wiefek, C. Reenock, R. Shudak, C. L. Jocoy, P. Pascale, and C. G. Knight, Weather and climate extremes, climate change, and planning: views of community water system managers in Pennsylvania's Susquehanna River Basin, *Journal of the American Water Resources Association,* 35(6), 1411-1420, 2000.

Pagano, T., H. C. Hartmann, S. Sorooshian, and R. Bales, Use of climate forecasts for water management in Arizona: a case study of the 1997-98 El Niño, *Climate Research*, forthcoming.

Pagano, T. C., H. C. Hartmann, S. Sorooshian, and R. Bales, Advances in seasonal forecasting for water management in Arizona: A case study of the 1997-98 El Niño, *HWR Report No. 99-040,* Department of Hydrology and Water Resources, University of Arizona, Tucson, AZ, 1999.

Pulwarty, R. S. and T. S. Melis, Climate extremes and adaptive management on the Colorado River: Lessons from the 1997-98 ENSO event, *Journal of Environmental Management*, 2000.

Pulwarty, R. S. and K. T. Redmond, Climate and salmon restoration in the Columbia River Basin: The role and usability of seasonal forecasts, *Bulletin of the American Meteorological Society,* 78(3), 381-397, 1997.

Ryan, G., Comments at the workshop "Climate Change and Potential Impacts on Great Lakes Water Levels," March 2001, Proceedings appear on the Great Lakes Regional Assessment web page.

Schilling, K. E. and E. Stakhiv, Summary and Commentary, *Water Resources Update*, Issue No. 112, pp. 1-5, 1998.

Snover, A. K., E. Miles, B. Henry, OSTP/USGCRP Regional Workshop on the Impacts of Global Climate Change on the Pacific Northwest, NOAA Climate and Global Change Program, Special Report 11, 1998.

Snover, A. K. and E. L. Miles, Introduction: The integrated assessment of the impacts of climate variability and change on the Pacific Northwest, in *Rhythms of Change: Climate Impacts on the Pacific Northwest,* edited by E. L. Miles, A. K. Snover, and The Climate Impacts Group, in review, 2001.

Sousounis, P.J., Executive Summary, in *Climate Change in the Upper Great Lakes Region: A Workshop Report*, edited by Sousounis, P. J. and G. Albercook, pp.3-9, University of Michigan, Ann Arbor, MI, 1998.

Sousounis, P.J., A Glimpse of What's to Come...? in *Preparing for a Changing Climate: The Potential Consequences of Climate Change and Variability, Great Lakes Overview*, edited by P. J. Sousounis and J. Bisanz, University of Michigan, Ann Arbor, MI, 2000.

Southwest Regional Assessment Group, *Preparing for a changing climate: the potential consequences of climate variability and change,* Institute for the Study of Planet Earth, The University of Arizona, Tucson, Arizona, 2000.

Stern, P.C. and W.E. Easterling, *Making climate forecasts matter.* Panel on the Human Dimensions of Seasonal to Interannual Climate Variability, Committee on the Human Dimensions of Global Change, National Research Council, National Academy Press, Washington, DC, 1999.

U.S. Environmental Protection Agency and Environment Canada, *The Great Lakes: An Environmental Atlas and Resource Book,* 3rd Edition, 1995.

Whitaker, P., H.C. Hartmann, R. Bales, and S. Sorooshian, Seasonal water supply forecast performance: issues and evaluations, *EOS Transactions-American Geophysical Union,* 81(19S), 205, 2000.

Winkler, J.A., and J.A. Andresen, personal communication, 2000.

Woodard, G.C., and C. Horn, *Effects of weather and climate on municipal water demand in Arizona,* Report prepared for the Arizona Department of Water Resources and Tucson Water, Division of Economic and Business Research, College of Business and Public Administration, The University of Arizona, Tucson, Arizona, August 1988.

Janet L. Gamble and John Furlow, U.S. Environmental Protection Agency, 1200 Pennsylvania Ave NW, Mail Code 8601D, Washington, D.C. 20460; gamble.janet@epa.gov

Amy K. Snover, JISAO/SMA Climate Impacts Group, University of Washington, Box 354235, Seattle, WA 98195

Alan F. Hamlet, Civil and Environmental Engineering, University of Washington, Box 352700, Seattle, WA 98195

Barbara J. Morehouse, Institute for the Study of Planet Earth, University of Arizona, 715 N. Park Ave (2nd floor), Tucson, AZ 85721

Holly Hartmann aand Thomas Pagano, Hydrology and Water Resources, University Arizona, Bldg. 11 N. Campus DRive. Tuscon, AZ 85721

17

Community Participation and Spatially Distributed Management in New York City's Water Supply

Steven E. Wolosoff and Theodore A. Endreny

1. INTRODUCTION

The U.S. Environmental Protection Agency (USEPA) Surface Water Treatment Rule (SWTR), a 1986 amendment to the original Safe Drinking Water Act (SDWA), mandates a set of criteria that unfiltered public drinking water supplies (DWS) must meet in order to avoid filtration. Under the SWTR surface fed DWSs may obtain a waiver from filtration if they meet standards for turbidity, fecal and total coliforms, provide adequate disinfection, and avoid waterborne disease outbreak, through watershed functions. The SWTR essentially focuses on non-point source (NPS) pollution, as opposed to more readily targeted point sources. The New York City (NYC) drinking water supply area, managed by the NYC Department of Environmental Protection (NYCDEP) can only meet these updated SDWA rules by implementing watershed-based management, where pollution control is spatially distributed throughout the contributing area. This approach will require the coordination of three main stakeholder groups, 1) scientists/engineers, 2) policymakers/managers, and 3) communities/interest groups. In this paper, we review obstacles and opportunities associated with such coordination.

Previous work has pointed to a dichotomy between the professional cultures of watershed scientists/engineers and watershed policymakers/managers, which present obstacles to the implementation of spatially distributed watershed management. This dichotomy inhibits the incorporation of updated scientific knowledge, as illustrated by Adams [1993] in the case of delayed spatially distributed forest resource management aimed at maintaining sustainable fishery populations in the US Pacific Northwest. In the case of drinking water, policymakers/managers

Water: Science, Policy, and Management
Water Resources Monograph 16

10.1029/016WM21

charged with protecting unfiltered surface DWSs are unprepared to implement these new SDWA rules due to training in approaches that worked for controlling point source pollution and managing more uniform large sized land parcels. Scientists/engineers are also unprepared, having not yet established governing rules for controlling NPS pollution loading in heterogeneous and suburbanizing watershed areas. Wolosoff and Endreny [2002] looked at suburban water quality issues in NYC's Croton water supply area and the Twin Cites, MN to demonstrate the response type and response time differences between these two stakeholders. In short, policymakers/managers were often rushed to find broad policies that could be applied across an entire, homogeneous, geopolitical region, whereas scientists undertook multi-year research projects to understand the complex interactions occurring within heterogeneous catchments. This account, however, neglected the role of a third stakeholder, communities/interest groups, in watershed-based management of unfiltered surface DWSs.

Letey [1999] documents limitations in watershed management stemming from this dichotomy between scientists/engineers and policymakers/managers, but through case study shows that communities/interest groups can often play a more significant role in decision-making processes than scientific research findings. In unfiltered surface DWSs responding to the SWDA amendments, where private land ownership and property rights are significant, as is the case in NYC's three multiple reservoir systems, a new paradigm arises. This paradigm places the three stakeholders on equal footing in the initiation of watershed management plans that seek to reduce NPS pollution through watershed functions and sustain the vitality of watershed communities. Understanding this paradigm of community shared control on what had been an 'expert' dominated field may help other watershed management plans progress more swiftly than the NYC example, which was delayed by communities/interest groups that wanted more involvement during the watershed protection planning process. While the SWDA has required input from scientists/engineers to policymakers/managers, it does not incorporate communities/interest groups. Yet watershed property rights might force this coordination in developing and enacting spatially distributed watershed management plans.

The involvement of communities/interest groups in watershed management is not a new concept, but prior to the 1986 SDWA amendments it was generally a directive set into legislation by policymakers/managers and implemented as a result of predetermined watershed plans. This was the case in the 1954 PL 566 legislation for Small Watershed Programs and in the Flood Control Act of 1960, which included provisions for, and funded local water projects involving, communities/interest groups in watershed-based management objectives [*Black,* 1982]. In a review of flood hazard management in the United States, the Federal Interagency Floodplain Management Task Force [1995] observed a 1960s shift in policymaker/manager response type from a reliance upon federally implemented dams and disaster relief toward land use management and stewardship of watershed functions implemented on a local level involving communities/interests groups in a river-corridor approach. The updated SDWA rules, however, do not mandate community/interest group participation, as do the above accounts. Watershed property rights, however, force the involvement of communities/interest groups in developing a watershed-based plan that is able to achieve the specified filtration avoidance criteria, through the use of watershed functions.

Several watershed management approaches have been identified that might work toward incorporating communities/interest groups in watershed planning. Riley [1998] describes the difference between a comprehensive and community based approach to watershed planning, where watershed plans are developed and implemented from either the top-down or bottom-up, respectively. Strong [1975] used the Brandywine Watershed Plan, intended to reduce flood damage caused by increasing suburban development outside Philadelphia, Pennsylvania USA, as a case study to show a comprehensive watershed management approach that did not give adequate consideration to community interests, and therefore failed. The Brandywine Conservancy, a non-profit organization, was later able to effectively mitigate environmental damage due to flooding by working within smaller watershed units, providing landowners with scientific knowledge of watershed functions such as water quality buffer zones, and developing incentives for community involvement and participation.

In NYC's case, policymakers/managers schooled in comprehensive planning progressed smoothly until the implementation phase, where it became apparent that communities/interest groups have a make or break influence, and do not blankly endorse centrally planned management. Hence, spatially distributed watershed management under the new paradigm encourages a shift from comprehensive to community-based planning to more quickly arrive at plans that will be implemented. This shift came about early on in NYC's response to the updated SDWA rules and was not initiated voluntarily, but instead was the reaction to threats of lawsuits from some watershed communities/interest groups.

Goal setting and participation of the community/interest group stakeholder in watershed planning, however, will not be straightforward. For example, downstream water supply consumers are primarily concerned about their human health risk and advocate increased watershed protection, while communities within upstream contributing areas are more affected by land use development pressures and often perceive additional controls as unnecessary. This traditional downstream/upstream division occurs where watershed management does not affect consumer's property rights, but directly affects those of watershed residents. NYC was forced to respect this division within the community/interest group stakeholder, because of the need for a community-based management approach. Parcelization of timberlands into smaller, fragmented, privately held pieces, as observed in New York State (NYS), will also affect watershed management. Germain and LaPierre [2001] observed increasing land parcelization within NYC's unfiltered surface DWS contributing areas, which corresponds to an increasing number and diversity of vested community interests as evidenced by the participants present at the Memorandum of Agreement (MOA) negotiations.

This paper uses a NYC DWS watershed case study to demonstrate specific instances in SDWA health risk management where stakeholder coordination, including all three groups, is working and where it is likely in need of improvement. We examine specific watershed-based management issues where the NYCDEP policymaker/manager group ultimately funded a) scientists/engineers to research watershed functions in relation to SDWA rules and b) communities/interest groups to coordinate management practices in the watershed. In this report, we review findings from scientific research on the pollutant removal capacity of watershed functions occurring within onsite wastewater treatment and disposal systems (OSTDS), including both large galley systems and smaller septic systems and urban stormwater best management practices (BMP). The scientific understanding obtained from these studies is presented alongside related community/interest group agendas. The following section provides a background to appreciate the NYC DWS system, while the third section focuses on three stakeholder coordination issues.

CASE STUDY: BACKGROUND ON NYC'S SURFACE DRINKING WATER SUPPLY MANAGEMENT

NYC's Water Supply Areas

Three hundred and seventy-five years of population growth in NYC, from a few hundred in 1626 to the 8 million counted in 2000, has required that the city progressively expand its drinking source area from local wells and ponds into larger watersheds in upstate NY. Water delivered to NYC residents is now supplied by three different systems, each with multiple reservoirs and controlled lakes and associated threats to water quality from processes occurring within their respective contributing areas, (Figure 1). Water quality threats include, but are not limited to, pollution from stormwater discharges, seepage from OSTDS, chlorine disinfection resistant pathogens, and suspected carcinogenic DBPs.

Source area expansion has been paralleled by growth of a significant suburban footprint in the Croton supply area, with parts in Westchester, Putnam, and Dutchess counties north of NYC. Likewise, the presence of agricultural land within water supply contributing areas stemmed from source area expansion west of the Hudson River, into the Catskill/Delaware systems. Moreover, tourism in this region has spurred additional economic growth and created new development pressures within the watershed, and has also introduced new, human generated, pollutant sources into NYC's unfiltered surface DWS. The combined impacts of the above mentioned land use types within all three of NYC's water supply basins poses a threat to the health of consumers and a challenge for scientists/engineers, policymakers/managers, and communities/interest groups (Table 1).

Figure 1. Map of New York City's water supply system [NYCDEP home page URL: http://www.ci.nyc.ny.us/html/dep/home.html].

The risk to human health from an unfiltered surface public DWS, addressed by the USEPA SWTR, forced NYC to obtain a Filtration Avoidance Determination (FAD) in each of its three water supply areas (Croton, Catskill, Delaware), or to design, build and operate a water filtration facility [*NYCDEP,* 1989]. Interestingly, both approaches require stakeholder coordination, where communities/interest groups influence land access and zoning. Protection of water quality by managing NPS pollution through watershed functions as opposed to building a costly water filtration plant requires improved stakeholder coordination. Increasing development pressures and the stricter water quality standards, however, are confounding the inherent obstacles attributed to interdisciplinary differences, in developing an improved watershed management plan for the NYC DWS. Thus far, NYCDEP has adopted a dual track approach to the new USEPA water quality regulations [*NRC,* 2000]. This is shown by the simultaneous design and siting of potential filtration facilities and demonstration of the watershed's potential for pollutant attenuation via watershed functions and improved management practices.

Table 1. Key stakeholder groups active in the management of NYC three water supply areas (not a complete list).

Scientists / Engineers	Policymakers / Managers	Communities / interest groups
National Research Council (NRC)	NYC Department of Environmental Protec tion	Coalition of Watershed Towns (CWT)
SUNY-ESF	NYS Department of Environmental Conservation	Hudson Riverkeeper
Syracuse University	US Environmental Protection Agency	Je rome Park Reservoir group
Cornell University	NYS Department of Health	Watershed Protection and Partnership Council (WPPC)
Upstate Freshwater Institute (UFI)	Westchester county	Watershed Agricultural Council (WAC)
US Geological Survey (USGS) Water Resource Division in Troy, NY	Putnam county	Catskill Watershed Corporation (CWC)
NYC Department of Environmental Protection	New York City	Manhattan Club
NYS Department of police Environmental Conservation	New York State	Croton Watershed Clean Water Coalition (CWCWC)
	NYCDEP water supply	New York Public Interest Research Group (NYPIRG)
		Catskill Center for Conservation and Development
		Sierra Club
		New York League of Conservation Voters
		Trout Unlimited

Communities/Interest Groups

Communities/interest groups in NYC have identified and resolved water quality problems in a manner that is distinct from the disciplinary methods of scientists/engineers and policymakers/managers. It should be noted that these community groups are many and diverse, and the examples provided in this case study would not represent the actions of all the groups. The diversity of community/interest groups is apparent in a division between support of more versus less watershed regulation and protection. Some groups, as represented by the NY based legal research non-governmental organization (NGO) River Keeper, are pursuing more stringent oversight and enforcement by the NYCDEP, while other groups, such as the Coalition of Watershed Towns (CWT) is more interested in fewer rules and more freedom for economic growth and self-rule. This difference may be attributed to the common upstream vs. downstream debate over external control within watershed lands of a surface DWS. Therefore, a community-based watershed management plan for NYC would not easily follow the empowerment of this stakeholder group as a whole. Instead the division between water supply residents and consumers outside of the water supply basin with regard to the level of control further complicates the operational differences between the other two stakeholders, scientists/engineers and policymakers/managers.

Watershed Land Acquisition

NYC began to plan for a more comprehensive ownership and management of watershed land in the upstate region in response to new USEPA filtration avoidance rules. Investment in watershed lands to allow for natural filtration of pollutants, as opposed to a filtration facility, was projected to save NYC as much as $6 to $8 billion [*Chichilnisky and Heal,* 1998], or $4 to $6 billion according to more recent analysis [*Brown,* 2001]. While watershed experts were excited and challenged by identifying and protecting watershed functions to the tune of billions of dollars, the notion of NYC managing private lands hit trouble. There was deep resentment among Catskill/Delaware landowners, who viewed NYC land acquisition and onerous rules, similar to practices within the early and middle 1900s when NYC removed entire villages for the use of Catskill/Delaware water, as unwelcome and intrusive. Public ownership of Catskill/Delaware critical watershed lands is only 26%, far less than the other major US cities with unfiltered surface DWSs, where close to 100% ownership is common [*World Resources Institute,* 2001] (Figure 2). In the case of the NYC unfiltered surface DWS, were most of the watershed lands are privately owned, bottom-up or community-based watershed planning may be the only means of capturing watershed functions and reducing health risks for consumers. Some methods of community based planning

Figure 2. Public ownership of critical watershed areas in the five major US cities with an unfiltered water supply [World Resources Institute, 2000: 211]

proposed by [*Riley,* 1998] include, 1) advocating individuals with a significant stake in the outcome to develop their own plan, 2) coming to a consensus when all stakeholders, ranging from government agencies to community interest groups are counted as equal representatives in plan development, or 3) planning centered around resolving specific conflicting interests to reach a consensus.

Memorandum of Agreement

Ultimately, five entities, 1) the upstate communities, organized into a unified CWT, 2) environmental organizations, led by the Hudson River Keeper, 3) NYS, 4) NYCDEP, and 5) the USEPA entered into a MOA that clearly delineated NYC watershed regulation powers, community autonomy and financial security for the Catskill and Delaware water supply basins. Although these MOA negotiations were only initiated after community threats of lawsuits against NYC, they represent an attempt by all three stakeholders to move toward a consensus approach to watershed planning. According to the MOA, land acquisition would be limited according to site specific criteria (Table 2) on further delineated "priority areas" and only be allowed on vacant or working landscapes, such as farms or within villages and hamlets [*Principe et al.,* 2000]. While many stakeholders were actively involved, the farming community was not present during MOA negotiations, although farmers could lose nearly 25% of tillable land under the proposed Watershed Rules and Regulations [*NRC,* 2000]. The Watershed Agricultural Council (WAC) was developed as a result of the MOA, and now the farming community is more involved in decisions made under a voluntary based watershed agricultural program (WAP) [*USEPA,* 2001].

The latest FAD for NYC's Catskill/Delaware system (Table 3) was negotiated as part of the MOA, which extended the determination from January 1997 to at least April 2002. The FAD requires that NYC implement a new set of Watershed Rules and Regulations, previously updated in 1953, to satisfy USEPA concerns that management of watershed functions alone provides the DWS with an effective pollutant filter. In place of including agricultural activity within the regulated MOA, the NYCDEP and USEPA endorsed the creation of the WAP to create Whole Farm Plans and manage for nutrient control on demonstration and volunteer subscribed farms. Milestones achieved and reported by the WAP will be considered in renewal of the FAD. The WAP resembles flood prevention management and PL 566 programs, where community/interest group participation comes out of a predetermined watershed plan, as opposed to being empowered during the negotiation of the MOA. NYC must also consider 1) the design and implementation of new scientist/engineer recommended BMPs on agricultural and residential lands, and 2) new and innovative alternatives suggested by policymakers/managers for NYC's watershed lands that incorporate updated scientific findings. Without giving consideration to community/interest groups, however, these

Table 2. Criteria for land acquisition in priority areas, increasing from A to C and 1 to 4, as described in the MOA [USEPA 1997]

Priority	Croton	Priority	Catskill/Delaware	
A	New Croton, Croton Falls and Cross River Reservoirs.	1A	Parcels > 1 acre in sub-basins within 60-day travel time and near intakes.	
B	Muscoot and portions of Amawalk and Titicus Reservoirs within 60-day travel time.	1B	Parcels > 5 acres in sub-basins within 60-day travel time not near intakes.	
C	Remaining reservoir basins and sub-basins beyond the 60-day travel time.	2	Sub-basins within terminal reservoir basins, not in priority areas 1A or 1B and,	At least partially located within 305 m (1000 ft) of a reservoir.
		3	Sub-basins with identified water quality problems, not in priority areas 1A, 1B or 2 and,	At least partially located within the 100-year flood plain.
		4	Remaining sub-basins in non-terminal reservoir basins and,	At least partially located within 91 m (300 ft) of a watercourse.
				Contain a federal jurisdiction or a NYSDEC wetland.
				Contain ground slopes > 15%.

tasks will not necessarily lead to implemented watershed protection for NYC's unfiltered surface DWS.

Catskill/Delaware vs. Croton Watersheds

The Catskill/Delaware systems and the Croton system are different hydrologically and have had different experiences with regard to obtaining a FAD. The Croton water supply system was classified as too densely developed to rely on natural filtration so watershed functions alone cannot clean suburban runoff. The alternative is to construct a filtration plant, estimated to cost $687 million [*NRC,*

Table 3. FAD provisions agreed to by NYC to satisfy EPA concerns [NRC 2000]

FAD COMPONENTS
Compliance with objective criteria of the SWTR
Design of filtration facility for the Catskill/Delaware system
Land acquisition in hydrologically sensitive areas
Data gathering and GIS development for watershed lands
Multi -tiered water quality modeling, from terrestrial to reservoir systems
Maintenance of the WAP
Kensico Reservoir modeling and remediation efforts
Nonpoint source pollution control
Whole community planning
Repair, replacement, and upgrade of septic systems (OS TDS)
Upgrade of wastewater treatment plants
Active disease surveillance

2000]; to filter supplies from 12 reservoirs and 3 controlled lakes. NYCDEP is in the process of designing, siting and constructing a filtration facility prior to US Department of Justice and NY State deadlines, as outlined in the US consent decree and the final Environmental Impact Statement (EIS) for the proposed water treatment plant (WTP) [*NYCDEP,* 1999], while continuing to research better ways to control pollution at its source, again endorsing a dual-track approach. The new paradigm that affected the NYCDEP response to the SDWA amendments in the Catskill/Delaware systems, motivated the inclusion of diverse community/interest group concerns during the development of a "Croton Plan," as part of the Watershed Rules and Regulations [*NYCDEP,* 1997].

The inability for the Croton supply system to obtain an FAD as opposed to the Catskill/Delaware systems can be attributed to differences in hydrologic properties and processes, and water quality concerns and management opportunities. The Catskill and Delaware water supply reservoirs exist in a predominantly agricultural and forested landscape, and are grouped together in this report. The science utilized to understand and manage the Catskill/Delaware versus the Croton systems emphasizes different processes as illustrated by the terrestrial process models used to simulate the quantity and quality of water entering supply reservoirs. In the Catskill/Delaware system, the NYC water authority has used two models, which are the Export Coefficient Model of Reckhow [*Reckhow et al.,* 1980] and the Generalized Watershed Loading Functions Model (GWLF) of Haith [*Haith et al.,* 1992]. These models simplify the watershed based on assumptions of forested and agricultural cover, with a few exceptions for residential and urban systems. Neither model explicitly handles the issues of setbacks, galley's, or stormwater loads. The Croton system water quality has been most recently modeled with a different set of tools, including SWMM and HSPF [*Wang et al.,* 2000]. These models offer limited explicit simulation of setbacks and galley systems, but do offer some improvements on stormwater retrofits through the incorporation of algorithms for suburban hydrologic processes such as the use of storm or sanitary sewer systems, BMPs such as detention ponds, and more complex nutrient/pollutant routines. The greatest distinction between the Catskill/Delaware system and the Croton systems is the result of the advanced degree of urbanization within the Croton system.

The Croton system water quality data has demonstrated that a transition from agriculture to residential/commercial land use negatively impacts the NYC DWS [*Iwan,* 1987]. Stormwater runoff is substantially affected by suburban land development as shown by sedimentation, nutrient loading, channel scouring, flooding, drought, summer low flows, higher temperatures, and pollution above pre-development conditions [*NYSDEC,* 1993]. The impacts of a continually suburbanizing contributing area upon the Croton water supply reservoirs stem from transport mechanisms, contaminant types, and flow pathways that are often more complex than in the Catskill and Delaware systems (Figure 3). The remainder of this report will focus on watershed management issues in the more developed 1000 km^2 Croton system.

Watershed management in the Croton water supply area requires an improved understanding of spatially and temporally varying human modifications to natural systems, such as lawn fertilizer/pesticide applications, additional watering, septic system leachate, pet waste, household product disposal as well as other NPS pollution. Moreover, the Croton system is continuously altered by new development and land parcelization, resulting in changes to watershed functions used in NPS pollutant attenuation, as well as a diversification of affected communities/interest groups. Improved understanding of pollutant attenuation through these dynamic and spatially varying watershed functions must be developed by scientists/engineers in an incremental and logical manner, communicated to communities/interest groups and policymakers/managers so that all three stakeholders are involved in the development of a watershed-based management plan that strikes a balance between controlling NPS pollution and private property rights.

NYC's Perceived Human Health Risks

Risk to human health can be measured by the number of reported waterborne disease outbreaks, which the Center for Disease Control and Prevention (CDC) defines as occurring when: a) two or more persons experience a similar illness after consumption or use of water intended for drinking, and b) epidemiologic evidence implicates the water as a source of illness. According to CDC estimates, disease outbreaks peaked near 25 per year in the early 1980s and are now fewer than

10 per year. The 265 reported outbreaks from 1974 to 1997 [*USEPA*, 1999e] represented different categories of infection (Figure 4) and risk for disease. This method considers all diseases equally risky, regardless of the total number infected. Another measure of water quality risk, however, might track the total number infected, which in the 1993 Milwaukee, WI *Cryptosporidia parvum* contamination of Lake Michigan resulted in nearly 50 deaths and 400,000 illnesses, soared two orders of magnitude above the 1980's outbreaks (Table 4) [*USEPA*, 1999e]. With Geographical Information System (GIS) technologies, risk will be tracked to its spatially distributed origins, which might also shift with storms.

According to storm and waterborne disease data analysis of 2,105 US watersheds from 1971 to 1994 [*Rose et al.*, 2000], increased incidences of extreme precipitation correlate with increased incidences of waterborne disease outbreak. One consequence of this pattern is that if a changing climate creates more extreme weather, we could begin seeing more waterborne disease outbreaks. Beck and Finney [1987] argue for a change from focus on time-invariant average water quality targets toward operational water quality management for wastewater treatment that would better mitigate elevated risk during such temporally varying extreme events. Policymakers/managers will look to

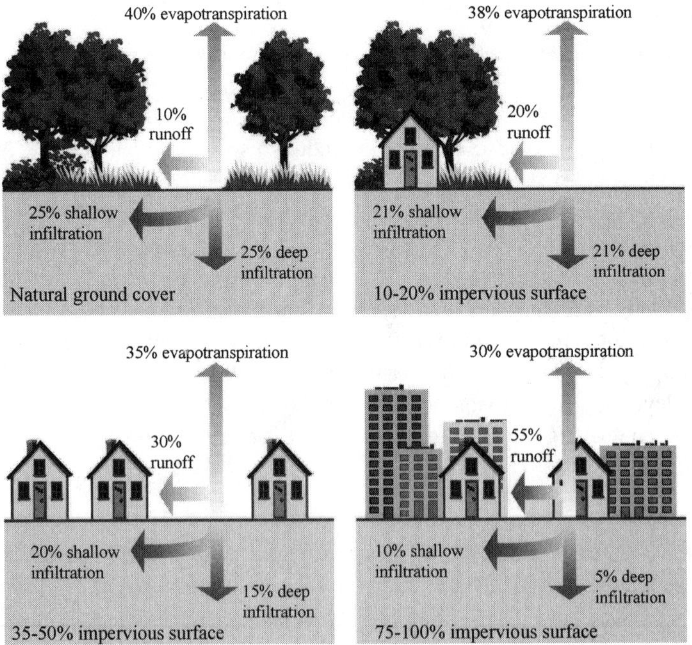

Figure 3. Water budget alterations as a result of urban/suburban land development [FIS-RWG 1998].

scientists/engineers, to describe both spatially and temporally varying pollutant transport mechanisms and for new technologies or management alternatives, such as decision support systems, to protect unfiltered surface DWS.

Community/interest group perceptions of risk, which are sometimes influenced more by personal experience and popular news media than actual risk calculations [*Singer and Endreny,* 1993], may direct the path that NYC water resource policymakers/managers take toward risk control. In an analysis of irrigation water drainage in the San Joaquin Valley, Letey [1999] observed that public "perceptions and beliefs," although usually subject to change, strongly influence management decisions, sometimes beyond more static, valid scientific findings. This concern was also expressed by the National Research Council (NRC) NYC Potable Water Committee, which questioned whether NYC's resources were inappropriately monopolized by phosphorous and other nutrient control, at the expense of research on pathogen and Disinfection Byproduct (DBP) threats with lower visibility yet higher risk. A public survey of NYC drinking water consumers discovered that primary concerns where taste, color and odor, none of which correlate with health risk. This survey both illustrates consumer interest and concern, as well as lack of information.

Disinfection Byproducts

NYC's principle disinfection technique is chlorination, which in the water distribution system can both serve to improve water quality by oxidizing harmful

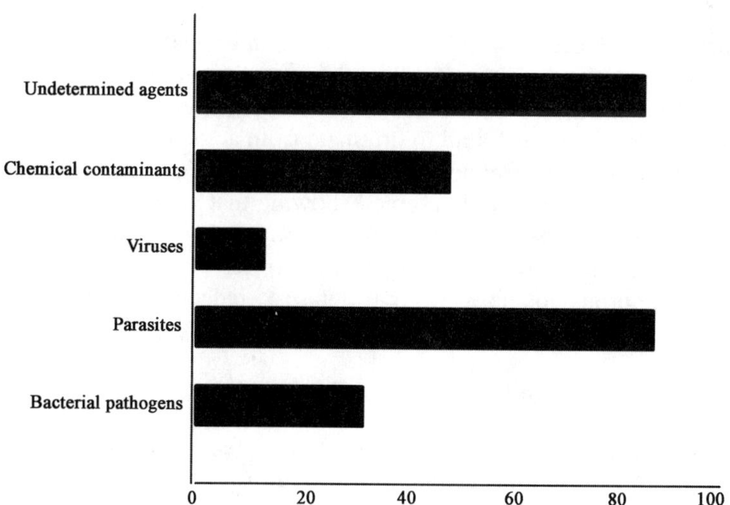

Figure 4. Number of reported outbreaks in the United States over a 25-year period (USEPA 1999e).

Table 4. Number of people affected by waterborne disease [USEPA 1999e].

Year	State	Cause of disease	Number of people directly affected
1985	MA	Giardia lambia (protozoan)	703 illnesses
1987	GA	Cryptosporidia parvum (protozoan)	13,000 illnesses
1989	MO	E. coli O157:H7 (bacterium)	243 illnesses, 4 deaths
1993	WI	Cryptosporidia parvum (protozoan)	400,000 illnesses, 7 deaths

contaminants or to create suspected DBP carcinogens such as Trihalo-Methane (THM) and chloroform by reacting with organic compounds in the water supply. Prevention of DBP formation requires spatially distributed management and a better understanding of their organic precursors, allochthonous and autochthonous organic carbon. Allochthonous organic carbon is loaded into reservoirs from terrestrial sources, while autochthonous organic carbon is generated within reservoirs. The relative abundance of these distinct precursors determines the best strategy for watershed managers to reduce total DBP formation, because they are controlled in two distinct manners [*NRC,* 2000].

Allochthonous carbon, usually responsible for most DBP formation throughout the year, poses a significant threat as runoff energy, volume and erosivity resulting from development increases. Improved watershed-based management in conjunction with land acquisition in critical areas, can provide reductions in the loading of allochthonous carbon from terrestrial sources to NYC's water supply reservoirs. On the other hand, two thirds of the total precursor mass was found to originate from autochthonous sources within the epilmnion of the Cannonsville Reservoir of NYC's Catskill water supply system from April to mid-summer, showing that autochthonous carbon sources must also be addressed [*Stepzcuck,* 1997]. Preventing algal blooms in the epilmnion of water supply reservoirs, which are often the result of phosphorous loadings from tributaries, reduces autochthonous organic carbon and associated DBP formation.

Pathogens Resistant to Disinfection

Specific pathogens with resistance to chlorine disinfection, regulated in the SDWA rules, are naturally occurring, and include *Cryptosporidia parvum*, *Giardia lambia*, and *Giardia parvum* protozoans. These pathogenic protozoa cysts are released from the intestinal walls of domesticated animals, livestock and wildlife and are then carried by watershed runoff into drinking supplies. Reducing spatially distributed sources of such waste is difficult, considering that the hosts, lifecycles, and transport mechanisms of these microbial pathogens are not completely understood. USEPA proposed that NYC construct an ultraviolet (UV) light treatment facility for the oxidation of chlorine resistant pathogens from the mixed Catskill/Delaware system at the Kensico Reservoir, estimated to cost $150 million, in lieu of designing a more costly WTP [*Brown,* 2001]. It is possible that the WTP will still need to be designed and built if the FAD is not renewed in 2002. The ineffectiveness of chlorine disinfection compounded with the spatially varying source area of these microbial pathogens challenges watershed policymakers/managers to develop new strategies to control these pollutants via watershed functions or water treatment systems that use new technologies, such as UV treatment

3. CASE STUDY: STAKEHOLDER COORDINATION IN THE NYC SURFACE DRINKING WATER SUPPLY

On-Site Treatment and Disposal Systems—Septic Systems

Residential development in the Croton water supply area is predominantly unsewered, where household wastewater is routed through an On Site Treatment and Disposal System (OSTDS) or septic system (Figure 5). Unsewered residential basins within the Croton water supply area contribute more nitrate, sulfate, boron, chloride and sodium to tributaries during baseflow than properly functioning sewered (e.g., piped to public wastewater treatment and disposal systems) residential, agricultural or undeveloped catchments [*Heiseg,* 2000]. The spatial siting of a septic system relative to receiving water bodies is an example of a setback distance, set to 30.5 m (100 ft) in the NYS Health Code. Under this code, new septic systems must also meet requirements for percolation rate and depth to water table in the effluent leachfield. In order to prevent the subsurface transport of contaminants in septic system effluent from reaching surface DWS, the USEPA mandated that the NYCDEP assess the effectiveness of this preceding septic system siting regulation. The resultant Septic Siting Study was initiated by researchers at the State University of New York College of Environmental Science and Forestry (SUNY-ESF) and provides a case study of the effectiveness of a preceding regulation in preventing pollutant leaching from OSTDS to surface water bodies, in all three of the NYC DWS basins.

Clusters of monitoring wells were installed 30.5 m (100 ft) downslope of six NYS Department of Health (NYSDOH) compliant OSTDS within all three of the NYC water supply areas and sampled routinely for two full years. Pathogen mimics were deposited into each of the septic systems studied two times, once in the fall and once in the spring, and then the wells were monitored for 75 days following the spike. Data demonstrated that the 30.5 m (100 ft) setback distance failed to contain septic pollutants. Pathogens were found in monitoring wells beyond 30.5 m (100 ft) from a septic leachfield in one third of the study sites.

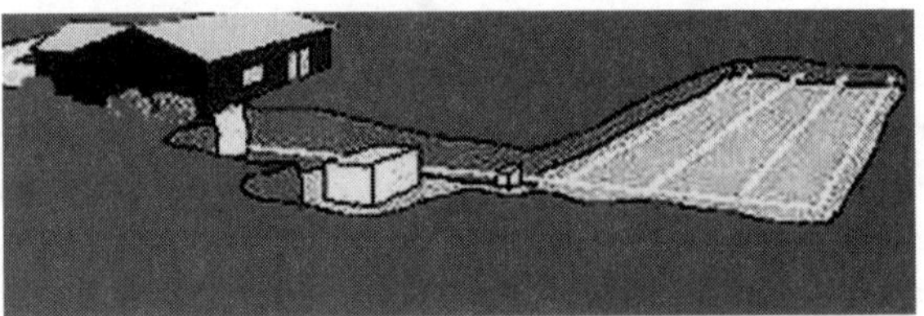

Figure 5. Schematic of a septic system
[URL:http://www.epa.gov/seahome/septics/src/page5.html].

Moreover, at some sites the investigators were uncertain that the downslope monitoring wells intersected the contaminant plume [*Curry,* 2000], thus it is possible that pathogens were transported greater than 30.5 m (100 ft) in more sites than were detected. The final report states that without further investigation there was not a strong enough argument that the preceding septic siting regulation was inadequate. The complex interactions between spatially and temporally varying site characteristics, including the flow path length, depth to the saturated zone, soil type, soil chemistry and the functioning of the septic system, was determined to be the source of uncertainty in the study conclusions [*Curry,* 2000].

Lessons for all three stakeholders can be extracted from the issue of septic siting for NYC's DWS risk management. One familiar lesson is that policymakers/managers are leery to incorporate spatial variability in scientific results into policy. Scientists/engineers, likewise, face the dilemma of synthesizing limited knowledge of spatially varying hydrologic processes into generalized theories helpful for policymakers/managers operating at the watershed scale. And finally, communities/interest groups may prohibit additional monitoring of septic sites for fear of detected failure and obligations for remediation.

Improved prediction of contaminant transport through scientific understanding of multiple parameters can empower policymakers/managers to utilize site-specific watershed characteristics in formulating distributed policies that directly address the hydrologic processes that are most responsible for impacts to receiving water quality. Such work was undertaken in the Catskill/Delaware water supply basins, where the Septic System Rehabilitation and Replacement Program altered its approach midway toward targeting potential septic failures within a 60-day travel time of a surface water body. Employing Darcy's Law, this buffer zone can be implemented for a variety of setback distances based on slope angle and saturated hydraulic conductivity (Figure 6). Septic siting policies focused upon this 60-day travel time buffer zone, rather than a static setback distance, would result in the generation of spatially varying septic system siting. This chart, when evaluated by policymakers/managers in the Croton water supply, provides a tool by which to identify potentially significant existing failures that can be targeted for rehabilitation of replacement and a set of criteria for new septic systems that is hydrologically based and therefore spatially distributed.

Without community/interest group participation, such an approach to reducing the effects of existing and future septic system failures is flawed. The Croton water supply can again gain from observing the accomplishments made through the Septic System Rehabilitation and Replacement Program that was implemented by a community group, the Catskill Watershed Corporation (CWC). The program provided economic support to interested property owners for septic system assessment, rehabilitation, and replacement in order to meet water treatment standards and was successful [*NYCDEP,* 2001]. Communities/interest groups are as spatially distributed as the septic systems themselves and therefore the empowerment of this stakeholder provides policymakers/mangers with an improved approach to controlling NPS pollution generated by contaminated septic system effluent.

On-Site Treatment and Disposal Systems—Galley Systems

As part of the Watershed MOA, the USEPA mandated that the NYCDEP assess the effectiveness of galley OSTDS, wastewater treatment systems used by multi-family communities, commercial operations and other entities consuming over 1000 gallons (3785 liters) of water per day. These systems are designed to treat higher wastewater flows than conventional septic systems and without a leach-field. The effectiveness of galley systems in treating wastewater and the potential impact upon surface water bodies is less certain. Neither the NYS Department of Environmental Conservation (NYSDEC) nor the NYSDOH provides any regulations for the design, installation, or operation of these systems [*Hassett et al.,* 2000]. Likewise, there is a limited amount of research into the environmental impacts of galley systems from the scientific/engineering stakeholder group, thus SUNY-ESF was funded by NYCDEP to determine the impact of galley systems upon water quality and human health risk in NYC's Croton DWS watershed.

Monitoring wells were installed 3.0 m (10 ft) downgradient and upgradient of three galley systems within the Croton water supply area and additional peizometers were installed 6.1 m (20 ft) and 45.7 m (150 ft) downgradient. Samples were gathered from monitoring wells and peizometers on a bi-weekly basis for one full year, and were analyzed for biological oxygen demand (BOD), chemical oxygen demand (COD), total suspended solids (TSS), dissolved oxygen (DO), temperature, nitrate, nitrite, ammonia, total Kjeldahl nitrogen, total phosphorous, total dissolved phosphorous, total coliform and fecal coliform [*Hassett et al.,* 2000]. Water quality analysis indicated that galley leachate had significantly elevated pollutant concentrations in all three sites (fecal coliforms often exceeded the maximum detection for contract labs) within a 3.0 m (10 ft) setback distance and as far as 61.0 m (200 ft) in sites where the main contaminant plume intersected a well.

The Galley Project related water quality and groundwater elevation data from monitoring wells to an extensive site characterization of three wastewater leach-fields. Interestingly, subsurface flow was observed to move in a direction transverse to the topographic gradient and along the interface of the native soil and overlain artificial fill layers [*Hassett et al.,* 2000]. The benefits of deeper percolation prior to such lateral flow may be significant in providing policymakers/managers with recommendations for improved regulations that can be applied to both septic and galley OSTDS that considers site-specific scientist/engineer knowledge of subsurface soil layering. Moreover, the presence of significant groundwater mounding underneath the galley systems was deduced based on non-natural changes in water level data and groundwater chemistry in upgradient wells. Topographically upgradient wells were found to be hydraulically downgradient in one of the study sites [*Hassett et al.,* 2000]. The groundwater mounding was attributed to the disposal of abnormal volumes of wastewater into thin and fine-grained soils overlying impermeable bedrock, typical of the Croton DWS water contributing areas.

This malfunction of galley OSTDS waste disposal confounds the variability in subsurface contaminant transport processes beyond that, which has been observed in conventional septic system effluent research. The Galley Project advised that the NYCDEP should place a ban on any galley systems for the NYC DWS. The study results and recommendations were reported to the NYCDEP, thus enabling policymakers/managers to make decisions based upon current scientific observations and interpretations of receiving water quality impacts of galley system effluent in the Croton water supply area. NYCDEP's funding of the Galley Project is an example where scientists/engineers were able to present knowledge of watershed functions to advise policymakers/managers, whom in turn are now advancing the study findings by working toward a ban on galley systems in the NYC DWS contributing areas. This effort will likely be approved by the NYSDOH as a result of strong support from community/interest groups such as the CWCWC.

Stormwater Pollution Prevention Using Best Management Practices

Nutrient enrichment and subsequent eutrophication within Croton DWS reservoirs was attributed to stormwater runoff from failed septic systems, lawns treated with fertilizers and pesticides, or parking lots with residual oil and anti-freeze deposits [*Borchert et al.,* 1996]. Even with a perfect stormwater plan, residential development still has a discernable impact on water quality, as work in the Croton by [*Heiseg,* 2000] showed with baseflow sampling for a variety of constituents.

Stormwater pollution prevention plans (SPPP) employ structural and vegetative BMPs in the design of new development, and retrofit, projects in order that runoff quantity and quality does not change from pre-existing levels. Applicability criteria for mandating the implementation of a SPPP within the NYC water supply, as described by the MOA, apply to new developments based upon slope angles and proximity to nearby water bodies [*USEPA,* 1997]. Prior to the MOA, less stringent criteria were based upon the NYSDEC state permit for stormwater discharge and elimination systems (SPDES). Whereas many urban BMPs are effective in maintaining pre-existing runoff volumes, the ability for these measures to control pollutant loads, including harmful pathogens, is less certain [*NRC,* 2000]. Data from studies examining the pollutant removal capacity of urban BMPs are widespread and values vary spatially depending upon the climate, geology, and land use where the device is functioning. NYCDEP funded research at SUNY-ESF that investigated the pollutant removal rates of BMPs, existing within NYC's Croton water supply area [*Green et al.,* 1997]. Average observed removal efficiencies from this and other investigations vary significantly between BMP types (Table 5).

Stormwater ponds and a variety of other urban/suburban BMPs are complex ecosystems with many factors affecting their pollutant removal capacity including rainfall, water temperature, turbulence, seasonality, and biological activity [*England,* 2001]. When considering this fact, policymakers/managers need to consult with sci-

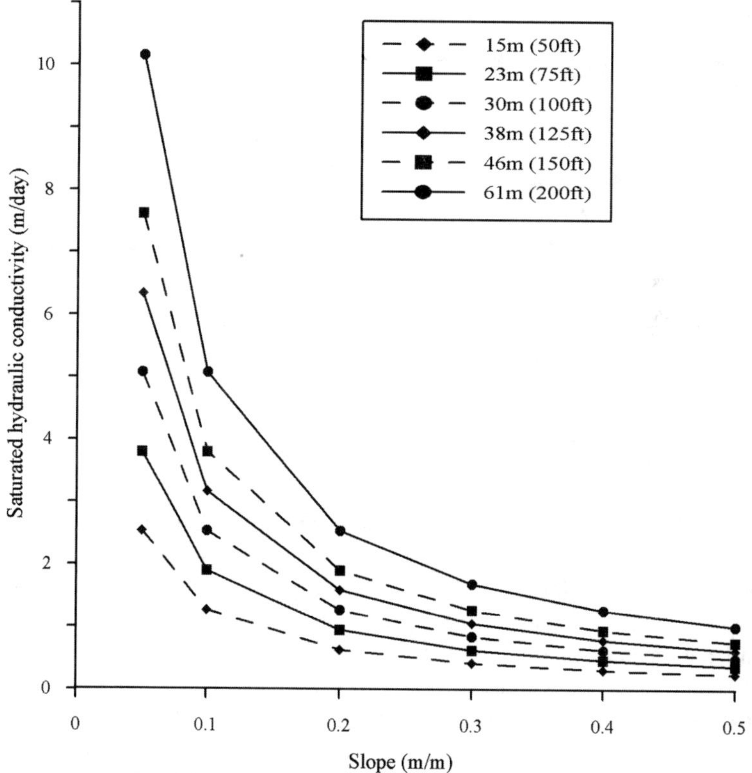

Figure 6. Series of setback distances that facilitate 60 days of residence time, prior to reaching a surface water body, when the soil porosity is 50%, based on slope angle and saturated hydraulic conductivity [modified from NRC 2000].

Table 5. Pollutant removal efficiencies for some residential/commercial BMPs

BMP Pollutant removal (%)	Wet pond (a)	Constructed wetlands (a)	Dry retention (b)	Wet detention with filtration (b)	Infiltration trench (c)	Sand filter (d)	Bioretention (e)	Vegetated swales (f)
Total phosphorous (TP)	68	< 0	61	61	60	33	70-83	9
Total nitrogen (TN)	85	< 0	91	0	60	21	68-80	38 Nitrate
Total suspended solids (TSS)	54	< 0	85	98	90	70	90	81
Biological oxygen demand (BOD)	*	*	92	99	75	70	*	67
Organics	*	*	*	*	90	48 Carbon	90	*
Metals	*	*	*	*	90	*	93-98	64
Fecal coliform	*	*	*	*	*	76	*	*

a) [Green et al., 1997]
b) [England 2000]
c) [USEPA 1999d]
d) [USEPA 1999b]
e) [USEPA 1999c]
[USEPA 1999a]

* = no data

entists/engineers that are now investigating pollution behavior within BMPs for better estimates of their removal capacity in specific ecological settings. The Evaluation of Non-Point Source Pollutant Removal by Best Management Practices study found that this variability occurring within a singular wet detention pond had significant repercussions upon the overall pollutant removal percentage during distinct sampling events [*Green et al.,* 1997]. Further research of standard BMP designs is needed to provide watershed policymakers/managers with more precise values for pollutant removal efficiency of BMPs depending upon watershed specific characteristics and dynamic pollutant pathways within the BMP [*Clary et al.,* 2001]. Scientists/engineers would then direct policymakers/managers toward appropriate guidelines for the design, construction and maintenance of specific permanent stormwater BMPs.

Landscape alteration employed to improve pollutant attenuation must be sensitive to communities/interest groups, otherwise implementation of a BMP will likely face opposition. One concern for communities/interest groups is the resultant insect breeding ground that is created by certain BMPs, often facilitating the spread of disease. In order to protect watershed residents from increased exposure to disease, such as West Nile Virus in the case of the NYC Croton DWS, alternative BMPs can be investigated by scientists/engineers and their effectiveness reported to policymakers/managers. One such BMP is a bioretention facility, or raingarden, that has a smaller footprint (10 – 100 m^2) than a stormwater wetland or wet detention pond. These systems would therefore target critical areas where NPS pollution generation occurs within small neighborhood sized suburban catchments. NYCDEP is currently funding research that is attempting to determine such critical areas or "hot spots" in suburban watersheds for stormwater pollution and potential treatment using small bioretention devices. Here, the extent of private land ownership in suburban watersheds forces the consideration of BMP approaches with a footprint that accommodates community/interest group needs.

4. DISCUSSION AND CONCLUSIONS

Updated USEPA SDWA rules have triggered a new paradigm in unfiltered surface DWS risk management, whereby NYC was coerced to move away from the traditional comprehensive approach, where policymakers/managers have central edict or acquire land and scientists/engineers instrument publicly owned watersheds, toward a bottom-up approach with increased community/interest group participation. This shift in management approach gave more responsibility to communities/interest groups, as shown by the many participants involved in the MOA negotiations. Public DWS consumers, charged with the threat of waterborne disease, perceive watershed protection and health risk reduction as the ultimate priority for policymakers/managers. Private landowners, including homeowners, farmers, and commercial businesses within the unfiltered surface DWS contributing area, however, elevate the value of property rights over more stringent controls for downstream drinking water risk reduction. This tradition upstream/downstream dichotomy exists within the community/interest group stakeholder and subsequently further complicated the development of a community-based DWS protection plan for NYC.

Watershed policymakers/managers are challenged to consider both public interests and scientific evidence when developing a protection plan for an unfiltered surface DWS. On one hand, watershed policymakers/managers are responsible for any technical determinations used in developing policies; therefore it is necessary that they consult scientists/engineers for credible solutions to water resource problems [*Jasanoff,* 1990]. Necessary empowerment of communities/interest groups in the MOA negotiations, however, led to a decrease in the weight applied to scientific findings during the development of a watershed-based management plan. Consideration of public perceptions and beliefs shed more attention upon the uncertainties associated with scientific research when community interests conflicted with preliminary scientific findings. Scientists/engineers need to educate and involve communities through site permissions and volunteer programs in research and other watershed protection projects that seek to manage diverse, spatially varying human activities and watershed functions within an unfiltered surface DWS. This will improve the coordination between the diverse agendas of communities/interest groups and the utilization of watershed functions for pollutant attenuation.

Scientific investigations attempting to unravel complex hydro-chemical processes in suburbanizing watersheds, such as pathogen transport in the subsurface or pollutant removal in stormwater BMPs, will almost certainly vary depending upon site specific hydrologic processes, thus requiring a spatially distributed DWS management approach. Moreover, increasing land parcelization has resulted in watershed communities/interest groups that are as diverse and spatially distributed as are the heterogeneities in watershed physical/chemical features. Many current policies, however, employ singular generalized values for spatially varying watershed functions within a heterogeneous mix of sub-catchments. This was shown in the Septic Siting study and the Galley study, which portray the challenge of properly regulating the siting of OSTDS due to spatial variability and complex interaction between hydrologic variables. The Evaluation of Non-Point Source Pollutant Removal by Best Management Practices study also demonstrates that our current knowledge of the impact of stormwater BMPs point to significant spatial and temporal variability, therefore requiring further scientific investigation in order to predict pollutant removal capacities of devices depending upon their design criteria and environmental setting. Scientists/engineers also need to characterize such spatially varying pollutant transport mechanisms in a manner that can be used by policymakers/managers to develop a watershed-based management plan that is directed at reducing NPS pollutant loadings.

This chapter reviewed selected risks to unfiltered DWS, new rules governing such supplies, and identified three stakeholder groups involved in addressing those risks. In the case of the NYC water supply, watershed policymakers/managers are challenged to develop appropriate rules and regulations that both reduce pollutant loadings in drinking water reservoirs below new USEPA water quality standards and satisfy communities/interest groups within reservoir contributing areas. The NYC unfiltered surface DWS is threatened by pollution from stormwater discharges, seepage from OSTDS, chlorine disinfection resistant pathogens, and suspected carcinogenic DBPs. Protection of water quality by managing NPS pollution through watershed functions as opposed to building a costly water filtration plant requires better stakeholder coordination. Increasing development pressures and stricter water quality standards are confounding the inherent obstacles attributed to the operational differences between scientists/engineers, policymakers/managers, and communities/interest groups, in developing an improved watershed management plan for the NYC DWS. Notwithstanding the facilitation of interdisciplinary communication in suburbanizing watersheds, the integration of site-specific scientific findings, environmental policies, and community empowerment, more often than not will require both a spatially distributed and bottom-up approach to watershed planning.

LIST OF ACRONYMS

BOD	Biological Oxygen Demand
BMP	Best Management Practice
COD	Chemical Oxygen Demand
CDC	Center for Disease Control and Prevention
CWC	Catskill Watershed Corporation
CWCWC	Croton Watershed Clean Water Coalition (CWCWC)
CWT	Coalition of Watershed Towns
DBP	Disinfection Byproducts
DWS	Drinking Water Supply
EIS	Environmental Impact Statement
FAD	Filtration Avoidance Determination
GWLF	Generalized Watershed Loading function
HSPF	Hydrologic Simulation Program in Fortran
MOA	Memorandum of Agreement
NRC	National Research Council
NYC	New York City
NYCDEP	New York City Department of Environmental Protection
NYPIRG	New York Public Interest Research Group
NYS	New York State
NYSDEC	New York State Department of Environmental Conservation
NYSDOH	New York State Department of Health
OSTDS	On-site Treatment and Disposal Systems
SDWA	Safe Drinking Water Act
SUNY-ESF	State University New York, College of Environmental Science & Forestry
SPDES	State Permit for stormwater Discharge and Elimination Systems
SPPP	Stormwater Pollution and Prevention Plan
SWMM	Stormwater Management Model
SWTR	Surface Water Treatment Rule
THM	Tri-halo Methane
TSS	Total Suspended Solids

UFI	Upstate Freshwater Institute
USEPA	United States Environment Protection Agency
USGS	United States Geological Survey
UV	Ultra-violet
WAC	Watershed Agricultural Council
WAP	Watershed Agricultural Program
WPPC	Watershed Protection and Partnership Council
WTP	Water Treatment Plant

Acknowledgments. We would like to thank the State University of New York College of Environmental Science and Forestry and the Edna Bailey Sussman fund for providing the funding which allowed for this analysis. We also would like to thank the researchers in each of the three case studies we have presented for their hard work and thoughtful scientific analysis of two very complex systems, the Catskill/Delaware and the Croton water supply areas.

REFERENCES

Adams, P.W., Closing the gaps in knowledge, policy, and action to address water issues in forests, *Journal of Hydrology, 150*, 773-786, 1993.

Beck, M.B. and B.A. Finney, Operational water quality management: problem context and evaluation of a model for river quality, *Water Resources Research, 23*(11), 2030-2042, 1987.

Black, Peter E., *Conservation of Water and Related Land Resources*, Praeger Publishers, New York, NY, 1982.

Borchert, Dale, Steven Marino, Suzanne Wechsler, Hannah Green and Jeffrey J. McDonnell, Evaluation of non-point source pollutant removal by best management practices, in *American Water Resources Association (AWRA) Symposium Proceeding; Watershed Restoration Management*, edited by Jeffrey J. McDonnell, Donald J. Leopold, James B. Stribling and L Robert Neville, pp. 37-47, AWRA Technical Publication SeriesTPS-96-2.

Brown, Jeff L., EPA to lift filtration plant requirement, *Civil Engineering, 71*(21), 2001

Chichilnisky, Graciela and Geoffrey Heal, Economic returns from the biosphere. *Nature, 391*(2), 1998.

Clary, Jane, Jonathan Kelley, John O'Brien, Jonathan Jones and Marcus Quigley, National stormwater best management practices database: a key tool to help communities meet phase II stormwater requirements, *Stormwater, 2*(2), 2001.

Curry, D., *The Final Report of the Septic Siting Study*, New York City Department of Environmental Protection (NYCDEP), Valhalla, NY, 2000.

England, Gordon, The use of ponds for BMPs, *Stormwater, 2*(5), 2001.

Federal Interagency Floodplain Management Task Force, Protecting Floodplain Resources; a Guidebook for Communities, Federal Emergency Management Agency Publication 268, 1995.

Federal Interagency Stream Corridor Restoration Working Group (FISRWG). Stream Corridor Restoration: Principles, Processes, and Practices. GPO Item No. 0120-A;

SuDocs No. A 57.6/2:EN 3/PT.653.

Germain, Rene and Seth LaPierre, Parcelization of non-industrial private forest lands in the New York City watershed, paper presented at the Wildland-Urban Interface Conference, Gainesville, FL, 2001.

Green, Hannah, Brenda Hill, Jeff McDonnell and Suzanne Wechsler, *The Evaluation of Non-point Source Pollutant Removal by Best Management Practices, Final Report,* State University of New York College of Environmental Science and Forestry (SUNY-ESF), Syracuse, NY, 1997.

Haith, D.A., R. Mandel and R.S. Wu., *Generalized Watershed Loading Functions Version 2.0 User's Manual,* Cornell University, Department of Agricultural and Biological Engineering, Ithaca, NY, 1992.

Hassett, James M., Don I. Seigel, Mark Sherlock and Albert Zumbuhl, *The Galley Project Final Report; a General Study of Galley Systems,* SUNY-ESF, Syracuse, NY, 2000.

Heiseg, Paul M., *Effects of Residential and Agricultural Land Uses on the Chemical Quality of Baseflow of Small Streams in the Croton Watershed, Southeastern New York,* Water Resources Investigation Report 99-4173, US Geological Survey, 2000.

Iwan, Gerald R., Drinking water quality concerns of New York City, past and present, *Annals of the New York Academy of Sciences, 502,* 1987.

Jasanoff, Sheila, *The Fifth Brach; Science Advisors as Policymakers,* Harvard University Press, Cambridge, MA, 1990.

Letey, J., Science and policy in integrated watershed management: a case study, *Journal of the American Water Resources Association, 35*(3), 1999.

National Research Council, *Watershed Management for Potable Water Supply,* National Academy Press, Washington, DC, 2000.

New York City Department of Environmental Protection (NYCDEP), *NYC's 2001 Watershed Protection Program Summary, Assessment, and Long-term Plan,* NYCDEP, Valhalla, NY, 2001.

NYCDEP, *Final Environmental Impact Statement for the Proposed Croton Water Treatment Plant, Notice of Completion,* NYCDEP, Valhalla, NY, 1999.

NYCDEP, *Final Watershed Rules and Regulations for the Protection from Contamination, Degradation, and Pollution of the NYC Water Supply and its Sources,* NYCDEP, Valhalla, NY, 1997.

NYCDEP, *New York City's Long-range Water Quality and Watershed Protection Program,* NYCDEP, Valhalla, NY, 1989.

New York State Department of Environmental Conservation (NYSDEC), *Reducing the Impacts of Stormwater Runoff from New Development, 2nd edition,* NYSDEC, Syracuse, NY, 1993.

Principe, Michael A., William N. Stasuik and Ira A. Stern. 2000. Protecting New York City's drinking water sources, in *Proceedings of the American Planning Association National Planning Conference,* edited by Kasson, Bill and Ray Quay, American Institute of Certified Planners, 2000.

Reckhow, K.H., M.N. Beaulac and J.R. Simpson, *Modeling Phosphorous Loading and Lake Response Under Uncertainty: a Manual and Compilation of Export Coefficients,* EPA-440/5-80-011, United States Environmental Protection Agency (USEPA), Washington, DC, 1980.

Riley, Ann L., *Restoring Streams in Cities,* Island Press, Washington, D.C., 1998.

Rose, Joan B., Scott Daeschner, David R. Easterling, Frank C. Curriero, and Jonathan A.

Patz, Climate and waterborne disease outbreaks, *Journal of the American Water Works Association, 92*(9), 2000.

Singer, Eleanor and Phyllis Endreny, *Reporting on Risk: How the Mass Media Portray Accidents, Diseases, Disasters and Other Hazards,* Russell Sage Foundation, New York, NY, 1993.

Stepczuk, C.L., A.B. Martin, P. Longabucco, J.A. Bloomfield and S.W. Effler, Allochthonous contributions of THM precursors in a eutrophic reservoir. *Lake and Reservoir Management, 14*(2-3), 1998.

Strong, Ann L., *Private Property and the Public Interest; the Brandywine Experience,* Johns Hopkins University Press, Baltimore, MD, 1975.

US EPA, Watershed protection programs. URL, http://www.epa.gov/r02earth/water/nycshed/protprs.htm#AG, 2001.

US EPA, *Stormwater Technology Fact Sheet, Vegetated Swales,* EPA 832-F-99-006, US EPA Office of Water, Washington, DC, 1999a.

US EPA, *Stormwater Technology Fact Sheet, Sand Filter,* EPA 832-F-99-007, US EPA Office of Water, Washington, DC, 1999b.

US EPA, *Stormwater Technology Fact Sheet, Bioretention,* EPA 832-F-99-012, US EPA Office of Water, Washington, DC, 1999c.

US EPA, *Stormwater Technology Fact Sheet, Infiltration Trench,* EPA 832-F-99-019, US EPA Office of Water, Washington, DC, 1999d.

US EPA, *25 Years of the Safe Drinking Water Act: History and Trends,* EPA 816-R-99-007, US EPA Office of Water, Washington, DC, 1999e.

US EPA, *Memorandum of Agreement for New York City's Surface Water Supply,* US EPA, Washington, DC, 1997.

Wang, J., J.M. Hassett, T.A. Endreny, and J.J. McDonnell, Criteria for Selection of Models for Water Quality Management in Urbanizing Areas, SUNY-ESF, Syracuse, NY, 2000.

Wolosoff and Endreny, Scientist & policymaker response types and times in suburban watersheds, *Environmental Management, 29*(6), 2002.

World Resources Institute, *World Resources 2000-2001; People and Ecosystems, the Fraying Web of Life,* Elsevier Science, Oxford, 2000.

Steven E. Wolosoff and Theodore A. Endreny, SUNY College of Environmental Science & Foresty, 211 Marshall Hall, 1 Forestry Drive, Syracuse, NY 13210-2778; swolosof@syr.edu.

18

Policy to Coordinate Watershed Hydrological, Social, and Ecological Needs: The HELP Initiative

Theodore Endreny, Benjamin Felzer, James W. Shuttleworth, and Mike Bonell

INTRODUCTION

The opening chapters have demonstrated the need for better coordination between the disciplines of water resources science and policy. This chapter presents some examples of how to proceed with forging this science and policy alliance. The Hydrology for the Environment, Life and Policy (HELP) program, a new international program crafted by the United Nations Educational Scientific and Cultural Organization (UNESCO) and the World Meteorological Organization (WMO), has been developed to coordinate hydrological, social, and ecological water resources issues. The chapter is organized to: present the motivation for the new HELP initiative, describe the structure and nature of the HELP initiative, outline the steps for participation, provide three case descriptions illustrating different basin themes from within the US, and reflect on challenges and opportunities for the HELP initiative. Within this chapter two themes blend together, but the reader should note that each has separate challenges. The first theme focuses on the coordination of hydrologic policy, management, and science, while the second concerns itself with the coordination of hydrology, sociology, and ecology.

Given that this chapter describes an international program, it should be mentioned at the outset why three US basins (some of which are partly in Mexico and Canada) were selected for the case descriptions. This choice is partly to illustrate the policy, management, and hydrologic variability between basins within a single political entity, and also partly the result of author proximity to the work in these basins, and hence ease of data access. Herein lies an important point addressed by the HELP initiative. By design, the HELP initiative will ultimately engage data sharing between all member basins, but currently data access to

Water: Science, Policy, and Management
Water Resources Monograph 16

10.1029/016WM22

many basins is extremely limited. Examples of coordinated policy, science, and management successes noted in the US basins are also evident in HELP basins located in New Zealand, South Africa, Europe, Asia, Central America, and elsewhere. At the same time, however, the reader should recognize that while the US basins represent some issues common to the international set of basins, each basin has a unique set of priorities and needs that are not represented in this chapter.

MOTIVATION TO COORDINATE POLICY, HYDROLOGY, AND MANAGEMENT

Global water resources are currently experiencing shortages in quantity and deficits in quality, with 1 billion people having no access to potable water, that jeopardize societal health [*Serageldin*, 2002]. The situation is predicted to worsen in many areas given climate changes and population growth (IPCC, 2001). A path forward must involve advances in the hydrological sciences that are directed by, and responsive to, these very real human water resource policy issues. Dialogue has been recognized as central to getting the team of water resource policy makers, managers, and scientists working toward watershed citizen or stakeholder needs [*Falkenmark et al.*, 1999].

Four central obstacles to implementation of dialogue between these three groups have been identified. Prior to setting a course for socio-ecological needs-driven watershed research, these obstacles must be addressed. The four obstacles to implementation of sustainable water management plans identified by UNESCO include: (1) a lack of global demonstration watersheds that show how hydrological practices can protect social and environmental resources (see Anderson (2001)

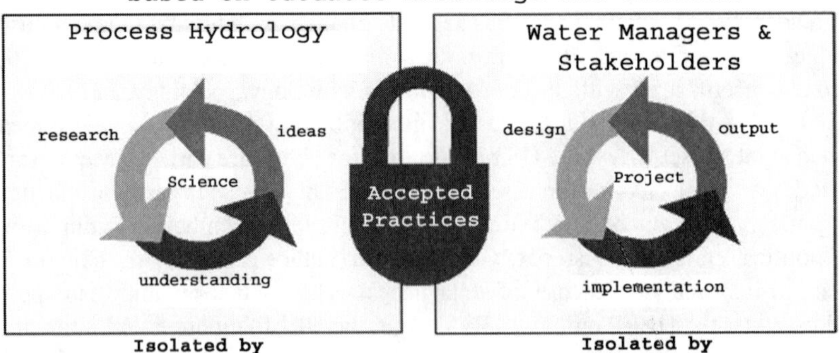

Figure 1. Illustration of science-policy paradigm lock, where scientists do not grasp or research the critical management questions and policy makers do not ask for the latest scientific theory that could aid management.

for a discussion of the many international programs that address the separate issues of sociology, ecology, and hydrology), (2) a lack of coordination in dialogue between countries and basins for sharing lessons learned (see *Sklarew* [2001] for an internet initiative addressing this concern), (3) a decline in national programs of field-oriented water quality and quantity monitoring and research (see *Fraser et al.* [2001] in their discussion of how the UN Global Environment Monitoring System (GEMS)/Water program has addressed this same trend), and (4) a paradigm lock that has traditionally delayed the implementation of hydrologic discovery to the benefit of society, due to scientists not grasping what water users and managers require, as well as managers not grasping what scientific alternatives are available (see Figure 1). The design of the HELP initiative is a direct response to these obstacles as well as the looming global water scarcity and quality crisis.

HELP Initiative Historical Origins

In response to these problems, the Hydrology for Environment, Life, and Policy Initiative (HELP) Initiative grew out of joint meeting of the United Nations Educational, Scientific, and Cultural Organization (UNESCO) and World Meteorological Organization (WMO) to help develop hydro-socio-ecological management plans. The origins of HELP stretch back to the creation of the International Hydrology Decade (1965 – 1974) to systematically study the hydrological environment and the five follow-up phases of UNESCO's International Hydrology Programme (IHP) [*Bonell and Askew*, 2000]. The HELP initiative was formally proposed at the 5th Joint UNESCO/WMO Conference on International Hydrology in February 1999, and was approved by the 28th Session of the IHP Bureau.

The HELP initiative was designed to combine advanced experimental hydrologic research with the most pressing sociologic and ecologic demands for water resource management and policy, creating for the first time a global hydrology program that was purely socio-ecologic needs driven. Formal preparatory work on the HELP initiative began with a task force meeting in Arizona in November 1999 dealing with the details of fitting HELP into a crowded UN agenda and refining the HELP objectives. By the spring of 2000 HELP documentation was in distribution and an open solicitation had been made for interested basins to apply and register their linked science-management-policy program.

In November of 2000, the HELP Interim Management Committee (IMC) met in Dublin and reviewed 24 basin applications and determined the need for additional basin attributes and application criteria. The revised basin nomination packets were requested of the nominees by mid February 2001. The HELP IMC reviewed the revised nominations, along with one additional nomination, in March of 2001, and by the end of May 2001 announced their findings on the 25 applicant basins, which are presented later in this document. The most recent meeting of a partial session of the HELP IMC occurred in Stockholm, April 2002, by invitation of the Swedish Research Council to coordinate a joint August 2002 symposium in Kalmar, Sweden directed toward increasing the dialogue between scientists, policy makers, and stakeholders. A central theme in the Stockholm meeting was the importance of not just policy, but of law, in providing a framework for approaching the dialogue on water use and water rights. Central themes of the Kalmar symposium included the needs of stakeholder involvement, dialogue between professional cultures, and establishment of funding, legal, and policy mechanisms to advance implementation of solutions.

STRUCTURE AND NATURE OF HELP

Goals, Outcomes, and Funding of HELP

Most fundamentally, the HELP initiative is about first meeting basic human needs through sustainable freshwater resource development, and second creating socio-economic and ecologic benefits to watershed residents and stakeholders through sustainable, integrated water use management, and advanced hydrologic research. To achieve these very important outputs, the HELP initiative intends to provide guidance for watershed teams that effectively collect new scientific data sets (e.g., needed meteorologic, hydrological, biological, and chemical parameters) and implement needed watershed management models (e.g., simulation tools for quantifying and visualizing management scenario benefits and constraints).

The architecture for HELP coordination has been carefully designed [*Bonell and Askew*, 2000], and updated at IMC meetings, to coordinate basins through regional secretariats that in turn work with the central international secretariat. Terms of reference for these relations and duties were drafted at the Stockholm 2002 meeting and are awaiting full IMC adoption. The HELP initiative does not budget operational costs for basin participation, but is funded to support the secretariat member activities, referred to here as task force operations. The task force will include directors of science, water policy, and water resource management, and this team of professionals is responsible for reviewing, overseeing, and partially coordinating HELP basin fieldwork as well as meeting with the technical advisory panels. Direct funding for establishing a watershed dialogue and implementing management plans within each HELP basin is expected to come from other donors and sources, likely within the host country.

Themes, Issues and Criteria For HELP

The HELP initiative initially solicits applications for participation from basins where existing or new advances in process-based hydrologic research (e.g., studies of cause and effect) will likely bring significant benefit to multiple issues of water law and policy, water resources management, and ultimately, stakeholder needs. This decision is made to decrease the initial time for results. The basins are expected to have unresolved hydrologic questions that connected to five themes of: (1) climate, (2) food, (3) pollution and human health, (4) environment, and (5) conflict. As long as these five themes for hydrologic-based research are integrated within the basin, the physical size of the basin is not important for HELP approval of basin enrollment in the initiative. In addition to the five themes, however, the HELP task force also identifies ten key issues, from international boundaries to population pressures (see Table 1), which are used to organize basin focus areas within the HELP initiative registry. It is recognized that the importance of these 10 issues will vary with global location, and hence they will form a means of organizing each HELP basin's policy, science, and management foci.

Once the basin application addresses the components in the above five themes and ten issues, the HELP task force further establishes five criteria that are to be satisfied for a basin to register and participate in the HELP initiative.

Table 1. Ten issues of concern that may be focus of new policy-science management coordination for the HELP basins.

Hydrological Issues of Concern
Threats to Sustainability
Impact of Global Scale Problems
Trans -Boundary Aspects
Long Term Trends
Ecological Damage
Social and Political Impacts
Economic Growth or Decline
Population Pressure
Risks to Human Health
Potential for Improved Management

1. Provide an opportunity to study a water policy or management issue for which hydrological process studies are needed,
2. Provide a cooperative relationship between relevant national and local agencies and the execution of the HELP program to ensure sustainability of technology transfer,
3. Provide adequate local capacity to participate in the program,
4. Provide monitoring of a minimum range of key variables and parameters, and
5. Provide data, information, and technological expertise that are shared openly and adhere to international data standards as well as quality assurance and quality control standards.

Criteria truly central to HELP goals are the second and fifth bullets, which address the issue of getting policy support to launch the HELP program, and then getting management and science support to share the results with other basins in the HELP initiative.

STEPS FOR PARTICIPATION IN HELP

HELP guidelines provide 5 themes, 10 issues, and 5 criteria for participation in this hydro-socio-ecological initiative, and an example of how to get started on the process is further provided. The HELP initiative identifies a set of seven steps (see Figure 2) that are perhaps best illustrated with work in the Motueka Watershed in New Zealand, where these steps are partly completed. There, the organizing and planning that links watershed environment and social needs preceded fieldwork by nearly two years [*Bonell and Askew*, 2000].

A critical step in the preparation of a HELP participating basin is to include watershed citizens, or stakeholders, in the initial dialogue that identifies the

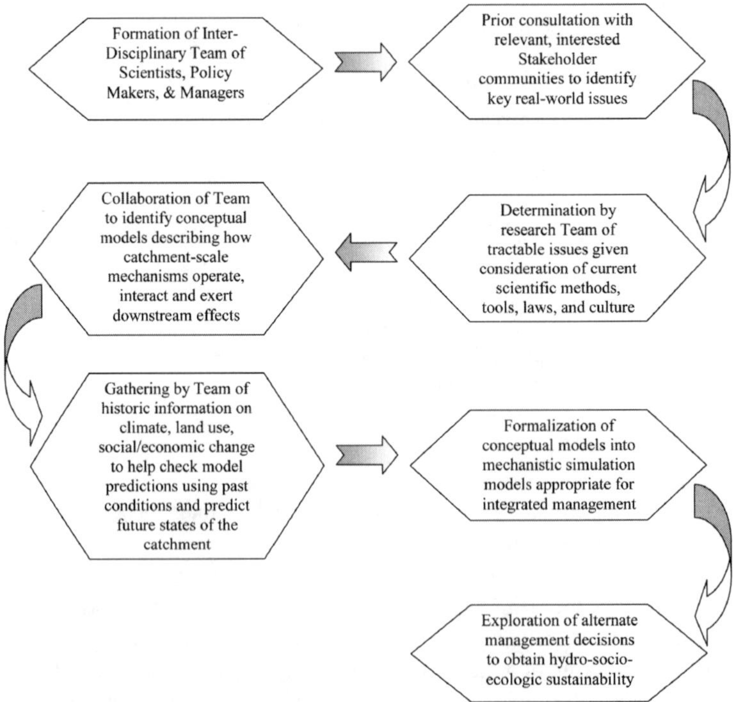

Figure 2. Suggested sequence of steps for developing an integrated, social needs driven, policy, science, and management plan for HELP initiative participation.

needs foci for coordinating policy, science, and management. In basins with large rural populations, as well as those with large urban centers, this step may pose the greatest challenge, as evidenced by other environmental programs using their constituents as the 'silent majority'. While a group may not identify itself or its needs, this does not signal that they have relinquished these needs. A related challenge for basins will be guiding stakeholders to distinguish between problems that coordinated hydrologic research can beneficially affect from those best left to other social programs. It is clear that not all basins will walk the same path toward dialogue, and the nuances for HELP implementation will vary across countries but possibly trend with socio-economic patterns and the social and ecological effectiveness of political and legal frameworks.

Distinct Environment, Life, and Policy Components

Determining whether a basin should consider participation in the HELP initiative may involve asking whether the hydrologic sciences can simultaneously, and in an integrated fashion, improve the watershed's environmental function, human standard of living, and water policy and management. These three components, environment, life, and policy, must lend themselves to cooperate with hydrology, prodding the hydrological scientists, to consider new monitoring, tools and models that address the complexity of sociologic and ecologic interactions within bounds of water policy and the principles of sustainable management. Further, the presence of a cooperative, inter-disciplinary group of water scientists, water policy makers, and water managers is the litmus test to examine whether the infrastructure is in place for HELP participation.

Initial Basin Lists, Classification, and Prospects

Some 25 HELP basins have formally applied for participation in the HELP initiative, yet the HELP IMC has the ultimate goal of creating an even larger network of experimental hydrological catchments representing a diverse set of climates, cultures, and hydrologic, ecologic, and sociologic needs. Table 2 lists the 25 basins, and their countries of origin, that submitted final applications to the HELP IMC for review in Wallingford, UK during March 2001, and provides the IMC ranking of the basin as reference, operational, evolving, or proposed. Figure 3 identifies each basin's geographic location on a world map, allowing for the observation that certain areas are possibly under-represented, such as central Asia and Africa. While these 25 basins represent the first wave of participants, an open invitation for new basin applicants will be extended in 2003, as the HELP initiative has not closed the door on future basin participants. Once a HELP member basin, the basin is expected to improve

Table 2. Listing of countries and basins that have submitted applications for inclusion in the HELP initiative.

Country or Countries	Basin	Preliminary Designation
Australia	Mount Lofty Ranges Basins	Operational
Australia	Murray Darling River Basin	Proposed / Reference[1]
Brazil	Jua and Branco River Basins	Evolving
Cambodia, Thailand, Viet Nam	Lower Mekong Basin	Evolving
China	Tarim River Basin	Evolving
Germany	Spree -Havel River Basin	Proposed
Germany and Austria	Upper Danube Basin	Evolving
Japan	Yasu	Proposed
New Zealand	Motueka River Basin & Tasman Bay	Operational
India	Subernanekta River Basin	Evolving
Panama	Pana ma Canal	Evolving
Peru	Jequetepeque River Basin	Evolving
Puerto Rico	Luquillo Mountain Basins	Proposed
South Africa	Olifants River Basin	Evolving
South Africa	Thukela River Basin	Evolving
Sri Lanka	Walawe River Basin	Evolving
Tajikistan, Uzbekistan, Turkmenistan, Kyrgyzstan, and Kazakhstan	Aral Sea Basin	Evolving
United Kingdom	Thames	Evolving
United Kingdom	Upper Severn	Evolving
United States & Canada	Lake Ontario Basin	Operational
United States & Mexico	San Pedro River Basin	Operational[2]
United States (NY & NJ)	Hudson River Basin	Proposed
United States (OK, CO, NM, KS, TX, MO, AR, & LO)	Red Arkansas River Basin (w/ Little Washita River focus)	Operational
United States (WA)	Yakima River Basin	Proposed

[1]The Murrumbidgee sub basin has been upgraded to Reference following the HELP IMC May 2001 listing.

[2]The San Pedro has been upgraded from Evolving following the HELP IMC May 2001 listing.

upon water policy, management, and science and move along the HELP classification, from proposed to evolving, evolving to operational, and from operational to reference basins.

The HELP IMC classification of basins in Wallingford assessed basins on, among other items, a demonstration of stakeholder involvement, identification of interdependent hydro-socio issues, clarification of research objectives, and verification of key agency support. Of the 25

HELP PILOT PHASE DRAINAGE BASINS

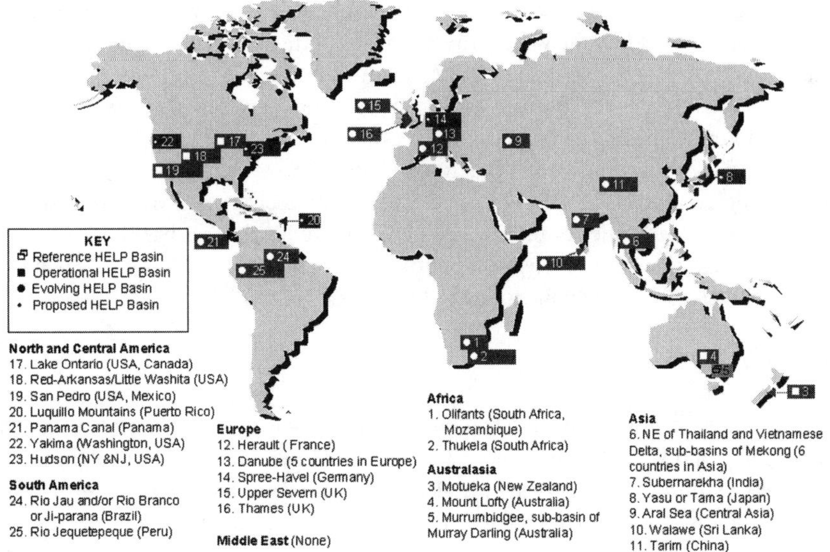

Figure 3. Global distribution of the 25 registered HELP basins, classified as Operational, Evolving, and Proposed. This is not a final list of participants, and additional basins may apply for registration.

nominated basins, none was initially considered a world "reference" basin, but four received an initial ranking of "operational," 14 were ranked as "evolving" and six were considered "proposed", indicating that additional information was needed. Since this ranking, one basin has moved into reference status and another to operational status. Reference basins are defined to functionally demonstrate HELP goals, and while they have areas for improvement, the most basic advances are needed within the other three classes.

Operational HELP basins are perceived to have adopted the HELP philosophy of implementing an active interface between management, science and society and is functioning to integrate a number of the HELP themes and issues. Evolving HELP basins are intended to express a clear intent to manage a basin following the HELP philosophy and have a comprehensive project plan that includes the participation of HELP stakeholder groups in basin management. Proposed HELP basins are perceived to have just begun the process of involving stakeholders and identifying key HELP issues. Examples of some of these basins illustrate how they are furthering the HELP process.

THREE CASE STUDIES WITHIN US HELP BASINS

The three basins described below are representative of how the HELP process can be used to build on the widely divergent strengths within existing basins. The Lake Ontario Basin, along the U.S./Canadian border, is a relatively wealthy region that has benefited from early cooperation between the US EPA and Environment Canada that helped initiate other cooperative management organizations. The Upper San Pedro Basin along the U.S./Mexican border is a much poorer region that has benefited from citizens volunteering their own time to discuss management issues of their watershed. The Red-Arkansas basin is an example of a region where extensive scientific data and analysis has not yet been paired effectively with applications in the basin.

Lake Ontario Basin

Numerous successful watershed management plans have been coordinated among US and Canadian hydrologists, water policy makers, and water resource mangers within the 90,000 km^2 Lake Ontario basin. The 18,000 km^2 Lake Ontario drains the upper four Great Lakes, or approximately 20% of the global freshwater volume, but receives watershed runoff from a much smaller area representing 1% of global freshwater. The Lake Ontario lake and land system has nurtured agricultural, fishing, industrial, and post-industrial economies (e.g., recreation, education, technology), which have caused pollution by metals, organics, chemicals, nutrients, sediments, and exotics. Further, the basin has benefited from advanced scientific methods, long periods of cross-border political stability, public participation, and relative financial wealth that has enabled advanced water resources planning and implementation [*Fuller and Harvey*, 2001]. The Lake Ontario basin historical experience has components common to other areas, and lessons learned should be shared within the HELP basin community. New problems will continue to arise, such as predicted water level changes during climate change (mostly lower levels), and invasive mollusk, fish, and plant species arriving in boat ballasts, which will require new management strategies and lessons, suggesting that the Lake Ontario might function as a learning and demonstration basin.

Cooperative international management of Lake Ontario reaches back to the beginning of the 20th Century, and in 1905 the International Waterways Commission was created to advise both Canada and the US about water levels and flows in the Great Lakes, with a focus on hydropower. In 1909 the Boundary Waters Treaty was signed, creating the International Joint Commission (IJC) management authority. The IJC's authority includes resolving trans-boundary water resource disputes through scientific and policy analysis performed by numerous affiliated scientists and governmental agencies [*Galloway and Clamen*, 2001]. Water quality management became a focus of IJC activity beginning in the late 1960s, yet it could be argued that Lake Ontario has only recently hosted comprehensive socio-ecological studies addressing the HELP socio-ecologic themes of sustainability. One example of this success are the Lakewide Management Plans (LaMPs), commissioned by the IJC [*IJC*, 1999], that begin to integrate hydrology, life, and environment, and could likely serve as a template for future HELP initiative projects. *Donahue et al.* [2001] discuss another potential integrative success in Lake Ontario, which is the creation of a new US-Canadian team to address how the management of water levels to meet human needs may impact ecological systems, and then potentially explore innovative hydraulic and hydrologic management alternatives.

Socially directed hydrologic research in the Lake Ontario Basin is not limited only to the larger Lake Ontario system, but also includes numerous citizen-driven watershed management projects for the numerous sub-basins [*FLOWPA*, 2000]. The FLOWPA report identifies two types of sub basins in the Lake Ontario system, those with direct connections to Lake Ontario, and the smaller internal watersheds within these sub-basins. In both basin types a stakeholder process was implemented to analyze whether hydrological research and action address local social concerns by reporting on the ecological impairment within the basin, summarizing agency reported concerns and priorities for restoration, and then providing a quantitative and graphical gap analysis, where gap is described as the congruence between water resource impairments and local priorities [*FLOWPA*, 2000]. Other HELP basins may find such gap analysis useful for quantitatively evaluating basin management plans, which serves the intent of HELP to share knowledge and methods between international basins.

Upper San Pedro Basin

The Upper San Pedro Basin (USPB) is a good example of a basin that has already undertaken a HELP-like process involving a wide-array of stakeholders, managers, and researchers in decision-making and planning [*Varady and Browning-Aiken*, in press; *Varady et al.*, 2000]. The San Pedro River originates in northern Sonora, Mexico, and flows north into Arizona, eventually joining the Gila River, which flows into the Colorado River. The San Pedro Riparian National Conservation Area (SPRNCA), an approximately 18,200-hectare area managed by the Bureau of Land Management, is located entirely within the USPB, which is one of the most ecologically diverse areas in the western hemisphere. Most of the water demand in the basin is for mining, municipal and domestic use, irrigated agriculture, and the riparian vegetation itself. Projected population increases in the U.S. Southwest will lead to increased water use for municipal and domestic purposes, while increased production of copper ore reserves in Mexico limits the groundwater supply there.

Water is managed by a complex array of federal, state, and local agencies on both sides of the U.S/Mexico border. In addition, a number of nongovernmental organizations, particularly environmental groups, are involved in water policy and water management discussions in the basin. Several research institutions conduct ongoing studies in the basin, including the Semi-Arid Land-Surface –Atmosphere Program (SALSA), the newly established NSF-supported Semi-arid Hydrology and Riparian Area (SAHRA) Science and Technology Center, and the US Department of Agriculture/Agricultural Research Service (USDA-ARS). One of the first applications of international environmental law under North American Free Trade Agreement (NAFTA) resulted in a study of excessive groundwater pumping within the USPB and its possible impacts on the San Pedro riparian system, whose loss would impact critical migratory bird populations in Canada, the U.S., and Mexico. The primary goal of water management in the USPB is sustainable development. Because the Upper San Pedro Basin is already estimated to be in a deficit, highly efficient water management on multiple fronts is required. In an extremely positive development, the Upper San Pedro Partnership (USPP) was formed in 1998 to meet water needs in the Sierra Vista sub-basin of the USPB and to sustain the viability of the San Pedro Riparian National Conservation Area.

A key component of the USPB is the use of citizen participation in research management and policy decision-making. The USPP, in particular, uses volunteers to help collect data on groundwater level and river flow. The Udall Center for Studies in Public Policy at the University of Arizona has taken a lead role by using citizens to help develop and administer a public-interest survey.

Red-Arkansas Basin

Parts of the Red-Arkansas basin have been extensively studied as part of the US Department of Agriculture research program and National Aeronautics and Space Administration (NASA) Earth Observing Program [*Sellers et al.*, 1992; *Starks and Humes*, 1996]. Much NASA research has direct applications for ecological and social systems [*Mohr et al.*, 2000], and can advance nicely the concepts of advanced hydrologic research for the HELP initiative. The Red and Arkansas rivers flow from the Rockies to the Mississippi River and include all of Oklahoma and parts of Colorado, New Mexico, Kansas, Texas, Missouri, Arkansas, and Louisiana. The Little Washita River sub-basin in Oklahoma has been extensively studied over the past 40 years and has a readily available network of meteorological and stream gage data [*Starks and Humes*, 1996]. There are 45 flood-retarding structures within the Little Washita River basin whose purpose is to control flooding in the low-lying alluvial areas through temporary storage of flood waters from upstream areas. Many of these structures are nearing the end of their design lifetime, so that decommissioning and remediation measures need to be developed and implemented. Water quality issues involving excessive concentrations of sulfate, chloride, nutrients, pesticides and suspended sediments limit water supplies. This project will also explore the potential linkages with other parts of the Red-Arkansas, such as involves the US Department of Energy (DOE) ARM/CART (Atmospheric Radiation Measurements/ Cloud and Radiation Testbed) sites and new sites associated with the Water Cycle initiative and the U.S. Bureau of Reclamation Fryingpan-Arkansas project. While these extensive data networks have been used in a wide range of scientific studies, there is still a need for better communication between the scientific and local communities to apply this research towards management decisions.

The Oklahoma Water Resources Board (OWRB) is the state's chief water agency. It monitors and studies drought and flooding, and it appropriates, distributes, and manages ground and surface water. It also supervises the Statewide Rural Energy and Water Conservation Program. Issues of water quality within these basins are managed by a combination of the USGS, USEPA, the Oklahoma Conservation Commission and the Oklahoma Department of Wildlife Conservation. In addition, the National Park Service Water Resources Programs, the Oklahoma GIS Council, the USGS and the Farm Services Agency have developed a GIS system on urban change and rural development, an environmental natural-resource management program, watershed and streamflow modeling, and mitigation and response measures to disasters. To resolve and prevent disputes over trans-boundary water shared with other states, Oklahoma participates in four interstate stream compacts. The Red River Boundary Commission handles problems associated with the river's frequent course changes and it resolves questions of land ownership, tax assessment, and jurisdictional problems. The HELP process can lead to better coordination of these activities and additional initiatives in other parts of the Red-Arkansas basin.

CHALLENGES AND OPPORTUNITIES IN THE HELP INITIATIVE

The challenges for the HELP initiative will arise directly from the goals that were defined at program inception. Fundamental to its program goals is solving the four obstacles to dialogue and progress, which are 1) lack of demonstration basins, 2) lack of coordination to share lessons, 3) decline in field data collection, and 4) the paradigm lock blocking policy maker and scientist communication. At this stage, the HELP initiative has established 25 member basins, and the framework for sharing lessons between basins will follow on the implementation of the terms of reference for global and regional secretariat functions, as mentioned above. Increasing the field data collection is a function of additional funding, and work on developing novel techniques for obtaining data, such as remote sensing tools, which are relatively inexpensive per unit area of coverage, might reverse this trend. Unlocking the paradigm lock is trickier, and may involve teaching the critical importance of dialogues to younger generations since that the old guard is often stuck in old ways. However, even this effort is sometimes a lesson lost as students grow [*Black*, 2001]. In terms of the four challenges, HELP has seen progress, and is still advancing.

The intent of the UNESCO and WMO HELP initiative, as explained above, is to improve the social and ecological condition of a hydrologic basin through needs based, innovative policy and science. The program goals target opportunities for improvement, and may appear particularly useful in nations that have not formally adopted constituent based environmental water resource protection legislation, yet some may consider a UN sponsored policy a disruption to local operations. As freshwater is a very dear and limited resource, actions that provide more for needy citizens may be those that upset pre-existing private and public power structures. Further, given that water is so central to a country's national health, and therefore defense, it is possible that any new initiative on water management, particularly one sponsored by an external organization, might be viewed with suspicion. In the US, where environmental legislation is rather advanced, and many water resource management programs have been well funded, the prospect of a foreign based water management initiative is doubly suspect due to its appearance as creating more work and complicating already complex operations. In short, the HELP initiative will be viewed as a potential threat if it were to supersede local control and create unfunded mandates.

Another level of challenge includes education of the policy makers, scientists, and managers that are either unaware of the HELP initiative or its intended benefits, as well as targeting programs for the education of watershed citizens looking to solve local water resource issues. The education of potential citizen groups to the HELP initiative has various levels, and is likely more challenging than breaking the paradigm lock between policy makers and scientists. For citizens, first there is creating awareness of the program, then informing them of the HELP initiative goals, then providing them with a reasonable set of expectations for outcomes from HELP, and, finally, encouraging them to enter into dialogue with local watershed policy makers, managers, and scientists.

Initial data limitations will create challenges for the new HELP initiative basins, as some basins reside in poorer countries without historical monitoring data, and others cross political borders and hence experience different levels of monitoring. An example of this data constraint is digital elevation model (DEM) data, which is available globally as GTOPO at a resolution of 30-arc seconds (approximately 1 km) and too coarse for detailed work. In areas where local sources are higher resolution, such as at 30 m in the US, the Lake Ontario and San Pedro basin must default to the coarser data of their neighbor country, which is 100 m for only part Canada, and 1 km for much of Mexico. Without these DEM data, basins are unable to make clear basin boundaries or delineate watershed study areas. Again, remote sensing provides a path forward, because it will provide necessary higher resolution data. For example, NASA Shuttle Ranging Topography Mission (SRTM) will issue 30 m data coverage over all these areas in late 2002. These scientific advances provide great opportunities for the HELP initiative. Another challenge for the HELP initiative is the development of a comprehensive clearinghouse for pertinent hydro-socio-ecological research results and data. Further, as integrated HELP-based programs continue to operate, the question then arises of how to catalogue findings, whether it is by keywords in management, policy, science, or instead by the social and/or ecological need.

CONCLUSIONS

The HELP initiative has chosen to specifically address the issue of coordinating policy, science, and management to improve the water resource benefits for social and ecological uses. While the program has a strong foundation, many coordination and data challenges must be met in its implementation. Given the focus of integrating water resources across professional disciplines, the HELP initiative is an exciting opportunity for scientists and hydrologists interested in the input of an ecological and social user group. For the managers and policy makers, it is an opportunity to envision new solutions and tools by coordinating with advanced scientific research. The program presents hydrologists with a new audience of users, potentially breaks open the paradigm lock, and asks that they demonstrate how new technologies and theory can help achieve socio-ecologic sustainability. To communicate effectively with this new audience, policy makers, scientists, and mangers will need to agree upon a common language that is understood by the citizen group that they hope to serve.

Benefits of joining the HELP initiative, beyond the prospect of achieving a more citizen-based and sustainable management plan include altruistic goals, such as working toward improvements in social and ecological systems and sharing lessons learned with other needy basins. Benefits for enrollment in the HELP initiative can also be more self-serving, such as (1) using the HELP endorsement to compete for additional funding for novel research initiatives, (2) benefiting from the outside review of UNESCO or WMO as an independent assessor on the sustainability of a watershed management plan, (3) accessing the HELP forum to learn helpful lessons from other basins that face a similar problem, and (4) revitalizing efforts to include all stakeholders in watershed management planning.

Acknowledgments. The authors thank the watershed citizens and programs that have supported coordinated policy, science, and management in the many HELP basins. Thanks also goes to the following agencies for generous funding support: the Great Lakes Research Consortium, the Semi-Arid Hydrology for Riparian Areas program, the Semi-Arid Land-Surface-Atmosphere program, the Ford Foundation, the Morris K. Udall Foundation, and the National Atmospheric and Oceanic Organization.

REFERENCES

Anderson, F. 2001. The Internationalizing World of Water Management. *Water Resources Impact*, 3(2): 3-8.

Black, P.E. 2001. Bridging the Paradigm Lock. *Water Resources Impact*, 3(5): 41.

Bonell, M. and Askew, A., 2000. *The Design and Implementation Strategy of the Hydrology for Environment, Life and Policy (HELP) Initiative: HELP Task Force*, United Nations Educational Scientific & Cultural Organization, Paris, France.

Donahue, M., Crane, T. and Manninen, C. 2001. A Water Resources Management Decision Support System for the Great Lakes: International Water Issues on the U.S. - Canadian Border. *Water Resources Impact*, 3(2): 16-19.

Falkenmark, M., Andersson, L., Castensson, R. and Sundblad, K. 1999. *Water: A Reflection of Land Use - Options for Counteracting Land and Water Mismanagement.* Swedish National Science Research Council, Stockholm.

FLOWPA, 2000. *The State of the New York Lake Ontario Basin: A Report on Water Resources and Local Watershed Management Programs*, Finger Lakes-Lake Ontario Watershed Protection Alliance, Penn Yan, NY.

Fraser, A., Dobarts, R. and Hodgson, K. 2001. The United Nations Environment Programme Global Environmental Monitoring System / Water Programme. *Water Resources Impact*, 3(2): 26-28.

Fuller, K. and Harvey, S., 2001. *The Great Lakes: an Environmental Atlas and Resource Book.* EN40-349/1995E, U.S. Environmental Protection Agency, Great Lakes National Program Office, Chicago, IL.

Galloway, G.J. and Clamen, M. 2001. The International Joint Commission: A Model of Cooperation in Dealing with Boundary Water and Transboundary Environmental Issues. *Water Resources Impact*, 3(2): 12-15.

International Joint Commission (IJC), 1999. *International Joint Commission Lake Ontario Stage 1 Lakewide Management Plan Review*, International Joint Commission, Washington, DC.

Intergovernmental Panel on Climate Change (IPCC), W., 2001. *Climate Change 2001 Impacts, Adaptation, and Vulnerability.* Third Assessment Report, WMO & UNEP Intergovernmental Panel on Climate Change, Working Group II.

Mohr, K.I., Famiglietti, J.S., Boone, A. and Starks, P.J. 2000. Modeling Soil Moisture and Surface Flux Variability with an Untuned Land Surface Scheme: A Case Study from the Southern Great Plains 1997 Hydrology Experiment. *Journal of Hydrometeorology*, 1: 154-169.

Sellers, P.J., Hall, F.G., Asrar, G., Strebel, D.E. and Murphy, R.E. 1992. An Overview of the First International Satellite Land Surface Climatology Project (ISLSCP) Field Experiment (FIFE). *Journal of Gephysical Research*, 97(D17): 18345-18371.

Serageldin, I. 2002. World Poverty and Hunger - the Challenge for Science. *Science*, 296: 54-58.

Sklarew, D., Annis, S., Mendler, J. and Hamid, M. 2001. Forging a Global Community to Address International Water Crises: The IW LEARN Project. *Water Resources Impact*, 3(2): 20-25.

Starks, P.J. and Humes, K.S., 1996. *Hydrology Data Report Washita '94.* NAWQL-96-1, US Department of Agriculture, Agricultural Research Service, Durant, OK.

Varady, R.G. and Browning-Aiken, A. In Press. *The Birth of a Mexican Watershed Council in the San Pedro Basin in Sonora.* Planeacion y Cooperacion Transfronteriza en la Frontera Mexico-Estados Unidos.

Varady, R.G., Moote, M.A. and Merideth, R. 2000. Water Allocation Options for the Upper San Pedro Basin: Assessing the Social and Institutional Landscape. *Natural Resources Journal*, 4(2): 223-235.

Theodore A. Endreny, SUNY College of Environmental Science & Forestry, 211 Marshall Hall, 1 Forestry Drive, Syracuse, NY 13210-2778; swolosof@syr.edu

Benjamin Felzer, NCAR, UCAR, Boulder, CO; bfelzer@msn.com

Mike Bonell, UNESCO, Division of Water Sciences, Paris, France

James Shuttleworth, University of Arizona, Department of Hydrology & Water Resources, Harshbarger Building, AZ 85721.

Acknowledgments

The editors and authors this book gratefully acknowledge the following individuals and organizations.

Susanna Eden and Richard Lawford thank the University Corporation for Atmospheric Research and the Climate Change Science Program (CCSP) Office, and CCSP agencies (particularly the National Oceanic and Atmospheric Administration) for their support.

David Ahlfeld and Weston Dripps appreciate the financial support of the Research Center for Groundwater Remediation Design and thank the several reviewers whose comments improved the manuscript.

Robert Varady and Barabra Morehouse are grateful for assistance from Robert Merideth and Leah Stauber at the Udall Center. Some of the ideas are extensions of previous work, co-written with Helen Ingram, Lenard Milich, Vera Pavlakovich-Kochi, and Doris Wastl-Walter. Additionally, W. James Shuttleworth, Michael Bonell, and James Wallace, formulators of the HELP initiative; and Aaron Wolf and Roger Pulwarty have, perhaps unknowingly, contributed influential insights. Finally, we acknowledge the Ford Foundation, the Morris K. Udall Foundation, the National Oceanic and Atmospheric Administration, and the Science and Technology Center for the Sustainability of semi-Arid Hydrology and Riparian Areas (SAHRA) for their past and present support for investigations on river-basin management.

Eric Reichard and Robert Raucher are grateful for the helpful technical and editorial comments of Richard Bernknopf, Steven Phillips, Clark Londquist, Gerald Woodcox, Anthony Buono, and the anonymous reviewers. We thank Theodore Johnson, Wayne Jackson, Jeff Micko, and Steve Bachman for providing information for the case studies. Phil Contreras prepared the illustrations.

Katharine Jacobs and Roger Pulwarty express appreciation for the assistance of Kenneth Seasholes of the Arizona Department of Water Resources, Nancy Beller-Simms of the Office of Global Programs, Sharon Megdal and Barbara Morehouse of the University of Arizona and the suggestions of additional reviewers.

Neal Lane, Rosina Bierbaum and Mark Anderson initially prepared their chapter while assigned to the White House, Office of Science Technology Policy. They are grateful to the U. S. Geological Survey and the National Oceanic and Atmospheric Administration for assistance with the illustrations. They also thank Peter Backlund, Rick Gold, Steve Longsworth, Ken Hollett and Carla Anderson for their helpful suggestions on the manuscript.

Richard Lawford, Jurate Landwehr, Soroosh Sorooshian and Martha Whitaker gratefully acknowledge the generous support provided by the University Corporation for Atmospheric Research, the NOAA Office of Global Programs, the Global Water Cycle Program Office, the USGS National Research Program, the program for Sustainability of semi-Arid Hydrology and Riparian Areas (SAHRA, Grant No EAR9876800), the Global Energy and Water Cycle Experiment (GEWEX) of the World Climate Research Programme (NASA, Grant No NAG5-8502), and NOAA Grant No NA16GP1577.

Russell Walker and Ellen Douglas gratefully acknowledge Dr. Richard Vogel and James Limbrunner for their review, and for their suggestions which helped make this a better document.

Holly Hartmann, Allen Bradley and Alan Hamlet gratefully acknowledge the support provided by the following grants from NOAA's Office of Global Programs: NA86GP0061 and NA16GP1577, as well as NASA Grant NAG5-8503 and NSF-STC Grant EAR-9876800.

Hope Mizzell and Venkat Lakshmi thank Dr. Michael Helfert and Melt Brown for their support and encouragement. They also express appreciation to Ray Newcome for his editorial guidance.

Janet Gamble, John Furlow, Amy Snover, Alan Hamlet, Barbara Morehouse, Holly Hartmann and Thomas Pagano thank Claudia Nierenberg, Susan Julius and other reviewers for their comments and suggestions. In addition, Amy Snover and Alan Hamlet wish to acknowledge the support of the Joint Institute for the Study of the Atmosphere and Ocean (JISAO) under NOAA Cooperative Agreement No. NA67RJ0155, Contribution #892.

Steven Wolosoff and Theodore Endreny would like to thank the State University of New York College of Environmental Science and Forestry and the Edna Bailey Sussman fund for providing the funding which allowed for this analysis. We also would like to thank the researchers in each of the three case studies we have presented for their hard work and thoughtful scientific analysis of two very complex systems, the Catskill/Delaware and the Croton water supply areas.

Theodore Endreny, Benjamin Felzer, James Shuttleworth and Michael Bonell thank the watershed citizens and programs that supported coordinated policy, science and management in the many Hydrology for Environment, Life and Policy Basins. They also express appreciation to the following organizations for funding support: the Great Lakes Research Consortium, the Sustainability of semi-Arid Hydrology for Ripirian Areas program, the Semi-Arid Land-Atmosphere program, the FORD Foundation, the Morris K. Udall Foundation, and the National Oceanic and Atmospheric Administration.

The editors would also like to thank the following agencies for their support in the form of employment, office space, and administrative support during the preparation of this book: University Corporation for Atmospheric Research, Office of Global Programs in the National Oceanic and Atmospheric Administration, Climate Change Science Program Office, University of New Mexico, University of Arizona, and American Geophysical Union. In addition the editors would also like to thank the following individuals who took the time to provide thoughtful reviews for various chapters of this book:

David Ahlfeld
Alice Auriel
Stewart Cohen
William Cosgrove
Teresa Culver
Shannon Cunniff
Peter Gleick
David Goodrich
Harvey Hill
Wayne Huber
Poul Hurremoes
Helen Ingram
Kathy Jacobs
L. Douglas James
Kenneth Kunkel
Jurate Landwehr
Dennis Lettenmaier
Harry Lins
Diane MacKnight
Jim Mjelde
Janet Neuman
Sandra Postel
Joan Rose
John Schaake
Richard Smardon
Soroosh Sorooshian
Eugene Stakhiv
Juli Trtanj
Paul Try
William van der Schalie
Robert Varady
Donald Wilhite
(with apologies for anyone whose name has been missed)